EFFECTS OF STRESS ON PHOTOSYNTHESIS

ADVANCES IN AGRICULTURAL BIOTECHNOLOGY

In preparation:

Akazawa T., ed: The New Frontiers in Biotechnology

Gottschalk W. and Müller H.P., eds: Seed Proteins: Biochemistry, Genetics, Nutritive Value. ISBN 90-247-2789-8

Komamine A., ed: Proceedings 5th International Congress of Plant Tissue Culture

Effects of Stress on Photosynthesis

Proceedings of a conference held at the 'Limburgs Universitair Centrum'
Diepenbeek, Belgium, 22—27 August 1982

edited by

R. MARCELLE

Laboratory of Plant Physiology, Research Station of Gorsem,
B—3800 Sint-Truiden, Belgium

H. CLIJSTERS and M. VAN POUCKE

Department SBM, Limburgs Universitair Centrum
B—3610 Diepenbeek, Belgium

1983

MARTINUS NIJHOFF / DR W. JUNK PUBLISHERS
THE HAGUE / BOSTON / LONDON

Distributors

for the United States and Canada

Kluwer Boston, Inc.
190 Old Derby Street
Hingham, MA 02043
USA

for all other countries

Kluwer Academic Publishers Group
Distribution Center
P.O.Box 322
3300 AH Dordrecht
The Netherlands

Library of Congress Catalog Card Number: 82-24535

ISBN-13:978-94-009-6815-8   e-ISBN-13:978-94-009-6813-4
DOI: 10.1007/978-94-009-6813-4

CONTENTS

PREFACE

This volume contains the papers, presented during a conference, organized jointly by the "Opzoekingsstation van Gorsem" and the "Limburgs Universitair Centrum", Belgium from 22 to 27 August 1982. For this third meeting, the chosen topic was the effect of different stresses on photosynthesis. Most of the research in this field is realized on water stress and temperature stress; this situation is reflected in the conference programme. However, the importance of the other factors such as light, CO_2, salinity, anaerobiosis, ... was also emphasized especially during the important discussion sessions.

We express our gratitude to Drs. J. Gale, P. Jarvis, G.H. Krause, P.E. Kriedemann and P.S. Nobel for their excellent leadership during the discussion sessions. Particular thanks are also due to Dr. H.W. Woolhouse who gave us an excellent inaugural address and whose erudition largely contributed to the interest of the discussions.

For the first time in our experience of editors, we decided to use camera ready copies in order to publish more rapidly the proceedings and at a lower price. For a lot of reasons (among other things the bad choice of type of letter to be used and the choice of instructions to authors which were not perfectly followed by the authors), the technical presentation of this book will appear as non homogeneous; we accepted this lack of homogeneity with the hope that the publication time would be shorter in spite of the fact that, some authors delivered their manuscript with delay.

We especially thank Prof. Dr. L. Verhaegen, rector of this University, and Prof. Dr. A. Soenen, director of the station, for their continuing support in organizing such meetings. The material organization was in the hands of the Limburgs Universitair Centrum; we thank the whole staff of the general administration for their help. Mrs. G. Berx-Vliegen did an excellent job as secretary during the meeting. We express also our thanks to our assistants and collaborators for their help.

Financial support of the "Nationaal Fonds voor Wetenschappelijk Onderzoek" and the "Ministerie van Nationale Opvoeding en Nederlandse Kultuur" is gratefully acknowledged. We very appreciate the interest, paid to this conference by the "Instituut tot aanmoediging van het Wetenschappelijk Onderzoek in Nijverheid en Landbouw".

R. Marcelle, H. Clijsters and M. Van Poucke

THE EFFECTS OF STRESS ON PHOTOSYNTHESIS

H W Woolhouse

John Innes Institute, Colney Lane, Norwich, England

1. INTRODUCTION

There are five main environmental factors which may limit the growth of plants through effects on the photosynthetic apparatus; these are light, temperature, water, nutrients and carbon dioxide. None of these factors affect photosynthesis uniquely, so that the stress response which we observe in a whole plant is usually an integral of effects on many of the facets of metabolism, of which photosynthesis is just one. It is also evident that over a wide range of habitats, factors such as high temperature and water stress may often be correlated; when this happens interpretation of the plant response may be a very complicated problem.

2. SOURCES OF STRESS

2.1 Light

The majority of land plants which have been studied show a response of photosynthesis to photon flux density (PFD) of the form shown in the lower three curves (Fig.1). The relationship between photosynthetic rate and PFD is linear at low values of PFD; at high PFDs the rate of photosynthesis gradually levels off, at which point photosynthesis is said to be light saturated. The upper curves in Fig.1. show that at higher temperatures the PFD at which photosynthesis is saturated is increased until at 20^0c or above, photosynthesis does not saturate over the normal environmental range of PFD. Non-saturating kinetics of this kind are characteristic of C_4 species. This family of curves obtained for single attached leaves of Spartina anglica emphasises how a particular stress, in this case sub-optimal amount of light, may be dependent upon other factors. It should also be emphasised that quantitatively different conclusions concerning the light requirement of the species would be obtained if photosynthesis were measured on the whole plant, thus introducing the complexities of mutual

shading between leaves and respiratory fluxes from non-photosynthetic parts
of the shoots.

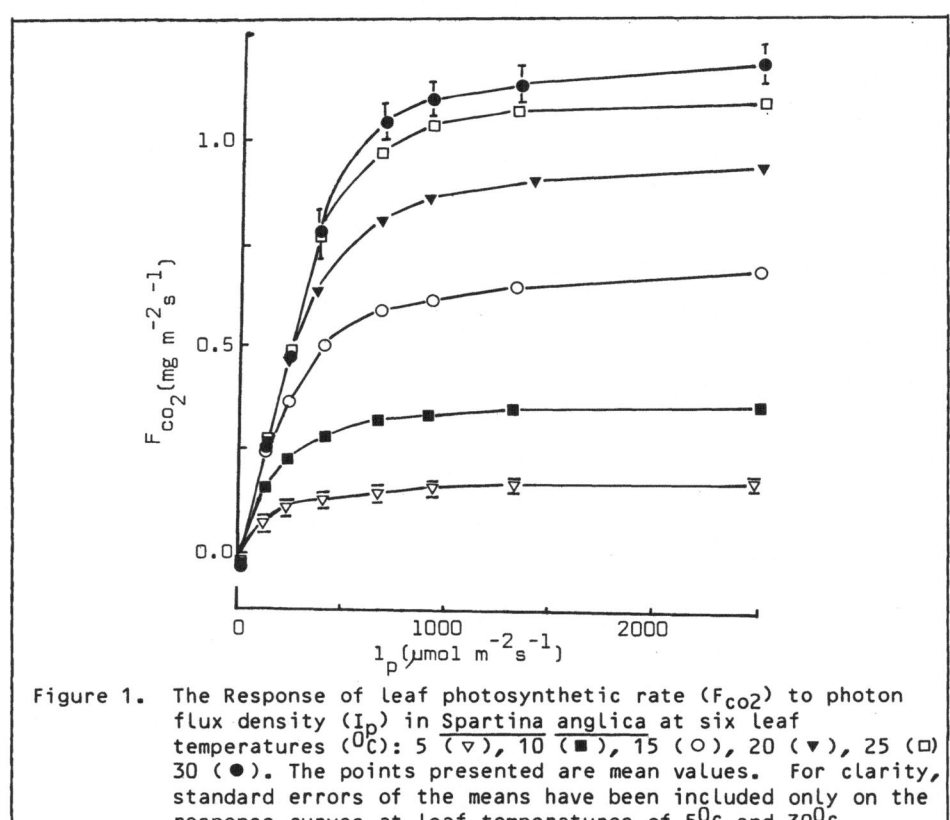

Figure 1. The Response of leaf photosynthetic rate (F_{CO_2}) to photon
flux density (I_p) in <u>Spartina anglica</u> at six leaf
temperatures (^{0}C): 5 ($\triangledown$), 10 ($\blacksquare$), 15 ($\circ$), 20 ($\blacktriangledown$), 25 ($\square$)
30 ($\bullet$). The points presented are mean values. For clarity,
standard errors of the means have been included only on the
response curves at leaf temperatures of 5^0C and 30^0C.

Stress may also be induced in some species by exposures to high PFDs.
Species which grow naturally in shade are particularly susceptible to
damage at high PFDs but other species may be affected. Evergreens such as
<u>Pinus</u> sp. may undergo photobleaching when exposed to high PFDs at sub-zero
temperatures [1].

2.2   Temperature

Stress responses to temperature vary greatly according to species,
temperature prehistory and cultural conditions. Many tropical and
sub-tropical species show a sharp decline in rates of photosynthesis below
15^0C (Fig.2) [2]. Long <u>et al</u> 1975 compared tropical and temperate
species. The rate of photosynthesis in seedlings of <u>Zea mays</u> declined

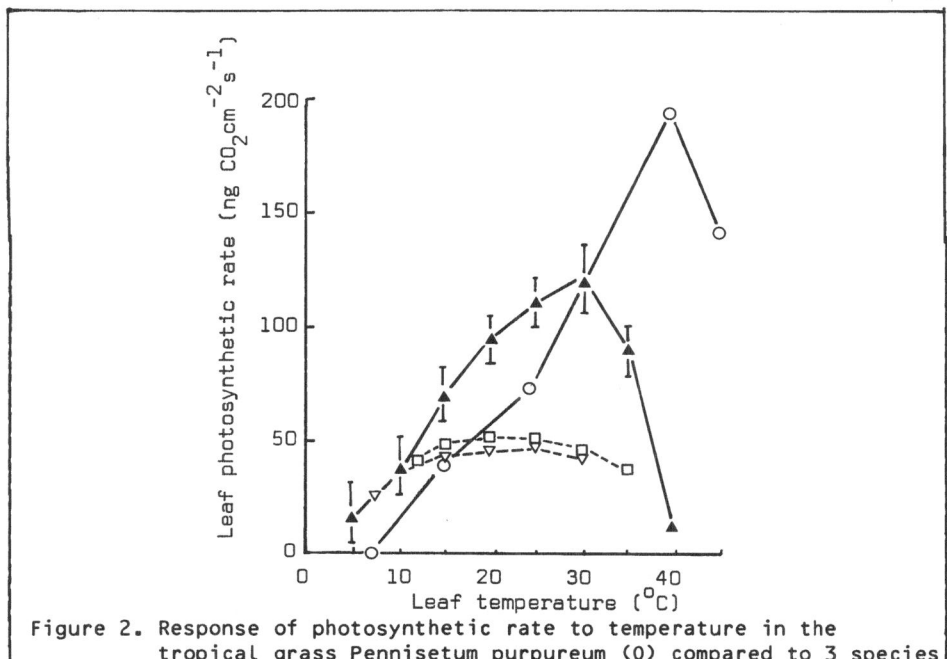

Figure 2. Response of photosynthetic rate to temperature in the tropical grass <u>Pennisetum purpureum</u> (O) compared to 3 species from temperate regions, <u>Spartina anglica</u> (▲), <u>Festuca arundinacea</u> (□), and <u>Sesleria albicans</u> (▽). Measurements made at incident photon flux density of 2,000 m mol $m^{-2}s^{-1}$.

continuously at a temperature of 13^0C [3]. Exposure to low temperatures for a period may lead to permanent damage, affecting performance when favourable conditions are resumed; exposure of maize seedlings to a temperature of 10^0C for three days decreased the rate of photosynthesis by 80% when they were returned to a temperature of 25^0C [4].

Plants vary greatly in their sensitivity to damage by high temperature. Although responses to heat damage are well documented [5], many of the earlier studies of effects on photosynthesis did not include rigorous control of the vapour pressure deficits (VPDs) so that the results obtained may be confounded by effects of water stress. Recent work makes it clear however that heat inactivation of photosynthesis may result from damage to the chloroplasts [6,7]. The respiratory system in the mitochondria is generally found to be less susceptible to damage by extremes of temperature than is photosynthesis in the chloroplasts.

2.3   Water

Availability of water is perhaps the commonest limitation to plant growth, but it is also one of the most difficult to study. It is abundantly clear that water stress leads to decreased photosynthetic rates but the

4

separation of effects on stomata, as opposed to metabolic changes in the mesophyll cells, are more difficult to unravel.Simultaneous measurements of CO_2 and water vapour fluxes may be used to calculate the stomatal and mesophyll conductances [g_s and g_m] as a function of leaf water potential but the results vary enormously. A survey of the literature on this subject reveals endless contradictions. Most of the difficulties are probably methodological. For example, it seems that one important consideration which has been frequently overlooked is the rate at which the stress is imposed. Where short sharp water deficits are induced, g_s may be the first to change, but often when the onset of water deficit is more gradual, there may be concomitant decreases in both g_s and g_m [8]. The finding of simultaneous effects of water stress on g_s amd g_m have raised such questions as: is stomatal conductance dependent on the rate of photosynthesis and does water stress affect both the guard cells and the mesophyll cells in a similar manner [9]? Gas exchange measurements show that photorespiration decreases in the course of water stress in wheat [10] and sunflower [11], suggesting that the effects of water stress on net photosynthesis are not due to increased photorespiratory efflux of CO_2.

2.4   Nutrients

Since almost all of the elements essential to plant growth are involved in some aspect of chloroplast metabolism, it is inevitable that most stresses caused by nutrient deficiencies will have substantial effects on photosynthesis. Toxic effects of excessive levels of nutrients such as Zn^{++} have also been ascribed to inhibitory effects on chloroplast metabolism. The magnitude of the fertiliser industry in the modern world testifies to the scale of potential nutrient stress in crop-producing areas. On a wider scale, plant growth over vast regions of the earth is limited by availability of nutrients, particularly nitrogen and phosphorus. The interaction of nutrient stress with water supply has many manifestations. Nutrient deficiencies leading to poor root growth can reduce the ability of the plant to extract water from the soil. High concentrations of nutrients may impose osmotic stress when water is in short supply. Direct inhibition of photosynthesis can arise from osmotic dehydration in the absence of effects on stomatal conductivity [12]. Species classified as hygro, meso and xerophytes differ significantly in the magnitude of such effects [13] (Fig 3).

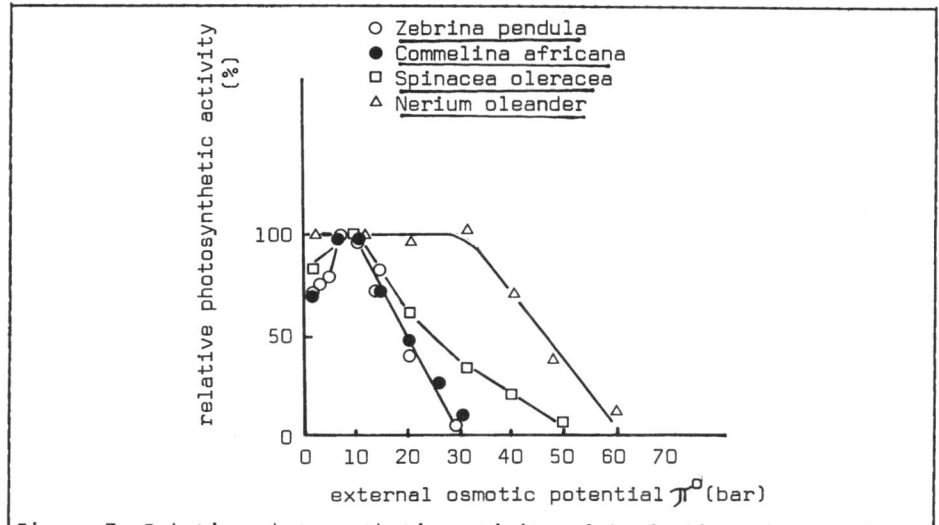

Figure 3. Relative photosynthetic activity of leaf slices from various plants at increasing external osmotic potential (π) with sorbitol as osmoticum. Control rates (100%) in μmol CO_2 fixed/mg chlorophyll^{-1}h^{-1}:Zebrina pendula 39, Commelina africana 45, Spinacia oleracea 108, Nerium oleander 75.

2.5   Carbon dioxide

It has long been known that the rate of photosynthesis under normal atmospheric conditions can be increased if the concentration of CO_2 is increased above the ambient level. Also of long standing is the knowledge that atmospheric concentrations of oxygen are inhibitory to photosynthesis, the so-called Warburg effect. In fact, the increase in rate of photosynthesis brought about by enhancement of CO_2 concentration can be simulated by lowering of the oxygen concentration. The kinetics of the oxygen inhibition suggested a direct competitive effect on CO_2 fixation. The competition of oxygen with CO_2 at the catalytic site of CO_2 fixation was subsequently established (16 and 17). At the same time that these facts were established, it also became clear that a distinctive feature of photorespiration, when compared to dark respiration, is the differential response of the two processes to oxygen concentration (18 and 19). From the association of these facts has developed the concept of the oxygenase reaction of ribulose biphosphate carboxylase/oxygenase (Rubisco) as the source of the photorespiratory carbon oxidation cycle [PCO] (20) (Fig.4).

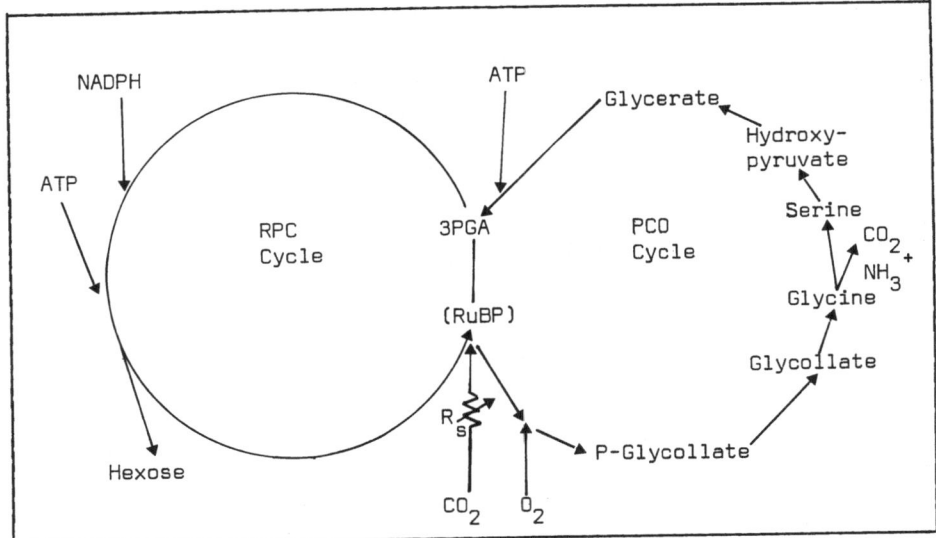

Figure 4. The photosynthetic carbon oxidation (PCO) cycle, coupled to the reductive pentose phosphate cycle through the competitive carboxylase/oxygenase reactions (ϕ). The coupling of the two cycles is completed as phosphoglycollate is metabolised through the sequence of intermediates shown in the diagram leading to the regeneration of 3PGA, with the release of CO_2 and ammonia. The variable resistance R_S represents the diffusive resistance to CO_2 uptake afforded by the stomata; P-glycollate, phosphoglycollic acid.

3. MECHANISMS OF STRESS EFFECTS ON PHOTOSYNTHESIS

It is convenient for purposes of discussion to distinguish effects of stress on stomatal conductance, electron transport and CO_2 assimilation. The potential complexities of such a simplification have already been noted in the case of relationships between g_s and g_m; similarly, there are reasons for not taking too simplistic a view of the light and dark reactions of photosynthesis as discrete and separate systems. Figure 5 is a diagram of the photosynthetic carbon reduction cycle (PCR) in which the enzymes requiring light for activation are emphasised. The photoactivation process is closely linked to the photosynthetic electron transport chain through the ferredoxin/thioredoxin system (21)(Fig.6), so that the state of enzymic activation may be determined by the rate of electron transport. Thus, at low PFDs the extent of activation of enzymes of the PCR cycle may limit the rate of photosynthesis in addition to the rate of electron transport per se.

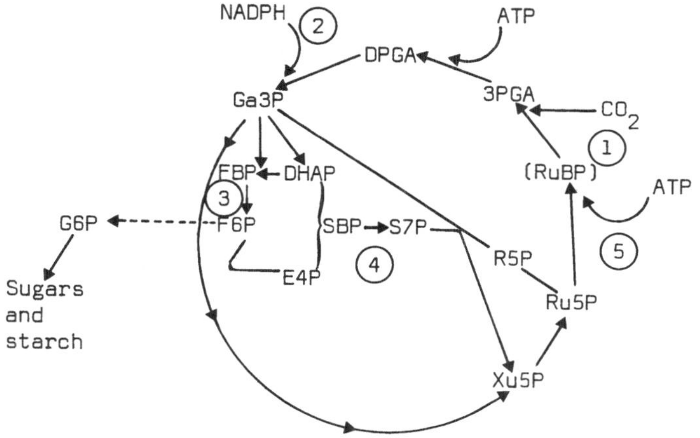

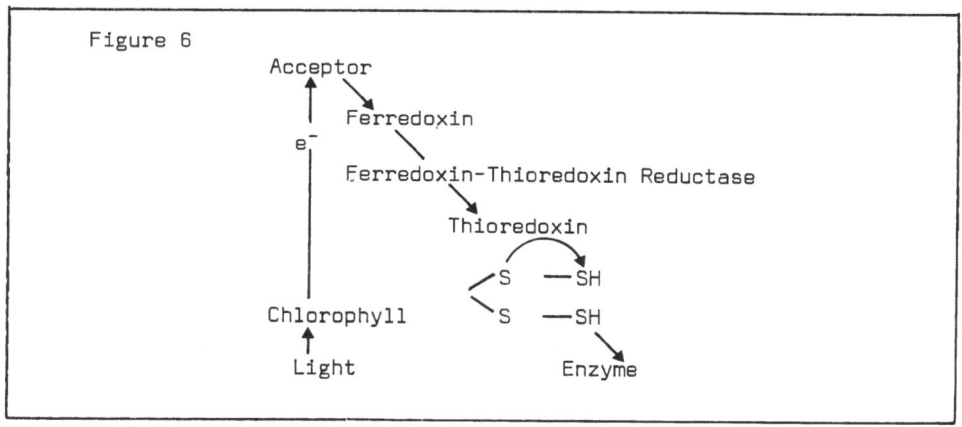

Figure 5. The reductive pentose phosphate cycle (RPC) of photosynthesis.
Note that a central feature of the scheme is the consumption
of ribulose 1,5 bisphosphate (RuBP) in CO_2 fixation and its
regeneration via a sequence of interconnected reactions. The
heavy arrows ——→ 1-5 indicate enzymes of the cycle
which are activated by light; they are :
1. ribulose 1,5 bisphosphate carboxylase;
2. glyceraldehyde 3-phosphate dehydrogenase;
3. fructose 1,6-bisphosphate phosphatase;
4. sedoheptulose 1,7-bisphosphate phosphatase and
5. phospho-ribulokinase.
Abbreviations : RuBP, ribulose 1,5-bisphosphate;
3PGA, 3-phosphoglycerate; DPGA, 1,3-diphosphoglycerate;
GA3P, glyceraldehyde 3-phosphate; FBP, fructose
1,6-bisphosphate; F6P, fructose 6-phosphate;
DHAP, dihydroxyacetone phosphate; G6P, glucose 6-phosphate;
E4P, erythrose 4-phosphate; SBP, sedoheptulose
1,7-bisphosphate; S7P, sedoheptulose 7-phosphate;
Xu5P, xylulose 5-phosphate; R5P, ribose 5-phosphate;
Ru5P, ribulose 5-phosphate.

Figure 6

3.1   The effects of stress on g_s and the consequences for photosynthesis.

Stress induced by low PFD, low temperatures, water shortage or nutrient deficiencies all give rise to decreased stomatal conductances. It does not follow, however, that this implies an increase in the extent to which the stomata are limiting photosynthesis; thus, if these further stresses are having even greater effects on the biochemistry of photosynthesis, then the relative contribution of the stomatal conductance may even be reduced. A simple coefficient (l) to describe the relative magnitude of stomatal limitation has been proposed (22);it takes the form:-

$$l = \frac{A_o - A}{A_o}$$

where A is the rate of photosynthesis observed and A_o is the rate of photosynthesis which would occur if resistance to CO_2 diffusion was zero; but see also (32).

If this analysis is applied to leaves maintained under limiting PFD partial closure of the stomata may occur but the extent to which they actually contribute to the limitation of photosynthesis is actually decreased [23]. Lowering of temperature will induce stomatal closure in many species, but in some, at least, this may be accompanied by an increase in the CO_2 concentration in the intercellular spaces of the leaf as the biochemistry is also slowed down, so that again the relative significance of the stomata is decreased [24]. Under water stress, the question of stomatal limitation is more complex. An analysis of published data by Farquhar and Sharkey [22] shows that in some species the relative contribution of the stomata to decreased photosynthesis may decrease under water stress whilst in fact the biochemical contribution may be increased. It may be that these differences are related to the extent to which abscisic acid synthesis is induced by water stress, since this appears to be an important factor contributing to stomatal closure in many species. In this context, it would be of interest to analyse the effects of K^+ stress on the relative stomatal contributions, since this ion plays a major role in the turgor changes associated with movements of the guard cells.

3.2   Effects of stress on the photochemical reactions of photosynthesis.

Photosynthetic electron transport in the thylakoid membranes of chloroplasts is sensitive to extremes of temperature. It is generally found that in species which are able to function normally at high temperatures, the photochemical reactions are inhibited and photosynthesis is consequently slowed down at low temperatures; the converse is true for species which are able to photosynthesise normally at low temperatures. The effects of low temperatures have been attributed particularly to a phenomenon referred to as lateral phase separation [25] in which the thylakoid lipids pass from the normal fluid to a gel condition [26]. There are probably several reasons why such phase changes should decrease the rate of electron transport and so reduce photophosphorylation and NADPH production. Transition to a mixed fluid-gel phase will increase the permeability of the thylakoids to ions and hence diminish their capacity to sustain the proton gradients needed to drive the synthesis of ATP. Nobel (1974) [27] obtained evidence of differential changes in the permeability of chloroplasts from chilling-sensitive and chilling-resistant plants when subjected to chilling treatments. Liquid-gel phase changes will also reduce the lateral diffusion of the electron-transporting protein complexes of the photosystems within the thylakoid membranes [28] on which the rate of electron transport will depend [29]. There is strong circumstantial evidence that slowing of the photosynthetic rate after exposure of leaves to high temperatures may also derive in part from damage to the electron transport system. Schreiber and Berry [30] found that changes in the chlorophyll fluorescence emission from intact leaves induced by high temperatures was correlated with damage to the electron transport system of the chloroplasts. Pearcy et al [7] found a progressive interruption of electron transport in isolated chloroplasts correlated with a reduction in the quantum yield of photosynthesis in the intact leaves from which they were extracted.

3.3   Effects of stress on the dark reactions of photosynthesis.

Temperature extremes are known to affect certain soluble enzymes of the PCR cycle but the evidence for the in vivo significance of these effects is less clear. The enzyme pyruvate orthophosphate dikinase, which regenerates phosphoenolpyruvate (PEP), the substrate for CO_2 fixation in C_4(PEP) species, appears to consist of hydrophobically bonded sub-units which become labile at low temperatures [35]. Exposure of plants to low temperatures led to reductions in the rate of photosynthesis which were correlated with the temperature sensitivity of the enzyme extracted from leaves [36].

Some enzymes of the PCR cycle become unstable to heat in vivo at temperatures close to those at which the rate of photosynthesis is depressed; these are, however, light-activated enzymes (NADP glyceraldehyde -P dehydrogenase, ribulose 5P kinase and NADP malate dehydrogenase) and the possibility cannot be ruled out that the effects were caused by the action of temperature on the electron transport-linked light activation mechanism. O'Toole et al [37] reported a correlation between decreased mesophyll conductance and ribulose bisphosphate carboxylase activity during water stress. The mechanism of this effect, if indeed it is a causal relation, remains to be investigated.

These few examples will suffice to emphasise that although there are substantial indications of effects of major environmental stresses on the photosynthetic capacity of leaves, the evidence is rarely definitive and rests on correlations which may not denote causal relations. Perhaps the most promising technique for making progress in this matter is the measurement of chlorophyll fluorescence at ambient temperatures which can be carried out on intact leaves on which photosynthesis can also be measured.

4. ADAPTATIONS OF THE PHOTOSYNTHETIC APPARATUS TO STRESS

4.1 Some general considerations

There is a vast amount of accumulated evidence to show that plant species differ in the temperature responses of photosynthesis. It is also evident that the pattern of these responses to temperature correlates broadly with the temperature regimes in the habitats where the species occur. We could at this point plunge into descriptions of a wealth of variations in the chemical composition of individual components of the photosynthetic apparatus, drawing inferences concerning the ways in which they modify tolerance of the system to temperature fluctuations and its potentialities under particular temperature regimes. It is well to recognise, however, that there is a substantial philosophical difficulty underlying this enterprise. We are concerned with the licence involved in attaching the label "adaptation" to any particular variation which happens to be the subject of a measurement. In the literature of photosynthesis, almost every measurable constituent of a leaf would appear to have been the subject of an "adaptive change" in the view of one author or another.

To take the matter a step further, one would require to say that the observed change contributes to the fitness of the plant, which will be measured in terms of the numbers of offspring it produces. We are not aware of any of the so-called adaptations of photosynthesis which have really been analysed at this level. What the variations in biochemical composition and fine structure of the photosynthetic apparatus in species from different environments amount to, as far as our present knowledge of them is concerned, are features which appear to achieve optimal functioning under particular sets of conditions.

It does not follow from this that specific attributes are involved directly with survival or the production of offspring since any one such attribute or character makes its impact not in isolation but integrated with all the other features of the whole plant. But if our variant feature of thylakoid composition, let us say, has to be recognised as an optimal solution to the functioning of the chloroplasts, against the background of all the other components of the system, we face an impossible task. Lewontin [38] provides a particularly stark statement of this difficulty. "In order to make the argument that a trait is an optimal solution to a particular problem, it must be possible to view the trait and the problem in isolation, all other things being equal. If other things are not equal, if a change in a trait as a solution to one problem changes the organism's relation to other problems of the environment, it becomes impossible to carry out the analysis part by part, and we are left in the hopeless position of seeing the whole organism as being adapted to the whole environment." This would appear to set severe limits on the value of optimisation theory as an approach to adaptation.

A long term but more solidly based programme for the evaluation of adaptation is to proceed from measurements of physiological responses to the environment to an analysis of the underlying biochemistry, thence to the genetic control of the biochemical characteristics and so to classical population genetic studies of the genes involved [39]. Amidst the complexities of the distinct yet inter-connected genomes of the chloroplast and nucleus, this may also be held to be an unrealistically complex programme, but it has the theoretical advantage of assessing the variations which are measured, ultimately in terms of the frequencies of the alleles by which they are determined.

The ideal approach to investigating the genetic determination of adaptive changes in the photosynthetic apparatus would be to develop isogenic lines differing only in a single gene basis affecting a defined attribute of the photosynthetic system. Assuming that the gene in question was not unduly pleiotropic in its expression, it should then prove possible to assess its effects in comparative studies of performance and fitness against the isoline carrying the wild type allele. One may suppose that such procedures will eventually be adopted by physiologists with concern for a sounder contribution to breeding programmes, but for the moment, there does not appear to be any example of such a study in respect of the metabolic aspects of photosynthesis.

4.2   Plasticity

Species which occur naturally in relatively uniform habitats tend to have their photosynthesis and other physiological processes geared to functioning optimally at around the median of the prevailing conditions. For the case of photosynthesis, thermal effects are the most thoroughly analysed. A major contribution to this field is that of Bjorkman and his colleagues at the Carnegie Institute [5], who have studied temperature acclimation in species from a wide range of habitats. It is a hazardous business to attempt to draw general conclusions at the present state of knowledge. Work with tropical cereals such as maize and sorghum suggests that in low temperature acclimation the actual ability to synthesise the photosynthetic apparatus is often a limiting factor [40]. Studies with plants of similar genotype grown at different temperatures, suggest that plants can alter the properties of the thylakoids to optimal functioning at the growth temperatures [41]. Plants raised at one temperature, then tend to function less well when transferred to a new temperature regime. In general, it would seem that species native to habitats subject to wide fluctuations in temperature have a greater propensity for modifying the properties of the photosynthetic apparatus according to the prevailing conditions. We know with reasonable confidence that thylakoid lipid changes are one of the important factors in this adaptation but more detailed information is needed on the regulatory properties of the enzymes involved in chloroplast lipid biosynthesis before we can define precisely how this thermal adaptation works.

4.3   Genetically determined adaptations of photosynthesis: the C_4
mechanism

4.3.1. Factors favouring plants possessing the C_4 mechanism

We shall conclude this discussion with a much more complicated
example of genetically determined modification in the photosynthetic
apparatus, which most physiologists would agree was adaptive in character,
notably the phenomenon of C_4 photosynthesis. This pathway presents
challenges not only from the standpoint of discovering how it may have
evolved but also because it raises in an acute form the problems to which
we have already referred concerning the assessment of adaptive significance
that is, the contribution which it makes to the fitness of the plant.

The key to an understanding of C_4 photosynthesis is photorespiration
in the sense that if photorespiration had not existed the C_4 mechanism
would never have evolved. A model equation describing the relative flows
of carbon through the PCO and PCR cycles ( ) takes the form:

$$0 \quad \frac{V \text{ oxygenase}}{V \text{ oxygenase}} = \frac{K_m O_2}{K_m CO_2} \times \frac{V_{max} CO_2}{V_{max} O_2} \times \frac{[CO_2]}{[O_2]}$$

where $V_{oxygenase}$ and $V_{carboxylase}$ are the rates of the oxygenase and
carboxylase reactions, $K_m O_2$ and $K_m CO_2$ are the Michaelis constants of
Rubisco for oxygen and CO_2 respectively and $V_{max} O_2$ and $V_{max} CO_2$ are the
maximum velocities of the enzyme with these substrates. From this equation
we see that the crucial variables are the relative concentrations of O_2/CO_2
at the catalytic site of Rubisco, the V_{max} of Rubisco with respect to
oxygenation and carboxylation and the relative affinity of Rubisco for CO_2
and oxygen.

We may consider these components of the equation in relation to some
of the main environmental variables that we have identified as sources of
stress affecting photosynthesis and see how these may have influenced the
evolution of the C_4 syndrome.

Temperature

The Q_{10} for photorespiration (Pr) is generally found to be greater
than 2, the normal value for a thermochemical reaction. The reasons for this
are that elevated temperatures have a differential effect on both the
relative solubilities of O_2 and CO_2 and possibly also on the affinity of
Rubisco for O_2/CO_2. Consequently, in most C_3 species there occurs a temperature, usually somewhere between 18° and 30^oC, where the net influx of

CO_2 in photosynthesis reaches a maximum and above this temperature the rate declines as the gradient of CO_2 concentration gets shallower due to increased Pr.

The C_4 mechanism serves to concentrate CO_2 in the bundle sheath of the leaf so that the carboxylation reaction of Rubisco in these cells is approximately saturated and the oxygenation reaction, which is favoured at these high temperatures, is thereby reduced. Water loss by transpiration is also reduced in these species because the very high affinity of the primary carboxylating enzyme, PEP carboxylase, for CO_2, permits photosynthesis to continue at much lower intercellular concentrations of CO_2 than in C_3 plants. Thus, the extent and frequency of stomatal opening can be reduced. It is for this reason that the water use efficiency of C_4 plants is roughly twice that of C_3 plants (Table 1). So, at high temperatures, C_4 plants generally have higher net rates of photosynthesis than C_3 plants and lose only half as much water per unit carbon gained.

It is not surprising, therefore, that we find a maximum of C_4 species in savannah, semi-desert communities and particularly amongst herbaceous forms where the additional heat load accruing from back-radiation is maximal.

This conclusion is reinforced at the level of general correlation by the work of Teeri and Stowe (43,44). They carried out a stepwise multiple regression analysis of the distribution of C_3 and C_4 grasses in relation to a number of environmental variables in twenty-seven geographical zones of North America. The proportion of C_4 grasses was highly correlated with the July minimum temperature but not with any other factors. When the study was extended to the dicotyledons (44) over the same region, distribution correlated with rates of summer pan evaporation and dryness ratio of the habitat and not with minimum July temperature. It seems probable that, in the dicotyledons, the total energy load is greater than for the grasses, because the leaves are generally broader and this brings the greater water use efficiency aspect of the C_4 mechanism into greater prominence as a component in selection.

It must be emphasised that these studies, whilst interesting, present only a broad correlative approach and may overlook many of the subtleties of adaptation peculiar to a particular species. The work of Caldwell et al.,[45] on a community comprising two shrubby species <u>Atriplex confertifolia</u> (C_4) and <u>Ceratoides Lanata</u> (C_3) in the Great Basin of Utah

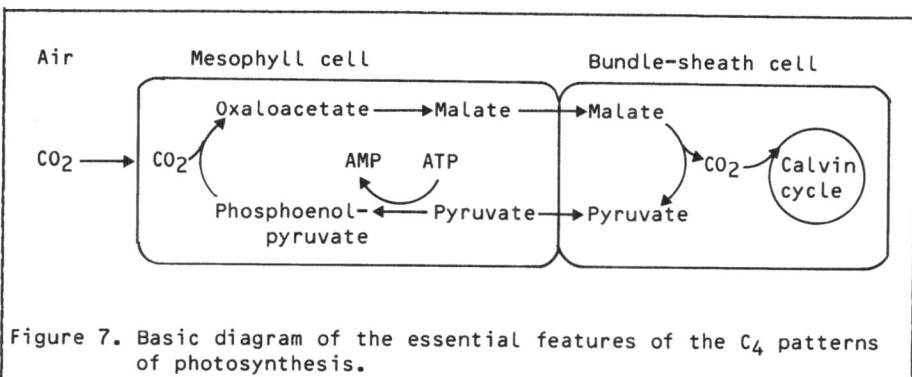

Air   Mesophyll cell   Bundle-sheath cell

Figure 7. Basic diagram of the essential features of the C_4 patterns
of photosynthesis.

Table 1. The relative water use efficiency of a selection of C_3 species
[g transpired water per g dry matter produced].

C_4 plants	
Maize	370
Millet	300
Amaranthus	300
Portulaca	280
C_3 plants	
Rice	680
Rye	630
Oats	580
Wheat	540
Barley	520
Alfalfa	840
Bean	700
Crimson Clover	640
Potato	640
Sunflower	600
Watermelon	580
Cotton	570

provides an illustration of this problem. Caldwell et al compared the various aspects of the carbon balance of the two species throughout the year, along with measurements of the allocation of carbon to the different organs and estimations of the water use. Both species were capable of photosynthesising in the early part of the year, April to July, though the rate of photosynthesis was higher in the C_3 species at this time. In the heat of August and September, however, the C_3 species ceased photosynthesis while the C_4 continued, albeit at a reduced rate. Over the season as a whole, the carbon and water economy of the two species was not significantly different but they were, in effect, exploiting different parts of the season. It appears that the different thermal optima of photosynthesis which show up in the seasonal responses of these two species may also influence their relative performances on a longer timescale. Thus quaternary studies show that in this region of Utah the relative proportions of Atriplex and Ceratoides have fluctuated, favouring the C_4 in warmer and the C_3 in cooler periods (B.N. Smith, pers. comm.).

The danger of drawing conclusions about the adaptive significance of the C_4 mechanism from broad generalisation is further illustrated in a recent publication concerning the shrubby C_4 vegetation of the deserts of Central Asia [46]. Vast areas in the region are covered by C_4 species of such genera as Anabasis, Haloxylon and Salsola (Chenopodiaceae) and Calligonum (Polygonaceae). Commenting on the relative absence of grasses in this region, the author notes that it is "probably due to extremely low temperatures during winter. The cold sensitivity of many C_4 grasses is well established and is in contrast to the cold-tolerance of many C_4 dicotyledons investigated here and elsewhere." There are serious objections to this argument. It is an example of aiming at generalisation which, by ignoring specific cases, leads to notions about the intrinsic attributes of a plant group or a particular physiological mechanism, which are not justified. For example, C_4 grasses, such as Distichlis sp. in brackish marshes in Eastern USA are often frozen solid. Spartina pectinata on the St Lawrence may be frozen into a metre of ice in some winters. Eragrostis curvula, a range grass of the Western USA, survives months of sub-zero temperatures. The author also ignores other work which does not accord with his proposition; almost all of the tropical C_4 dicotyledons including species in the Asteraceae, Nyctaginaceae and Euphorbiaceae are just as sensitive to cold as are the tropical C_4 grasses. The author also fails to recognise that in these Asiatic regions there are other factors which favour shrubs rather than grasses and C_4 rather than C_3 species. The low and erratic rainfall, combined with a supply of phreatic water at depth of 3-10 metres is beyond the reach of fibrous-rooted grasses but well within the compass of deep-rooted shrubs. The pattern of temperatures is also critical, fluctuating between very hot by day in summer, in the temperature range favouring C_4 species, to very cold at nights and in winter, when it is not favourable to significant growth in C_3 or C_4 species. The transition of the seasons is also relatively rapid in these continental regions so that the spring and autumn seasons when one might expect intermediate temperatures favouring C_3 plants, are of relatively short duration.

4.3.2. The occurrence of C_4 species in atypical habitats. We have seen already that there are substantial difficulties in the matter of attributing the occurrance of some C_4 species in particular habitats in terms of adaptive advantages of their photosynthetic mechanism. Indeed, it seems almost certain that some species have migrated or been subject to a change in environmental conditions such that they now occur where it is extremely unlikely that the C_4 mechanism would have arisen de novo. Examples of C_4 species from shaded, aquatic and temperate salt marsh habitats will serve to illustrate this situation.

4.3.2.1. Shaded habitats. Shaded habitats are characterised not only by lower than average photon flux densities but are also usually cooler than adjacent open sites. It has been suggested that such habitats are disadvantageous to C_4 species because the mechanism requires more ATP per unit of carbon fixed than the C_3 pathway and temperatures are lower than in the open so that photorespiration is rarely a problem. For a C_3 species, in the absence of photorespiration, the energy requirement is 18ATP per molecule of hexose synthesised. For a C_4 species, in which photorespiration is taken as zero and all CO_2 fixation goes initially through a PEP carboxylation, the substrate for which is generated by the pyruvate, Pi dikinase enzyme system, the ATP requirement is circa 30 ATP per molecule of hexose synthesised.

If we can consider an elevated temperature at which photorespiration becomes significant, rising to, say, 40% of photosynthesis in a C_3 species and 10% of photosynthesis in a C_4 species, then the ATP requirements will rise to approximately 38 and 36 ATP per molecule rate of hexose synthesized

18

in the respective cases. Thus we can see that under many circumstances the quantum requirement for C_4 species would be expected to be higher and the quantum efficiency of photosynthesis correspondingly lower than for C_3. C_4 species could thus be at a disadvantage in low light - a proposition which has been summarised in the form of a simulation model [47].

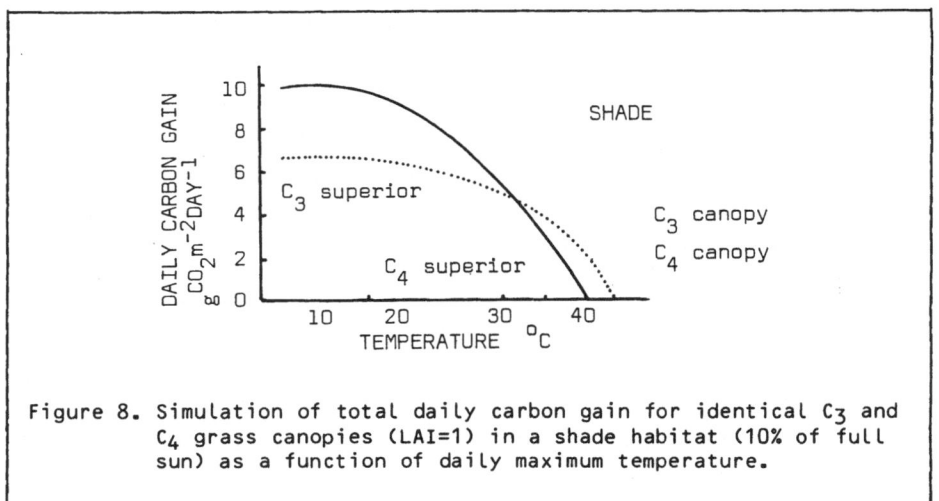

Figure 8. Simulation of total daily carbon gain for identical C_3 and C_4 grass canopies (LAI=1) in a shade habitat (10% of full sun) as a function of daily maximum temperature.

Figure 8 shows the application of this model to identical C_3 and C_4 grass canopies and it shows that only when the temperature in shade rises above 32^0C would the C_4 species have any conceivable advantage. The facts, however, do not support the model, since there are well documented cases in which C_4 grasses and Euphorbia species appear to be successful in shaded habitats [48,49,50]. The case of Euphorbia species in the flora of Hawaii will serve as an example [49].

There are fourteen native species of Euphorbia in Hawaii; all are C_4, all endemics in the same sub-genus and agreed by taxonomists to have a common ancestry. Some of the species are shrubs which occupy open ground; these pose little problem in the present context in the sense that one may suppose that in the early volcanic history of Hawaii the original habitat was open and well suited to the progenitor of these Euphorbias which arrived there and may already have possessed the C_4 mechanism. As the islands developed forests, some of the Euphorbias became adapted to the shaded habitat which was thereby created and evolved a quasi arborescent habit as components of the understorey.

Robichaux and Pearcy [49] studied one of these shade-adapted species, E.forbsii, comparing it with Cladoxylon sandwichense, a C3 species of comparable life form from the same forest habitat. Both species were found to have relatively low photosynthetic rates at light saturation, required low PFDs to saturate photosynthesis and had low rates of dark respiration. Quantum yields of photosynthesis were found to be similar in the two species at 22^0C, not 32^0C or above as indicated by the Ehleringer model.

Other examples of C4 species adapted to growth in shade are to be found in the grasses Paspalum conjugatum [50] and Microstegium vimineum [48]. Of particular significance in the shade-tolerant C4 species studied to date appears to be a lower rate of dark respiration, which offsets the potential carbon deficit imposed by the higher quantum requirements.

4.3.2.2. Aquatic Habitats.   The grass Vossia cuspidata and the sedge Cyperus papyrus are both species of tropical swamps which, under appropriate conditions, can even grow to form floating islands. They grow in regions of high insolation but the humidity and general availability of water render it unlikely that there are strong selection pressures favouring the selection of the C4 mechanisms in such areas.   Moreover, these species frequently occur with other grasses and sedges (Phragmites sp. Gania sp. etc) which possess the C3 mechanism of photosynthesis. Even more remarkable, perhaps, are the cases of species such as Spartina pectinata and Cyperus longus which also inhabit freshwater swamps but extend into cool temperate regions of Europe and North America [51]. Again, one is drawn to the conclusion that these species probably had antecedents native to hot dry regions in which the C4 mechanism originated and that the mechanism has been retained through subsequent periods of speciation and invasion into habitats where the mechanism is now selectively neutral.

4.3.2.3.   Salt marshes in cool temperate regions: the case of Spartina anglica.   Spartina anglica is a vigorous allotetraploid which arose in Britain from a hybridisation between the native S.maritima and S.alterniflora, an introduction from North America.   Details of the cytology and attendant nomenclature of the species are complicated and need not concern us here.   Suffice it to say that the native parent S.maritima has remained a scarce species, of relatively weak growth, which rarely sets seed in Britain, whilst the other parent, S.alterniflora, has never gained a foothold.   Spartina anglica, however, produces fertile seed, probably in part by apomixis, and now occupies over 12,000 hectares around the coast of Britain, often spanning salt marshes from the coastal to the landward side. It has Kranz anatomy and the gas-exchange characteristics of a C_4 species [2, 52, 53], as do both parent species. The productivity of the species is relatively high, of the order of 40 tonne $ha^{-1}a^{-1}$, of which an exceptionally large amount, over 50%, is invested in below-ground parts, most notably a massive rhizome system. Clearly, from the lack of fitness displayed by the parent species of S. anglica, it must be supposed that the key to its fitness does not rest primarily with the C_4 photosynthetic system. The reasons for the lack of fitness in S.maritima and S. alterniflora in Britain are not known; the lack of vigour in the former suggests that it may be a poor competitor, particularly on account of its greater light requirement than its C_3 neighbours; for S.alterniflora conditions are probably too far below the normal temperature requirements of the species. S.anglica may have inherited a measure of temperature adaptation from S.maritima in that it is able to photosynthesise at similar rates to neighbouring C_3 species, even at $5-10^0C$ [52,54,55]. There is still some suggestion of imperfect adaptation to temperature, on the other hand, when the seasonal pattern of growth is considered [56]. As noted, the vigour and productivity of the species is high. The huge below-ground system must impose a severe respiratory demand upon the plant but against this, the rhizome provides the abundant air spaces needed for oxygen transport in the anaerobic estuarine muds, whilst the stout anchor roots enable the plant to establish and maintain itself in the seaward side of the marsh. Indeed, in places where wave action is not too vigorous, it may be seen that the species is, in essence, creating its own habitat as it goes along, as the stout stems and rhizomes slow the movement of water and encourage silting.

Thus, the burden of the root system may be offset by the gaining of a new niche in which, there being no other species, the light-sensitive C_4 mechanism is spared the burden of competition from other species. Support for this view is provided by experiments in an area where Phragmites communis was invading Spartina marsh in Poole Harbour [57]. The Spartina became etiolated, then ceased to flower, then failed to produce tillers and finally died over 25m landward of the seaward limit of Phragmites. When the stems of Phragmites were cleared from a square metre of ground in the landward extinction zone of Spartina, the Spartina recommenced growth, tillering and the production of flowers.

Behind these general considerations, which must be considered in accounting for the rapid spread of S.anglica, there remains a basic issue concerning photosynthesis in this species. It is that although the significance of the C_4 mechanism, as a part of the totality of features which enable the plant to thrive, may be relatively small, it is nonetheless distinct from other C_4 species in being able to proceed at significant rates at low temperatures. It is of interest, therefore, to know how this is achieved. Reference was made earlier (Section 3.2) to studies suggesting that the phospholipid composition of the thylakoids in relation to liquid/gel phase transitions in the membranes may be an important factor in the adaptation of the photosynthetic apparatus to low temperatures. Nothing is known concerning this aspect of the photosynthetic apparatus in Spartina. It was also noted (Section 3.3) that some enzymes of the carbon assimilation pathway in C_4 plants may be more sensitive to low temperatures; the NADP-malate dehydrogenase and the pyruvate, phosphate dikinase being particularly fragile in this respect.

Table 2 shows the optimised activities of enzymes of the C_4 and PCR cycles in extracts of leaves of S.anglica together with measurements of the chlorophyll and protein content and rate of photosynthesis. From these data it is clear that S.anglica possesses many of the enzymes necessary to catalyse CO_2 assimilation in C_4 species at levels of activity comparable to those found in tropical species. Also, it is a PEP-CK species; that is to say, it decarboxylates oxalo-acetic acid in the bundle sheath cells to form phosphoenol-pyruvate (PEP) and CO_2. It will be noted, however, that in Table 2 results are not given for NAD-malic enzyme or for pyruvate, phosphate dikinase which is usually regarded as a key enzyme of the C_4 cycle. These results are omitted because of great technical difficulty which we encountered leading to variability in the assays. Negative results are nowhere more suspect than in the case of labile enzymes; it is noteworthy, however, that despite elaborate cross-checks for inactivation using mixed extracts from leaves of maize and Spartina, we were never able to obtain levels of activity of the dikinase adequate to support more than 15% of the observed rate of photosynthesis. When this result is considered alongside the virtual absence of NADP-malic dehydrogenase in leaves of Spartina (Table 2), it tempts the hypothesis that there might be a short-circuiting of these enzymes in this species consequent on a direct transfer of the PEP generated in the decarboxylation reaction in the bundle sheath, back into the mesophyll [Fig.9].

Table 2. Activities of enzymes of the C_4 cycle and the reductive pentose phosphate pathway and rate of photosynthesis of leaves of <u>Spartina anglica</u>. Values are means ± S.E. of estimates from six different leaves unless shown otherwise in parentheses.

Enzyme	Activity $\mu mol\ min^{-1}\ g^{-1}$ fresh weight*
Phosphoenolpyruvate carboxylase	25.1 ± 3.8
Phosphoenolpyruvate carboxykinase	8.8 ± 0.9 (7)
NADP-malic enzyme	0.1 ± 0.02(7)
Pyruvate kinase	0.9 ± 0.08
NAD-malate dehydrogenase	80.9 ± 5.4
NADP-malate dehydrogenase	0.5 ± 0.07
Aspartate aminotransferase	33.2 ± 2.1
Alanine aminotransferase	17.6 ± 1.5
Ribulose-1 5-bisphosphate carboxylase	6.3 ± 0.7 (7)
NADP-gyceraldehyde-3-phosphate dehydrogenase	14.6 ± 2.1
Ribulose-5-phosphate kinase	26.2 ± 2.1
Rate of photosynthesis [Net CO_2] incorporation	3.8 ± 0.1 (43)

*Chlorophyll content [mg g^{-1} fresh weight] 1.06 ±0.07 (8)
Protein content   [mg g^{-1} fresh weight] 7.24 ±0.98 (7)

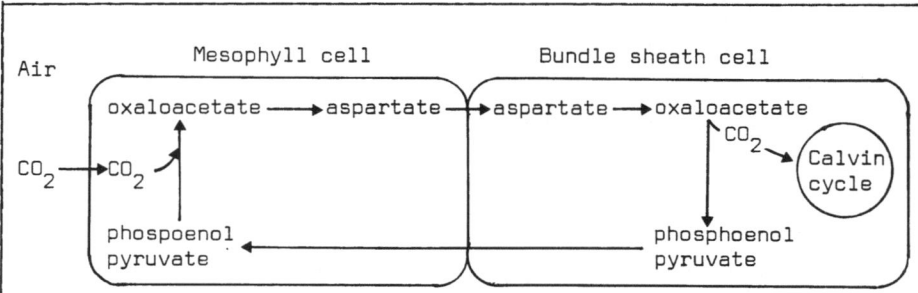

Figure 9.Diagram to illustrate an alternative pathway to the regeneration of PEP as a substrate for CO_2 fixation in the mesophyll cells of C_4 species of the PEP carboxinase type, without the involvement of the pyruvate-phosphate dikinase enzyme system (compare Fig.7).

This is an unresolved problem requiring further work, but it is of interest to note that this question of the pathway of regeneration of PEP in the C_4 cycle in PEP-CK plants is also problematical in other species. It has been suggested [58,59] that PEP generated by decarboxylation of OAA is converted to pyruvate in the bundle sheath either via pyruvate kinase or a PEP phosphatase. Pyruvate or alanine then moves to the mesophyll where pyruvate is converted to PEP by the pyruvate, phosphate dikinase. It has also been suggested however that PEP may move directly from bundle sheath to mesophyll without conversion to pyruvate [60] (Fig.9). Our results to date do not permit a resolution of the problem but it is evident that this presents an intriguing possibility for adaptation by reducing the dependence of photosynthesis on those enzymes of the C_4 cycle which appear to be most cold labile.

5. EVOLUTION OF ADAPTATIONS TO STRESS IN THE PHOTOSYNTHETIC APPARATUS

It will be evident from the span of this review that consideration of the evolution of adaptions to stress is a subject which could occupy whole symposia. For present purposes, we shall confine ourselves to C_4 photosynthesis as an adaptation and consider its possible origins. We have considered already (Section 4.3.1) the types of habitat and the selection pressures which seem to have been associated with the emergence of the C_4 mechanism; we now face the question of its antecedents. The following essential features of the mechanism would appear to be pertinent to this question. It is polyphyletic and represents a derived condition, C_4 species having arisen from C_3 progenitors. There is as yet no known example in which a C_3 species is suggested to have arisen from a C_4 progenitor. The C_4 mechanism is generally associated with Kranz anatomy and the possession of PEP carboxylase as the primary carboxylating enzyme. In the further details of these attributes however, as for example the presence or otherwise of an extra ring of cells between bundle sheath and mesophyll, the degree of chloroplast dimorphism and the pattern of enzymes associated with the CO_2 pump, there is considerable variation. It is also becoming increasingly evident that the spectrum of enzymes associated with C_4 photosynthesis in C_4 species is not unique to this group; all appear to be present in species with C_3 and Crassulacean acid metabolism-type photosynthesis. What is different between these groups are the relative amounts, tissue distribution and regulatory properties of these enzymes.

The suggestion has been made that many of the enzymes of the C_4 mechanism are intrinsic to the stomatal apparatus and that the CAM and C_4 systems represent varying patterns of redeployment of these enzymes [61]. It is also worthy of note in this connection that there is an increasing appreciation of the physiological and biochemical complexity of stomata. Physiological variation is found in response light, CO_2 concentration, temperature, vapour pressure deficits and hormone levels [62,63]. Biochemical diversity is seen in the finding that all guard cells appear to pump K^+ ions in changing turgor but vary in the extent to which malate or chloride forms the counter ion. Recent evidence suggests that guard cells also vary in whether or not they contain Rubisco [64]. If this finding is confirmed, then we may have, in effect, a progenitor for the differential expression of a part of the PCR cycle in one set of chloroplasts; in this case those of the guard cells relative to the rest of the leaf - analgous to the difference in expression of this part of the cycle between the chloroplasts of the mesophyll and bundle sheath in C_4 species. Seen in this light, the vast array of changes involved in the emergence of the C_4 syndrome may come to look somewhat less formidable. It may be that the initial key events involved changes in a relatively few major genes controlling the site of expression of this spectrum of enzymes as between guard cells, mesophyll and bundle sheath. Once this has occurred, there could arise modifications in the amounts and the regulatory properties of the enzymes towards an optimisation of metabolism in the cells in which they are now being expressed.

Rather than indulge in further speculation on possible genetic mechanisms involved in the evolution of the C_4 syndrome, we shall conclude by considering the possibilities for a further experimental analysis of the problem. The presence of species showing characteristics which are apparently intermediate between C_3 and C_4 has been reported from such genera as Panicum, Mollugo, Moricandia and Neurachne [65,66,67,68] and it is an attractive proposition to look at these more closely. It must be said, however, that at the present time there is a lack of definitive evidence in any of the reported cases to sustain the contention that the species in question are genuinely intermediate in character. The case of Alloteropsis alata, in which two ecotypes, one C_3, the other C_4, were reported [69] offered the promise of a system which might be amenable to genetic analysis. Recent work (Marks, unpublished) has shown however, that the C_3 "ecotype" is diploid, whilst the C_4 is hexaploid and must presumably now be regarded as a distinct species. Faced with the enormous complexity of the C_4 syndrome, it may be felt by some that the pursuit of the genetic analysis of adaptive changes in the photosynthetic apparatus would be better conducted for the present using simpler systems such as possible changes in one or two thylakoid lipids. As an exercise in "the art of the possible" this would have much to commend it, though the more complex challenge will remain to nag the curious.

REFERENCES

1. Tranquillini W. 1964. The physiology of plants at high altitudes. Ann. Rev. Plant Physiol. 15, 345-62.
2. Long S.J., Incoll L.D. and Woolhouse H.W. 1975. C_4 photosynthesis in plants from cool temperate regions with particular reference to Spartina townsendii. Nature 257, 622-4.
3. Scott D. 1970. CO_2 exchange in plants III. Temperature acclimatisation of three species. N.Z. Journal Bot.8, 369-379.
4. Taylor A.O. and Rowley J.A. 1971. Plants under climatic stress I. Low temperatures, high light effects on photosynthesis. Plant Physiol. 47, 713-18.
5. Berry J.A. and Bjorkman O. 1980. Photosynthetic response and adaptation to temperature in higher plants. Ann. Rev. Plant Physiol. 31, 491-543.
6. Bjorkman O., Troughton J. and Berry J.A. 1976. Comparison of the heat stability of photosynthesis, chloroplast membrane reactions, photosynthetic enzymes and soluble protein in leaves of heat-adapted and cold-adapted C_4 species. Carnegie Inst. Washington Yearbook. 75, 400-7.
7. Pearcy R.W., Berry J.A. and Fork D.A. 1977. Effect of growth temperature on the thermal stability of the photosynthetic apparatus of Atriplex lentiformis (Torr) Wats. Plant Physiol. 59, 873-78.
8. Redshaw A.J. and Meidner H. 1972. Effects of water stress on resistance to the uptake of carbon dioxide in tobacco. J.Exp.Bot. 23, 229-40.
9. Cowan I.R. 1977. Stomatal behaviour and environment. Adv. Bot. Res., 4, 117-228.
10. Lawlor D.W. 1976. Water stress induced changes in photosynthesis, photorespiration, respiration and CO_2 compensation concentration of wheat. Photosynthetica 10, 378-87.
11. Lawlor D.W. and Hock H. 1975. Photosynthesis and photorespiratory CO_2 evolution of water stressed sunflower leaves. Planta. 126, 247-238.
12. Plaut Z. 1971. Inhibition of photosynthetic carbon dioxide fixation in isolated spinach chloroplasts exposed to reduced osmotic potentials. Plant Physiol. 48, 591-95.
13. Kaiser W.M. 1982. Correlation between changes in photosynthetic activity and changes in total protoplast volume in leaf tissue from hygro-, meso- and xerophytes under osmotic stress. Planta 154, 538-45.
14. Warburg O. 1920. Uber die Geschwindigbeit der photochemischen Kohlensauezersetung in lebenden Zellen.II. Biochem.Zeit. 100, 188-217.

15. Ogren W.L. and Bowes, G. 1971. Ribulose diphosphate carboxylase regulates soybean photorespiration. Nature New Biol. 230, 159-60.
16. Bowes G., Ogren, W.L. and Hageman R.H. 1971. Phosphoglycolate production catalysed by ribulose diphosphate carboxylase. Biochem. Biophys. Res. Comm. 45, 716-22.
17. Bowes G. and Ogren W.L. 1972. Oxygen inhibition and other properties of soybean ribulose, 1,5-diphosphate carboxylase. J. Biol. Chem. 247, 2171-76.
18. Goldsworthy A. 1970. Photorespiration. Bot. Rev. 36, 321-340.
19. Jackson W.A. and Volk R.J. 1970. Photorespiration. Ann. Rev. Plant.Physiol. 21, 385-432.
20. Lorimer G.H. and Andrews T.J. 1980. The C-2 photo- and photo-respiratory carbon oxidation cycle. In: The Biochemistry of Plants, ed. M.D. Hatch, N.K. Boardman, 8, 329-74. New York Acad. Press.
21. Buchanan B.B. 1980. Role of light in the regulation of chloroplast enzymes. Ann. Rev. Plant Physiol. 31, 341-74.
22. Farquhar G.D. and Sharkey T.D. 1982. Stomatal conductance and photosynthesis. Ann. Rev. Plant Physiol. 33, 317-45.
23. von Caemmerer and Farquhar G.D. 1981. Some relationships between the biochemistry of photosynthesis and the gas exchange of leaves. Planta, 153, 376-87.
24. Drake B. and Raschke K. 1974. Prechilling of Xanthium strumarium L. reduces net photosynthesis and independently of stomatal conductance while sensitising stomata to CO_2. Plant Physiol. 53, 808-12.
25. Linden C.D., Wright K.L., McConnell H.M. and Fox C.F. 1973. Lateral phase separations in membrane lipids and the mechanism of sugar transport in Escherichia coli. Proc. Natl. Acad. Sci. SA 70, 2271-75.
26. Raison J.K. 1980. Membrane lipids - structure and function. Biochemistry of Plants 7, ed. P.K. Stumpf. New York Acad. Press.
27. Nobel P.S. 1974. Temperature dependence of the permeability of chloroplasts from chilling-sensitive and chilling-resistant plants. Planta. 115, 369-72.
28. Murata N., Troughton J.H. and Fork D.C. 1975. Relationships between the transition of the physical phase of membrane lipids and photosynthetic parameters in Anacystis nidulans and lettuce and spinach chloroplasts. Plant Physiol. 56, 508-17.
29. Hackenbrock C.R. 1981. Lateral diffusion and electron transfer in the mitochondrial inner membrane. Trends Biochem. Sci. 4, 151-154.
30. Schreiber N. and Berry J.A. 1977. Heat induced changes of chlorophyll fluorescence in intact leaves, correlated with damage of the photosynthetic apparatus. Planta, 136, 233-38.
31. Broyer J.S. and Bowen B.L. 1970. Inhibition of oxygen evolution in chloroplasts isolated from leaves with low water potentials. Plant Physiol. 43, 612-15.
32. Jones H.G. 1973. Limiting factors in photosynthesis. New Phytol. 72, 1089-94.
33. Moharty P. and Broyer J.S. 1976. Chloroplast response to low leaf water potentials. Quantum yield is reduced. Plant Physiol. 57, 704-9.
34. Keck R.W. and Broyer J.S. 1974. Chloroplast response to low leaf water potentials III. Differing inhibition of electron transport and photophosphorylation. Plant Physiol. 53, 474-79.

35. Shirahaski K., Hayakawa S. and Sugiyama T. 1978. Cold lability of pyruvateorthophosphate dikinase in the maize leaf. Plant Physiol. 62, 826-830.

36. Sugiyama T., Schmitt M.K., Ku S.B. and Edwards G.E. 1979. Differences in cold lability of pyruvate, Pi dikinase among C_4 species. Plant Cell Physiol. 20, 965-71.

37. O'Tool J.C., Crookston R.K., Treharne K.J. and Ozbun J.L. (1976) Mesophyll resistance and carboxylase activity. A comparison under water stress conditions. Plant Physiol 57, 465-68.

38. Lewontin R.C. 1977. Adaptation. In: The Encyclopaedia Einaudi, Torino Giulio Einaudi Edition.

39. Woolhouse H.W. 1981. Aspects of the carbon and energy requirements of photosynthesis considered in relation to environmental constraints. In: Physiological Ecology an Evolutionary Approach. ed. P. Calow and C.R. Townsend. Pub. Sinauer Assos. U.S.A. 51-85.

40. Slack C.R., Roughan P.G. and Basset H.C.M. 1974. Selective inhibition of mesophyll chloroplast development in some C_4 pathway species by low night temperature. Planta 118, 57-73.

41. Seemann J.R., Downton W.J.S. and Berry J.A. 1980. Field studies of acclimation to high temperatures: Winter ephemerals in Death Valley. Carnegie Institute Washington Yearbook. 78, 157-62.

42. Lloyd N.D.H. and Woolhouse H.W. 1978. Leaf resistances in different populations of Sesleria caerulea (L.) Ard.

43. Teeri J.A. and Stowe L.G. 1976. Climatic patterns and the distribution of C_4 grasses in North America. Oecologia 23, 1-12.

44. Stowe L.G. and Teeri J.A. 1978. The geographic distribution of C_4 species of the Dicotyledons in relation to climate. Amer. Natur., 112, 609-23.

45. Caldwell M.M., White R.S., Moore R.T. and Camp L.B. 1977. Carbon balance, productivity and water use of cold desert shrub communities dominated by C_3 and C_4 species. Oecologia 29, 275-300.

46. Winter. 1981. C_4 plants of high biomass in arid regions of Asia – occurrence of C_4 photosynthesis in Chenopodiaceae and Polygonaceae from the Middle East and USSR. Decologia 48, 100-106.

47. Ehleringer J.R. 1978. Implications of quantum yield differences on the distribution of C_3 and C_4 grasses. Oecologia 31, 255-67.

48. Winter K., Schmitt M.R., and Edwards, G.E. 1982. Microstegia vimineum, a shade adapted C_4 grass. Plant. Sci. Lett. 24, 311-18.

49. Robichaux R.H. and Pearcy R.W. 1980. Photosynthetic responses of C_3 and C_4 species from cool shaded habitats in Hawaii. Oecologia. 47, 106-109.

50. Ward D. and Woolhouse H.W. Unpublished.

51. Jones M.B., Hannon G.E. and Coffey M.D. 1982. C_4 photosynthesis in Cyperus longus L., a species occurring in temperate climates. Plant Cell and Environment. 4, 161-68.

52. Long S.P. and Woolhouse H.W. (1978) The responses of net photosynthesis to vapour pressure deficit and CO_2 concentration in Spartina x townsendii (Sensuo lato) J.Exp. Bot. 29, 567-77.

53. Long S.P. and Woolhouse H.W. (1978). The responses of net photosynthesis to light and temperature in Spartina x townsendii (Sensuo lato) J. Exp. Bot. 29, 803-14.

54. Long S.P. and Woolhouse H.W. (1979). Primary production of Spartina marshes. Proceedings of the 1st European Ecological Symposium. Ed. R.L.Jefferies and A.J.Davey. Blackwell Scientific Publications, Oxford. p.p. 333-352.

55.   Long S.P. and Incoll L.D. 1979. The prediction and measurement of photosynthetic rate of _Spartina_ x _townsendii_ (Sensuo lato) in the field. J. Appl. Ecol. 16, 879-91.

56.   Woolhouse H.W. 1980. Possibilities for the modification of the pattern of photosynthetic assimilation of CO_2 in relation to growth and yield of crops. Proc. 15th Int. Potash. Inst. Colloquium. Wageningen 1.

57.   Ranwell D.S. 1972. Ecology of Salt Marshes and Sand Dunes. 258pp. Chapman and Hall, London.

58.   Hatch M.D. 1979. Mechanism of C_4 photosynthesis in _Chloris gayana_: pool sizes and kinetics of $^{14}CO_2$ incorporation into 4-carbon and 3-carbon intermediates. Arch. Biochem. Biophys. 194, 117-127.

59.   Hatch M.D. and Kagawa T. 1976. Photosynthetic activities of isolated bundle sheath cells in relation to differing mechanisms of C_4 pathway photosynthesis. Arch. Biochem. Biophys. 175, 39-53.

60.   Rathnam C.K.M. and Edwards G.E. 1977. C_4 dicarboxylic acid decarboxylation in bundle sheath chloroplasts, mitochondria and strands of _Eriochloa borumensis_ Hack., a phosphoenolpyruvate-carboxykinase type C_4 species. Planta 133, 135-44.

61.   Cockburn W. 1981. The evolutionary relationship between stomatal mechanism, crassulacean acid metabolism and C_4 photosynthesis: Opinion. Plant, Cell and Environment. 4, 417-418.

62.   Rashke K. 1975. Stomatal action. Ann. Rev. Plant Physiol. 26, 309-340.

63.   Jewer P.C. and Incoll L.D. 1981. Promotion of stomatal opening in detached epidermis of _Kalanchoe daigremontiana_. Hornet et Perr. by natural and synthetic cytokinins. Planta. 153, 317-318.

64.   Soundararajan M. and Smith B.N. 1982. Localisation of ribulose biphosphate carboxylase in the guard cells by an indirect immunofluorescence technique. Plant Physiol. 69, 273-277.

65.   Rathnam C.K.M. and Chollet R. 1979. Photosynthetic carbon metabolism in _Panicum milioides_, a C_3-C_4 intermediate species. Evidence for a limited C_4 dicarboxylic acid pathway of photosynthesis. Biochem. Biophys. Acta. 548, 500-519.

66.   Sayre R.T. and Kennedy R.A. 1977. Ecotypic differences in the C_3 and C_4 photosynthetic activity in _Mollugo verticillata_, a C_3-C_4 intermediate. Planta. 134, 257-262.

67.   Apel P. 1980. CO_2 compensation concentration and its O_2 dependence in _Moricandia spinosa_ and _Moricandia moricandioides_ (Cruciferae) Biochem. Physiol. Pflazen. 175, 386-388.

68.   Hattersley P.W., Watson L. and Johnston C.R.   Remarkable leaf anatomical variations in _Neurachne_ and its allies (Poaceae) in relation to C_3 and C_4 photosynthesis.

69.   Ellis R.P. 1974. The significance of the occurrence of both Kranz and non-Kranz leaf anatomy in the grass species _Alloteropsis semialata_. S. Afr. J. Sci. 70, 169-173.

#

MOLECULAR ASPECTS OF PHOTOSYNTHESIS AT LOW LEAF WATER POTENTIALS

J. S. Boyer and H.M. Younis*
USDA/ARS, Department of Botany and Department of Agronomy, 289 Morrill Hall, University of Illinois, 505 S. Goodwin Avenue, Urbana, Illinois 61801, U.S.A.

ABSTRACT

Chloroplasts lose activity when leaf photosynthesis is inhibited at low water potentials. The losses often limit the rate of photosynthesis even though the stomata close at the same time. The chloroplasts show no structural degradation but the thylakoid lamellae are thinner than in leaves having high water potentials. The activity of chloroplast coupling factor, a subunit of these membranes, is inhibited when the protein is prepared from leaves having low water potentials, and the inhibition is associated with altered conformation of the protein and decreased binding affinity for ADP. The changes in photophosphorylating activity and in coupling factor activity can be simulated by exposure of the chloroplasts or protein to high Mg^{2+} concentrations (above 5 mM) prior to assay. Since Mg^{2+} concentrations are typically 1 to 3 mM in chloroplast stroma, ion concentrations above 5 mM could readily occur during dehydration of leaves and may be involved in the activity losses by the chloroplasts.

INTRODUCTION

Leaves generally lose photosynthetic activity as leaf water potential decreases (Boyer, 1976). The losses are attributable to stomatal closure and decreased activity of the chloroplasts (Boyer, 1976). The closure of the stomata occurs because of the failure of guard cells to retain solutes necessary for generating the osmotic potential and turgor required for stomatal opening (Ehret and Boyer, 1979). The chloroplasts lose activity particularly in membrane-associated reactions of the thylakoids. Photosystem II and photophosphorylation are often severely inhibited (Keck and Boyer, 1974) whereas

*Present Address: Department of Plant Protection, Faculty of Agriculture, and Biochemistry and Molecular Biology, University Science Center, University of Alexandria, Egypt.

Photosystem I and enzyme activity from the Calvin Cycle are somewhat less affected (Boyer, 1976).

The losses in chloroplast activity are not associated with chloroplast degradation or loss in structural integrity when viewed with the electron microscope (Fellows and Boyer, 1976). Rather, the thylakoid membranes undergo changes in conformation, appearing thinner than the controls when situated in leaves having low water potentials. These alterations in conformation can be detected after the chloroplasts are isolated and are correlated with the losses in activity that are also detectable upon isolation. When a protein subunit of the thylakoid membrane, coupling factor (ATP synthetase when situated in the membrane), is extracted from chloroplasts, most of the major biochemical characteristics (molecular weight, number of subunits, mobility on gels) are unaltered by low leaf water potentials (Younis et al., 1979). However, the conformation of the protein is altered and is associated with a loss in binding affinity of ADP to the protein (Younis et al., 1979). It consequently seems likely that the altered conformation of the thylakoid membranes reflects conformational changes in the membrane components and these, in turn, manifest themselves as altered activity of partial reactions of photosynthesis in the membranes.

The mechanism causing the loss in activity of photophosphorylation is not well understood. We have attempted to simulate the effects of low water potentials by altering ion concentrations known to affect photosynthetic partial reactions that are sensitive to low water potentials. Mg^{2+} is an important regulator for several of these reactions, and we therefore used Mg^{2+} in the simulation.

MATERIALS AND METHODS

The experiments were conducted with chloroplasts isolated from commercial spinach (*Spinacea oleracea* L.) purchased from a local market. The isolation of the chloroplasts and coupling factor, the dehydration of the leaves, and the measurement of leaf water potential were carried out as described by Younis et al. (1979).

Magnesium was supplied to isolated chloroplasts or coupling factor in a preincubation medium containing buffer (Younis et al., 1979) for 30 min, then the chloroplasts or coupling factor were transferred to assay medium for cyclic photophosphorylation or Ca^{2+}-ATPase as described (Younis et al., 1979). The assay medium diluted the preincubation concentrations of Mg^{2+} to 1% of the preincubation levels.

RESULTS

Table 1 shows that the activity of chloroplasts for cyclic photophosphorylation decreased when the chloroplasts were preincubated in Mg^{2+} concentrations as high as 10 mM. The activity of Ca^{2+}-ATPase of coupling factor was also decreased after preincubation of the protein in Mg^{2+} concentrations as high as 10 mM (Table 1). The effect was larger if the exposure to Mg^{2+} occurred after heat activation of the protein but was also observed prior to heat activation.

Table 1. Magnesium concentration during preincubation, and phosphorylating activity of isolated chloroplasts and Ca^{2+}-ATPase activity of chloroplast coupling factor during subsequent assay.*

Mg^{2+} Concentration During Preincubation	Photophosphorylating Activity	Ca^{2+}-ATPase Activity	
		Before Heat Activation	After Heat Activation
mM	μmole·h^{-1}·(mg chl)$^{-1}$	μmole·h^{-1}·(mg chl)$^{-1}$	
0	1500	21	21
5	1050	23	7
10	750	14	5

*Preincubation was carried out in the presence of buffer, and a small aliquot of the preparation was transferred to the assay medium after 30 min at room temperature. The assay medium diluted the Mg^{2+} concentration to 1% of the preincubation concentration. Details of methods are given in Younis et al. (1979).

Table 2 shows that photophosphorylation was inhibited when chloroplasts were isolated from leaves having low water potentials. The expected stromal Mg^{2+} concentrations were calculated from the water content of the tissue prior to chloroplast isolation (Table 2) and the stromal concentrations of Mg^{2+} (assumed to be 3 mM (Portis and Heldt, 1976; Portis, 1981)) prior to leaf dehydration. The data show that water loss by leaf tissue at low water potentials is sufficient to cause expected stromal Mg^{2+} concentrations as high as 9 mM. This is similar to the concentrations of Mg^{2+} used in the Mg^{2+} preincubation experiment (Table 1).

Table 2. Phosphorylating activity of isolated chloroplasts, approximate water content of leaves, and calculated stromal Mg^{2+} concentrations at various leaf water potentials*.

Water Potential	Photophosphorylating Activity	Approximate Water Content	Calculated Stromal Concentration of Mg^{2+}
bar	$\mu mole \cdot h^{-1} \cdot (mg\ chl)^{-1}$	% of turgid	mM
-2	1060	100	3
-15	740	55	6
-25	475	35	9

*Chloroplasts were isolated from spinach leaves that had been dehydrated to levels shown. Assays were conducted as in Younis et al. (1979). Stromal Mg^{2+} concentrations were calculated from water content of tissue assuming 3mM Mg^{2+} in chloroplast stroma of turgid leaves and no changes in compartmentation of the Mg^{2+}.

DISCUSSION

This experiment was designed to simulate low water potentials in leaves by using high Mg^{2+} concentrations in preincubation media. Thus, the chloroplasts were pretreated by dehydrating the leaves before isolating and assaying for photophosphorylating activity and, similarly, the chloroplasts were pretreated at high Mg^{2+} concentrations before assaying for phosphorylating activity. Both experiments required the chloroplasts to remain in the altered state after the pretreatment conditions had ended in order for an effect to be detected in the uniform assay media. The parallel simulation with coupling factor allowed the effects of the simulation to be tested on a subunit of the thylakoid membranes of the chloroplasts.

The similarity in Mg^{2+} concentrations (to 10mM) necessary to inhibit photophosphorylation and likely to be present in chloroplast stroma of dehydrated leaves suggests that Mg^{2+} could be a factor causing the inhibition of photophosphorylation at low water potentials. The high Mg^{2+} and low water potential caused persistent changes in chloroplast and coupling factor activities of similar magnitude for both pretreatments, which further suggests that high magnesium concentrations may have altered photophosphorylation *in vivo* as water potentials decreased (see Younis et al., 1979 for coupling factor activities at low water potentials).

However, although the Mg^{2+} simulation is consistent with the alteration in photosynthesis that is observed at low water potential, other ions could also be involved. Thus, the results reported here may reflect a general chloroplast response to high ion concentrations rather than a response specific for Mg^{2+}.

REFERENCES

1. Boyer JS. 1976. Water deficits and photosynthesis. In: Water Deficits and Plant Growth, vol. IV, T.T. Kozlowski, ed., Academic Press, Inc., New York. Pp. 153-190.
2. Ehret DL, Boyer JS. 1979. Potassium loss from stomatal guard cells at low water potentials. J. Exp. Bot. 30:225-234.
3. Fellows RJ, Boyer JS. 1976. Structure and activity of chloroplasts of sunflower leaves having various water potentials. Planta 132:229-239.
4. Keck RW, Boyer JS. 1974. Chloroplast response to low leaf water potentials. III. Differing inhibition of electron transport and photophosphorylation. Plant Physiol. 53:474-479.
5. Portis AR Jr. 1981. Evidence of a low stromal Mg^{2+} concentration in intact chloroplasts in the dark. Plant Physiol. 67:985-989.
6. Portis AR Jr, Heldt HW. 1976. Light-dependent changes of the Mg^{2+} concentration in the stroma in relation to the Mg^{2+} dependency of CO_2 fixation in intact chloroplasts. Biochim. Biophys. Acta 449:434-446.
7. Younis HM, Boyer JS, Govindjee. 1979. Conformation and activity of chloroplast coupling factor exposed to low chemical potential of water in cells. Biochim. Biophys. Acta 548:328-340.

INTEGRATION OF BIOCHEMICAL PROCESSES IN THE PHYSIOLOGY OF WATER STRESSED PLANTS

D.W. LAWLOR
Botany Department, Rothamsted Experimental Station, Harpenden, Herts. AL5 2JQ. U.K.

ABSTRACT

Crop productivity under water stress is determined by the way in which biochemical processes are integrated and change in relation to the environment. In the field, productivity is determined by the responses of the plant at the morphological or physiological process level of organisation e.g. leaf area or net photosynthesis. Their relative importance in determining crop growth is briefly considered. The manner in which stress affects photosynthesis, photorespiration and energy utilisation in the leaf and cell and the basic lesions to metabolism are outlined. ATP synthesis by photophosphorylation is inhibited whilst electron transport and reductant supply are relatively in-sensitive. Inbalance in these processes causes changes in secondary carbon and nitrogen metabolism involving amino acid and quaternary ammonium compounds and phospholipid synthesis; terpenoid metabolism may also be linked. Aspects of these in relation to carbon and nitrogen fluxes and cellular ener-getics are considered together with possible controls. Stress induced changes in metabolism of glycophytes are discussed in relation to the concepts of directed metabolism with useful biological control function, or to undirect-ed metabolism, resulting from stress, without important control function.

INTRODUCTION

Plant response to water deficits is partly determined by the degree and timing of water stress experienced in relation to growth stage, temperature, irradiance etc. [7]. Most important in determining growth and yield are the biochemical and physiological mechanisms of the plant and how these respond quantitatively to the deviations in environment from an optimum, as implied by the concept of stress. Practically, the aim is to improve crop production under stress by modifying cultural practices or otherwise manipulating the biol-ogical mechanism or by breeding, selection etc. to establish the advantageous features in the genome [11].

Uncertainty about the necessary modifications results from inadequate understanding of how the biochemical mechanisms (which together interact to produce the physiological responses at a higher level of organisation) are affected by water stress (4). A qualitative picture of the biochemical interactions responsible for growth and assimilation is emerging from the numerous studies of plant responses (4,10,13). Here I consider what processes are affected by water stress, their relative response to degree of stress, how they are related (to see if a common mechanism can be advanced for stress effects) and to assess, semi-quantitatively, the importance to growth of water stress in teracting with light and nutrient supply.

It is a truism that the plant is a highly integrated system, dominated in terms of energy capture and utilisation by carbon and nitrogen assimilation. In C_3 plants, changes in the fluxes of carbon, nitrogen and energy involve the photosynthetic carbon reduction cycle and glycolate pathway with RuBP carboxylase/oxygenase enzyme characteristics and the ammonium cycle linking them (8). Energy consumption is related to the relative size of the fluxes of material through both systems. Water stress decreases CO_2 assimilation by stomatal closure under mild to severe stress and this affects the utilisation of energy captured by the photosynthetic machinery of the chloroplast thylakoids. More severe stress damages biosynthetic systems and causes serious inbalance in metabolism which decreases the efficiency with which CO_2, light etc. are used and reduces the biological stability of the plant and decreases productivity of ecological and agricultural systems. Intermediates of the glycolate pathway appear to be precursors of compounds which accumulate in stressed leaves (15, 16).

In multistep systems, with interaction between pathways, common intermediates and rigid coupling of processes by adenylates and pyridine nucleotides (1), modification of one process may have non-linear effects in other parts of the system or on output. Alternate routes with different control mechanisms may serve to stabilise the system or maintain essential processes when conditions are suboptimal. Closer to the optimum, subtle interactions between processes, difficult to detect over short periods, determine the efficiency of crop production over time. One of the most challenging tasks of stress physiology is to explain crop yield in terms of metabolism over the spectrum of stresses.

In considering these individual processes and their integration under stress conditions the question of 'value' to the plant is always present (9,10,

11).   The biochemical process   or modification is often viewed as having a
direct advantage or disadvantage which can be exploited in breeding etc.   Such
discussions often stray into subjectivity and teleological argument, which
whilst stimulating understanding and further study may also cause misuse of resources because the concepts are based on inadequate understanding of the integrated metabolism.   Ultimately, the value of a process or its modification
can only be assessed in terms of the biological or economic productivity of the
plant, its long-term stability in a given environment, and the reproductive
success of the plant and the efficiency with which the change enables environmental inputs to be exploited.

DETERMINANTS OF CROP PRODUCTION
    Dry matter yield is determined by gross photosynthesis and the rates of
photo- and dark respiration per unit leaf area, total leaf area, and the losses
due to respiration.   Economic yield depends on the proportion of total assimilate diverted to the required organ.   Water stress affects all of these
processes or components, the degree depending on plant and conditions.   As yet
there is little quantitative analysis of the importance of individual processes in water-stressed crops under a range of conditions.   Cell expansion decreases with slight stress and determines leaf area upon which light absorption and total yield depend.   Leaf senescence is greatly increased by water
stress in all crops, however the causes are not understood and there is little
work specifically directed to water stress.   Stomatal conductance in the field
may decrease yield little under mild stress, possibly because prolonged
stomatal closure could impair metabolism.   Crop respiration linked to growth
and maintenance may affect yield differences between stress treatments, however little is known of its control.   Stress acting via gross photosynthesis,
photo- and dark respiration causes only a small decrease in assimilation rate
per unit of leaf area in many crops.   An experiment at Rothamsted on a barley
crop subject to water stress (17) which halved dry matter yield suggests that
40% of the decrease was caused by smaller leaf area duration, and less than 10%
by stomatal resistance with small effects of assimilation rate.   However,
quantitative estimates are too few to establish the major limitations in such
highly integrated physiological systems in the field.

PHOTOSYNTHETIC METABOLISM AND ITS CONTROL BY STRESS

Assimilation. Photosynthesis is central to the growth and productivity of
plants, the basic process of light reactions, electron transport and synthesis
of NADPH and ATP, the photosynthetic carbon reduction cycle and photorespiration are now well understood as are the links to nitrogen metabolism leading
to the photorespiratory ammonia cycle (8). As assimilation is greatly decreased by stress under controlled conditions the causes, or lesions in the
assimilation pathways have been analysed. The relation of carbon and nitrogen
assimilation to synthesis of amino acids, terpenoids, quaternary ammonium
compounds and to phospholipids although apparent have yet to be quantitatively evaluated; this will be considered here. These products influence growth during stress and recovery e.g. osmotic adjustment by solutes or production
of hormones such as abscisic acid.

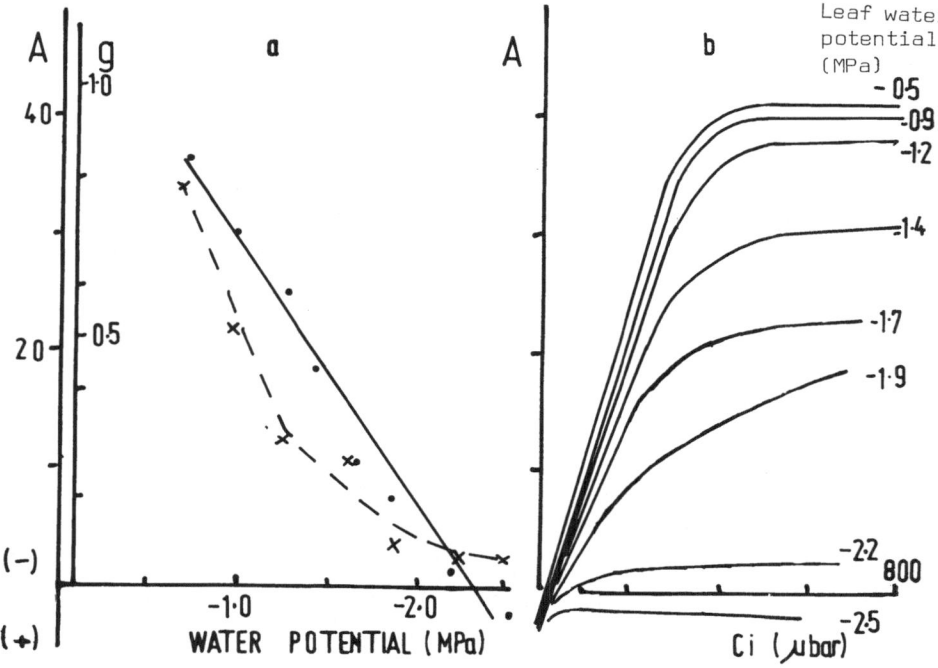

Figure 1a). Assimilation of CO_2 (A, μmol m^{-2}s^{-1}) and leaf conductance (g,
mol m^{-2} s^{-1}) related to leaf water potential at 330 μbar external CO_2;
b)A related to internal CO_2: both at 210 mbar O_2 and 1200 μmol quanta m^{-2} s^{-1}.

Carbon assimilation decreases with stress (Fig.1) as water potential decreases to severe wilting over a period of five days.   Water potential of O to -1.0 MPa may be regarded as 'mild' stress, -1.0 to -1.6 MPa as medium stress and below -1.6 MPa as very severe stress; the terms are approximate and depend on plant and growth conditions.   Most important is the progressive loss of turgor during the increase in stress, culminating in zero turgor and plasmolysis at severe stress.   Stomatal closure (conductance decreased from 0.8 to 0.1 mol m^{-2} s^{-1}) contributes to the decrease and is confounded with smaller turgor. Similar effects have been frequently observed (4,6,8) for rapidly stressed plants grown in osmotica, small soil volumes etc.   To distinguish the stomatal control from metabolic effects on assimilation the internal CO_2 concentration (Ci) is calculated from $Ci = Ca - \frac{A}{g}$ where A is assimilation rate (μmol m^{-2} s^{-1}), g is the conductance (mol m^{-2} s^{-1}) and Ca is the ambient CO_2 concentration (μbar); A is related to Ci at saturating light.   The slope, A/Ci, expresses the dependence of metabolism when CO_2 is limiting and is governed by the affinity of the enzyme RuBP carboxylase/oxygenase for CO_2 and O_2 (see 8,9). The plateau is a region where A is limited by regeneration of RuBP, this depends on ATP and NAD(P)H synthesis and on the capacity of the carbon reduction cycle enzymes.   RuBP carboxylase may also limit if the amount and characteristics of the enzyme are inadequate.   Decreasing water potential from -0.5 to -1.2 MPa does not significantly change the dependence of A on Ci and has a small effect on the plateau.   Stress from -1.2 to -1.8 MPa decreased the slope little; the small decrease is possibly because of physical restrictions to CO_2 supply from air spaces to enzyme reaction centres.   However, the RuBP regeneration region is much more affected.   With severe stress (below -1.8 MPa) slope and particularly the plateau are affected.   Evidently stress affects RuBP regeneration or enzyme capacity much more than the ratio of RuBP carboxylase to oxygenase activity.   Measurements of the ratio in vitro show no consistent differences with stress (Keys & Lawlor, unpublished).   Also, with rapid stress the total amount of enzyme is little changed.   Therefore, slow regeneration of RuBP is the probable cause of reduced assimilation.

Photorespiration and water stress.   The relative magnitude of the RuBP carboxylase to oxygenase reactions is determined by the ratio of CO_2 to O_2 at the active site (6,8).   Stomatal closure decreases CO_2 in the leaf more than O_2 because of the great difference in concentration in the atmosphere.   Flux of carbon into the glycolate pathway increases (Fig.2) relative to assimilation which decreases, and hence relatively more photorespiratory CO_2 is released compared to assimilation (4,6). Decreasing the O_2 concentration eliminates CO_2 release in C_3 plants between -0.5 and -1.6 MPa but not at severe stress, where CO_2 is produced by 'dark respiration' and compensation point increases greatly (4). Carbon flux through glycolate consumes NAD(P)H and may synthesise ATP from mitochondrial glycine decarboxylation, which produces 0.25 mol CO_2 per mol of carbon in glycine. With net photosynthesis of 2 μmol m^{-2} s^{-1} and 0.5 μmol photorespiration $\approx$ 110 ATP and 70 NAD(P)H (μmol m^{-2} s^{-1}) are consumed. With 4 μmol m^{-2} s^{-1} net photosynthesis and equal photorespiration about 40 ATP and 35 NAD(P)H μmol m^{-2} s^{-1} are used, a relatively increased consumption but still smaller than the unstressed plants. Stressed plants must therefore dissipate considerable excess energy or be damaged (photo-inhibition, 9). However, unstressed plants at CO_2 compensation point (21% O_2)m are not damaged but stressed are (9). Simple energy balance cannot explain this; more serious lesions must occur in metabolism.

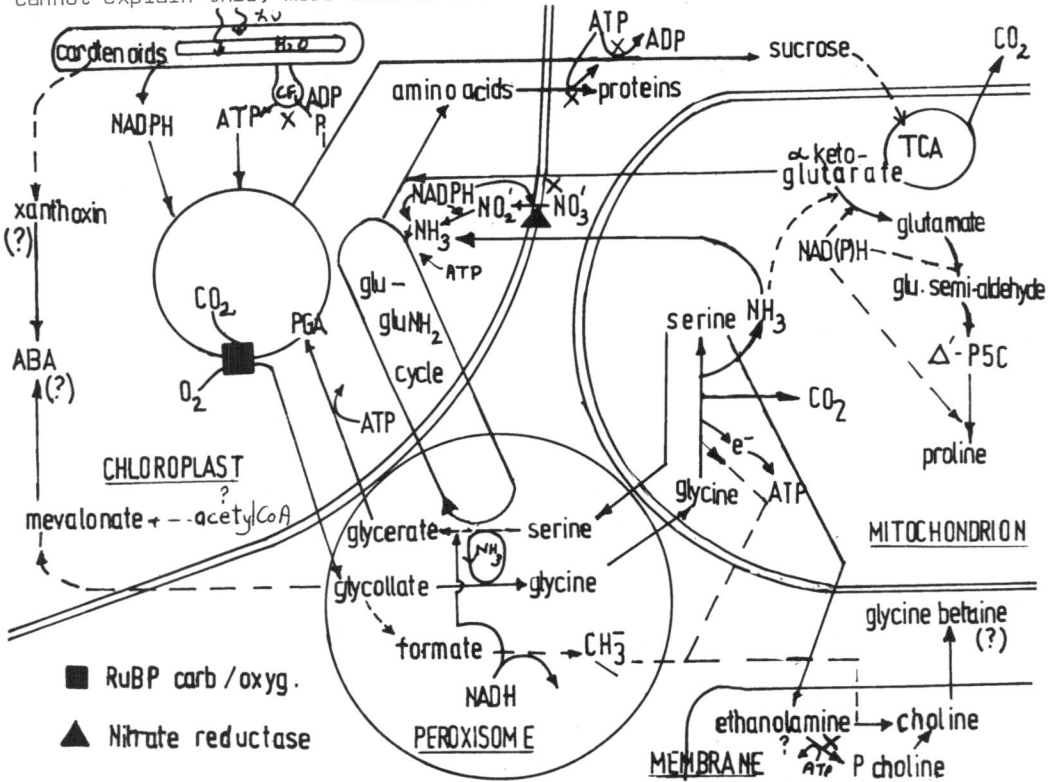

Figure 2   Carbon, nitrogen and energy interactions in photosynthetic metabolism, grossly simplified. Speculative routes of synthesis, - - - - -; uncertain location, (?) . Major lesions caused by stress, X.

Regeneration of RuBP, enzyme capacity, NAD(P)H and ATP synthesis.

Decreased CO_2 assimilation gives less 3PGA and therefore RuBP as the cycle is autocatalytic.   The decrease in RuBP is due either to enzymes or NAD(P)H or ATP supply.   Enzymes are inhibited only at extreme stress (3) and generally exceed the level for maximum assimilation.   Light harvesting and electron transport from water to NAD(P)H are little affected by stress (3) in the hydrophobic environment of the thylakoid membrane; slower rates may be due to methods of isolation etc, but evidence is that even with severe stress electron transport is substantial.   Thus in severely stressed cells, ferredoxin is probably reduced and the $NAD(P)H/NAD(P)^+$ ratio high.   Mitochondrial electron transport is probably also insensitive to stress.

ATP synthesis by photophosphorylation decreases with severe stress (3, 13) and the ATP/ADP ratios are smaller and change more slowly with illumination.   The function of CF_1, a complex of 5 different subunits, is altered by stress as shown by uncouplers, but it is not known how (13).   Oxidative phosphorylation is possibly little affected by stress.

CONSEQUENCES FOR METABOLISM

Over-reduction of pyridine nucleotides and low phosphorylation potential (1) which occur at mild to severe stress will greatly affect metabolism.   Insufficient ATP or other phosphorylated nucleotides in equilibrium will disrupt ribosomes, slow protein (4) and membrane lipid synthesis, affect ion pumping and change equilibria of enzyme reactions using ATP.   Inhibition of respiration in the light may be overcome and glycolysis stimulated, e.g. by the removal of ATP inhibition of phosphofructo-kinase; reactions using NAD(P)H will be stimulated.   Accumulation of some stress induced metabolites is briefly considered, in a speculative way (see Fig. 2) because in many cases their origin is unclear and it may serve to clarify how they relate to the major fluxes of materials and also function.

Nitrogen and amino acids.

During rapid, severe stress NO_3' may be reduced to NH_3 before inhibition of nitrate reductase (4), thus increasing the pool of glutamate.   The photorespiratory flux of NH_3 is much larger (8) than from NO_3' and with stress will consume a larger proportion of reductant and ATP, but too small to balance decreased CO_2 fixation, probably less than 15% of total energy.   Alternatively, with stress glutamate could be made from NH_3 released in the tissues and α-ketoglutarate and NAD(P)H, with sucrose (which is lost under rapid stress (5)) contributing to the C skeletons in mitochondrial re-

actions.   In sunflower glutamate increases 2-fold (rates 3 nmol cm^{-2} hr^{-1} with
stress (Fig. 3); glycine and serine accumulate at similar rates and are formed
in part, by the glycolate pathway (5,6).   Proline accumulates most (to 500
nmol cm^{-2} at rates of 10-20 nmol cm^{-2} hr^{-1}) but only at very severe stress and
low turgor (5).   This is not due to blocked protein synthesis but faster synthesis, slower oxidation and some protein breakdown.

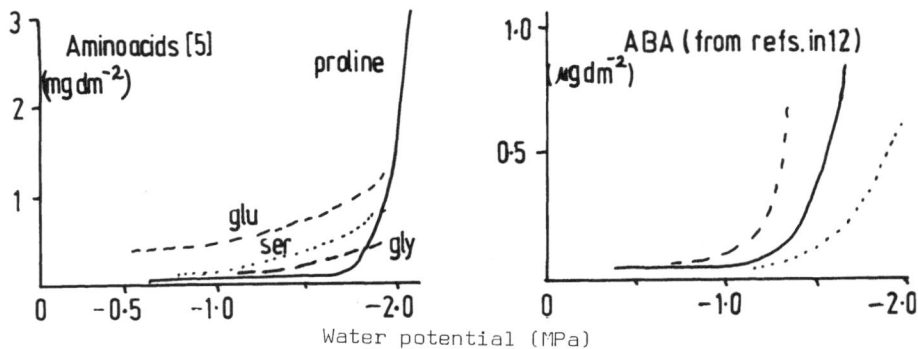

Figure 3.   Amino acid and abscisic acid accumulation.

Proline is synthesised from glutamate via Δ'-pyrroline-5-carboxylate, probably
in the mitochondria, and active glycolysis and TCA cycle and a carbohydrate
supply are needed.   Proline accumulation consumes trivial amounts of excess
reductant but may remove NH_3 and is mainly a response to metabolic inbalance;
its osmotic rôle is not a primary response to stress (2,11).   Nitrogen metabolism under stress must be viewed as an entity, and not only one amino acid
considered, if its control is to be understood.

Quaternary ammonium compounds (QAC'S).   Glycine betaine accumulates in halophytes and also in stressed plants (2), although the threshold water potentials
are not established.   It may act as counter ion and osmoticum; it protects enzymes and accumulates in cytosol and vacuole.   Synthesis is from serine to
ethanolamine, which requires $\overline{CH}_3$ groups derived by reduction from C_1 units (2)
possibly formate, and leads to phospholipid synthesis, e.g. phosphatedyl choline
via choline, which needs ATP or equivalent.   Glycine betaine is an end product of the pathway and is only slowly metabolised.   As the amount of betaine
synthesis is small, consumption of reductant is trivial (2) in relation to tota
energy supply; carbon is probably available in excess from glycolate pathway
serine and C_1 units, e.g. tetrahydrofolate in glycine decarboxylation or formate derived non-enzymically from glyoxalate.

Evidence of inositol accumulation and protein breakdown (5) and ion leakage suggests that severe stress causes membrane breakdown, which may supply choline for betaine synthesis.   Lipids turnover rapidly, so ATP would quickly affect cell compartmentation; why then is electron transport little affected? There is surprisingly little information on membranes under stress.   In conclusion, betaine synthesis could be regarded as a consequence of metabolic inbalance and damage, rather than directed synthesis with osmotic and protective roles.

Terpenoid metabolism.   Synthesis of components of the thylakoid membrane, carotenoids, plastoquinone etc. is 'linked' to abscisic acid (ABA) and farnesol, which are potent hormones, by synthetic pathway from mevalonic acid (MVA) and farnesyl pyrophosphate (12).   ABA accumulation and its effects are well documented, however the site of synthesis and the source of carbon and controls are uncertain. De novo synthesis of bulk ABA may require acetyl CoA, which would be available in mitochondria, and reductant for MVA formation would be ample, if NADPH level is indeed large.   However, the ATP requirement for MVA to farnesyl pyrophosphate is counter to the hypothesised ATP shortage. Alternative sources of carbon are the glycolate pathway (12) or from carotenoid breakdown under strong illumination but neither has been examined in stressed cells.   Absolute amounts of ABA are very small, hence substrate, reductant and perhaps even ATP may not be limiting to its synthesis.   ABA accumulates at very low water potentials (Fig. 3) where turgor is near zero (12) and inbalance in metabolism greatest, suggesting that it is a damage response.   Evidence of stomatal closure before ABA accumulation would support this.   Better correlation of stress, ABA and its effects during stress development is needed to resolve the question of time and importance of ABA production.

Quantum efficiency and photoinhibition.   Stress decreases efficiency by inhibition of CO_2 assimilation (4, 9) which uses perhaps 70% of absorbed light energy, so excess energy must be dissipated by other routes which will be overloaded under sudden, severe stress.   Thus carotenoids, which quench excited chlorophyll and high energy states of oxygen, could be broken down.   Electrons may pass to oxygen giving oxygen and -OH radicals, H_2O_2 etc. which destroy lipids and may 'swamp' such safety mechanisms as carotenoids, catalase etc. particularly if ATP shortage limits resynthesis of macromolecules.   Increased PSII fluorescence suggests that this occurs.   Thus, within a short time light may severely damage stressed leaves, despite photorespiration.

CONCLUSION

Water stress may affect crops in the field by influencing leaf area mainly, damage to metabolism is an extreme event only at severe stress. Disruption to metabolism is probably caused by inhibition of ATP synthesis and over-reduction of photosynthetic cells with consequent stimulation of synthesis of some compounds, e.g. amino acids, and inhibition to others. Increase in glycolate pathway metabolism is important in energy regulation and providin material for synthesis of compounds under stress.

Thanks are due to Professor Barry Osmond, of the Australian National University where the studies on A/C_i curves were made, and to Dr. John Lenton . for stimulating discussions on terpenoids.

REFERENCES

1. Atkinson DE. 1977. Cellular energy metabolism and its regulation. New York Academic Press.
2. Hanson AD, Nelsen CE. 1978. Betaine accumulation and ^{14}(C) formate metabolism in water stressed barley leaves. Plant Physiol. 62, 305-312.
3. Kaiser WM, Heber U. 1981. Photosynthesis under osmotic stress. Effects of high solute concentrations on the permeability properties of the chloro plast envelope and on activity of stroma enzymes. Planta 153, 423-429.
4. Lawlor DW. 1979. Effects of water and heat stress on carbon metabolism of plants with C_3 and C_4 photosynthesis. In:Stress Physiology in Crop Plants. ed. Mussell H and Staples RC. p. 304-326.
5. Lawlor DW, Fock H. 1977. Water stress induced changes in the amounts of some photosynthetic assimilation products and respiratory metabolites of sunflower leaves. J.exp.Bot. 28, 329-337.
6. Lawlor DW, Pearlman JG. 1981. Compartmental modelling of photorespiratio and carbon metabolism of water stressed leaves. Plant, Cell & Environment 4, 37-52.
7. Legg BJ, Day W, Lawlor DW, Parkinson KJ. 1979. The effects of drought o barley growth; models and measurements showing the relative importance of leaf area and photosynthetic rate. J.agric.Sci.Camb. 92, 703-716.
8. Osmond CB. 1980. Integration of photosynthetic carbon metabolism during stress. In: Genetic engineering of osmoregulation. p.171-185. ed. Rains DW, Valentine RC, Hollaender A. New York, Plenum Publishing Corp.
9. Osmond CB, Winter K, Powles SB. 1980. Adaptive significance of carbon dioxide cycling during photosynthesis in water stressed plants. In: Adaption of plants to water and high temperature stress. p. 139-154. ed. Turner NC, Kramer PJ. New York, John Wiley & Sons.
10. Singh TN, Paleg LG, Aspinall D. 1973. Stress metabolism. 1. Nitrogen metabolism and growth in the barley plant during water stress. Aust.J.Biol.Sci., 26, 45-56.
11. Stewart CR, Hanson AD. 1980. Proline accumulation as a metabolic response to water stress. See Ref. 9. p. 173-189.
12. Walton DC 1980. Biochemistry and physiology of abscisic acid. Ann.Rev.Plant Physiol. 31, 453-489.
13. Younis HM, Boyer TS, Govindjee. 1979. Biochim.Biophys.Acta. 548, 328-340.

WATER STRESS RESISTANCE OF PHOTOSYNTHESIS: SOME ASPECTS OF OSMOTIC RELATIONS

Hanno RICHTER and Sylvia B. WAGNER

Botanisches Institut der Universität für Bodenkultur, A-1180 Wien, Austria

1. ABSTRACT

The generally accepted tolerance-avoidance classification of resistance mechanisms is rather problematic in the case of water stress. This is mostly due to the multiple effects of water stress which inhibits several fundamental processes at different levels of tissue water potential. A less ambiguous evaluation is possible for mechanisms involved in the resistance of single processes, as for instance photosynthesis. In this paper, attention is concentrated on the components of total water potential which permit plants to resist the negative effects of water stress on photosynthesis. For several shrub species a close connection between turgor loss and increased stomatal resistance has been demonstrated; osmotic potentials in mature leaves of these species remain constant under water stress. Wheat and other cereals make use of different strategies: turgor loss and stomatal closure occur at different relative water contents, and osmotic adjustment leads to rapid responses under the impact of water stress. A method for inducing osmotic adjustment in detached leaves is described and the value of osmotic adjustment as an avoidance strategy for the relief of photosynthesis is discussed.

2. INTRODUCTION

Plant strategies for maintaining photosynthesis under conditions of water stress are numerous, including morphological and biochemical approaches. We shall concentrate here on the role of active changes in the solute content of cells, describing some phenomena and their possible contribution towards the water stress resistance of photosynthesis. A short discussion of stress terminology will be necessary to show the restricted meaning of terms like "avoidance" and "tolerance" for the description of resistance mechanisms in the case of water stress.

3. METHODS

3.1 Measurement of total water potential (ψ_t) for daycourses and PV curves

ψ_t was determined with standard commercial pressure chambers (Soilmoisture Equipment Corp., Santa Barbara, Ca., U.S.A.). Techniques followed RITCHIE and HINCKLEY (1975) for the field measurements, and HINCKLEY et al. (1980) and KARLIC and RICHTER (1982) for the establishment of pressure-volume (PV) curves in the laboratory. The most important steps of data acquisition for PV curves are:

a) the complete resaturation of detached leaves in a dark, humid chamber with the cut surface immersed in water;

b) the repeated weighing of the slowly transpiring leaf to follow the water loss closely;

c) the careful determination in the pressure chamber of water potentials corresponding to specific weight data; and

d) the release of the chamber pressure after each measurement at a rate slower than 0.02 MPa per second to avoid injury to the leaf tissue.

3.2 Construction and evaluation of PV curves

From saturated weight (SW), intermediate fresh weights (FW) and the weight after drying at 100 °C (DW), relative water contents (R) could be calculated:

$$R = \frac{FW - DW}{SW - DW}$$

Both "type I" and "type II" transformations (cf. TYREE and RICHTER 1981, 1982) were constructed and evaluated for general points of curve discussion such as acceptability of data points for the straight portion of the curve. Numerical values for osmotic potential and turgor potential were obtained from linear regression analysis on type II curves.

3.3 Stomatal resistance

A null-balance diffusion porometer (LICOR LI 1600) was used on detached leaves of wheat. The fully saturated leaves were brought into a humid chamber under a light bank (1000 μmol . m^{-2} . s^{-1}), until their stomata showed minimum resistances after about half an hour. The leaf bases were then taken out of the water; resistances and weights were followed until the stomata closed again under the impact of water stress after about 20 to 30 minutes.

4. RESULTS AND DISCUSSION

The information on water relations obtained in routine field measurements is essentially restricted to water potential and stomatal resistance, which are readily measured with fast-working, reliable and robust instruments. Such data are however often difficult to interpret. Stomata respond to a number of environmental variables in addition to water, and their opening state may depend more directly on internal parameters other than total water potential; turgor potential, ψ_p, seems to play a major role for many plants as a para- meter controlling stomatal opening. PV curve analysis shows osmotic and tur- gor potentials as functions of total water potential; daycourses of total water potential can therefore be "translated" into daycourses of turgor po- tential. Fig. 1 shows that two virtually identical lines for ψ_t, measured on young and old leaves of the same evergreen shrub, may conceal widely dif- ferent turgor relationships throughout the day (cf. also KARLIC and RICHTER 1982). Leaf growth and maturation is obviously accompanied by an increase in the solute content of vacuoles and, thus, improved turgor maintenance.

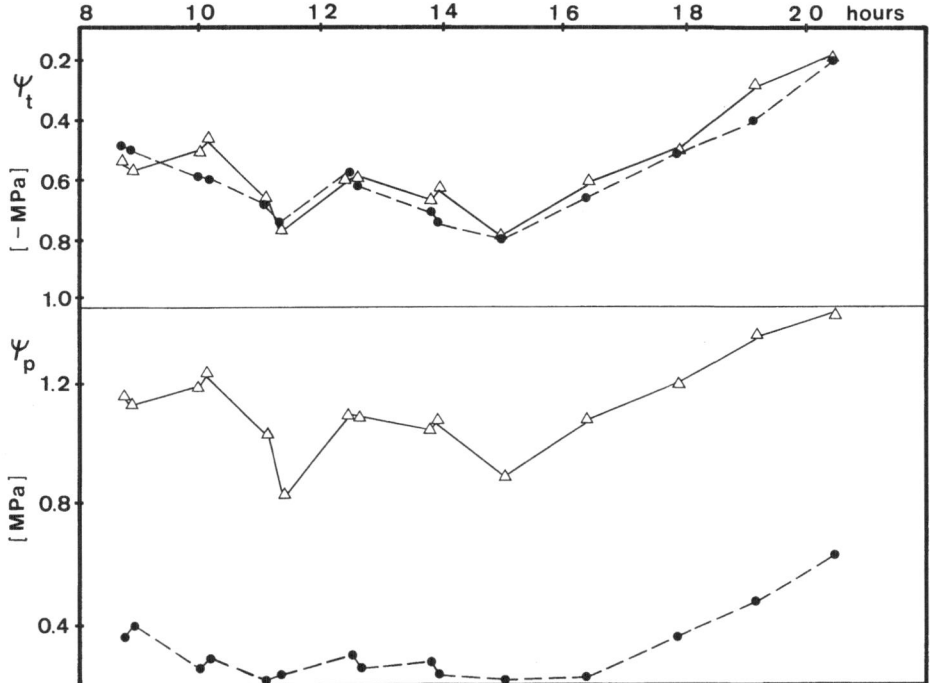

Fig. 1: *Ilex aquifolium* L., daycourses of total water potential (ψ_t) and tur- gor potential (ψ_p) for young (●) and old (△) leaves, 1978 - 07 - 03.

HINCKLEY et al. (1980) investigated several shrub species from dry sites. They showed that stomatal closure, when induced by water stress, occured in all cases whenever Ψ_t dropped to the immediate vicinity of the turgor loss point. Osmotic relations of mature leaves, as indicated by PV parameters, were remarkably stable for a given species not only at one site, but even in places differing widely in respect to precipitation and temperature. There was no indication of a short-time osmotic adjustment, which may be defined as active accumulation of additional solutes in response to water stress.

Thus we find a rather consistent pattern in woody perennials: stomata close when leaf potentials drop to the turgor loss point; the osmotic potential at this point changes only during the early (and, to a small degree, the very late) phases of leaf development, while remaining more or less constant in mature organs regardless of stress. However, such findings should not be generalized prematurely. Recent results from a detailed study on wheat, *Triticum durum*, indicate the existence of strategies which are quite different. These strategies involve both the variability of osmotic parameters and the connection between the loss of turgor and stomatal closure.

Wheat cultivars have been among the many crop plants intensively studied in the past ten years for their ability to perform osmotic adjustment (for literature see TURNER and JONES 1980). A decrease in the osmotic potential at full saturation was elicited in all these experiments by reducing the substrate water potential and thereby Ψ_t of intact rooted plants. TURNER and JONES (1980) present evidence that the rate of stress development has a major influence on the magnitude of the response: slow rates of Ψ_t reduction result in better adjustment than rapid ones. 1.2 MPa per day were the most rapid rate reported for a successful adjustment reaction.

It is therefore quite surprising to see that adjustment reactions may be triggered by a completely different and much less elaborate protocol (fig. 2). Measurement of PV curves involves the drying of leaves to values of R well below the turgor loss point. With wheat, the time needed to reach a relative water content of about 0.6 is around 3.5 hours at room temperature. This corresponds to a rate of 0.65 MPa per hour or 16 MPa per day. The linear part of the resulting PV curve is as straight and highly correlated as has been known from our studies on other species. After resaturating the same leaf again at room temperature in the dark, we find a value for the saturated weight which differs by not more than a few milligrams from the first value. A second set of PV data can then be obtained from this leaf by repeating the standard

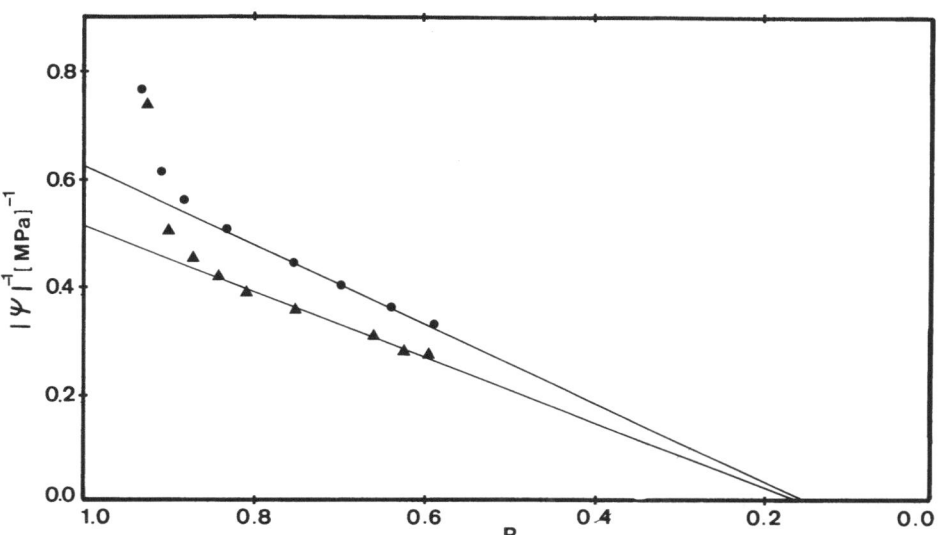

Fig 2: *Triticum durum* cv. Grandur, flag leaf of a field-grown plant at anthesis. PV curves from first (●) and second (▲) drying cycle.

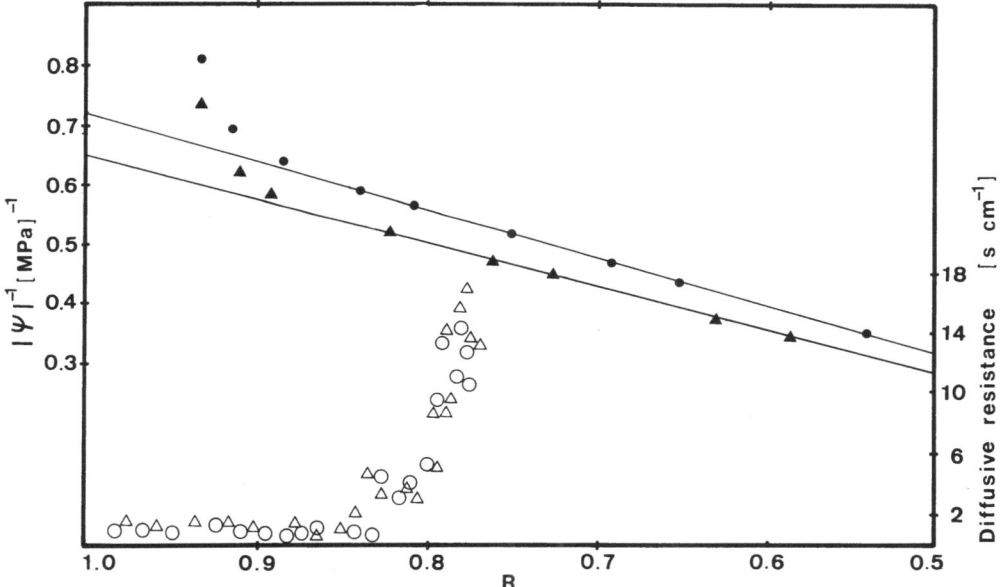

Fig. 3: *Triticum durum* cv. Grandur, PV curves and stomatal resistances of
flag leaves at late anthesis.● ○ : unstressed;▲△: predroughted.

procedure. The second curve shows a decrease in the osmotic potential of 0.1
to 0.4 MPa, which is in the range of adaptive responses reported for the leaves of rooted wheat plants. The magnitude of the adaptation depends on the
conditions during drying and resaturation as well as on variety, developmental stage and pretreatment of the plant (WAGNER and RICHTER, in preparation).
Similar effects have been observed in rye *(Secale cereale)*, but not in a
number of fruit tress such as cherry and plum. This rapid osmotic adaptation
in cereals could perhaps depend on the existence of fructanes as carbohydrate reserves.

Maintenance of stomatal opening and photosynthesis down to lower values
of Ψ_t are among the well-known effects of osmotic adjustment. They became
evident in some of our experiments (fig. 3). When stomatal resistance of rapidly drying leaves is plotted against R, there is exactly the same response
in pre-droughted and resaturated leaves as in those of unstressed controls.
Since the osmotic potential of pre-droughted leaves is lowered, this means
that the stomata will only close if the continuum factors enforce a lowering
of Ψ_t to this new value for Ψ_o. Another conspicuous feature in wheat besides
the existence of osmotic adjustment is the wide distance along the R axis
between the turgor loss point and the point of stomatal closure. This is very
unlike what we know from drought resistant shrubs (HINCKLEY et al. 1980), but
it explains the observation that field values for Ψ_t drop frequently below
$\Psi_{o(tlp)}$, the osmotic potential at the turgor loss point as estimated from a
PV curve (table 1). It would stress our imagination to assume a causal connection between leaf turgor and stomatal closure in a case like this.

Table 1. Ψ_o at the turgor loss point, minimum Ψ_t in the field, and time
for which Ψ_t stayed below the turgor loss point (flaccid time)
for two wheat varieties at different dates.

Date	Variety	$\Psi_{o(tlp)}$ (-MPa)	$\Psi_{t(min)}$ (-MPa)	flaccid time (hours)
1981 - 05 - 22	Grandur	1.65	1.77	3
	Unidur	1.74	1.68	-
1981 - 05 - 31	Grandur	1.90	2.45	6
	Unidur	2.02	2.29	6
1981 - 06 - 08	Grandur	1.98	2.33	5
	Unidur	1.88	2.18	5

It would now seem appropriate to assess osmotic adjustment and stomatal reactions in wheat as factors for photosynthesis under stress. This requires a short discussion of the terminology for water stress resistance first.

Mechanisms of drought resistance are often classified according to LEVITT´s tolerance – avoidance concept (LEVITT 1972). This approach was originally developed for the case of freezing resistance, where it has considerable advantages. When trying to use it in the classification of drought resistance phenomena we encounter however numerous problems. They are illustrated by continuing attempts to obtain coherent sets of definitions (cf. TURNER and KRAMER 1980, PALEG and ASPINALL 1981). The dilemma proves to be quite fundamental. It is only partly due to the ill-defined meaning of the term drought; most problems are instead posed by the complex nature of water stress which affects dozens of physiological and biochemical processes. Well-known review articles (e.g. HSIAO 1973) point out that certain adverse effects of water stress are fully developed at comparatively high values for total water potential, whereas other reactions and metabolic pathways become affected one after another whenever a new threshold value is reached. We must therefore conclude that a given combination of strategies can result in successful stress avoidance for a number of physiological processes, whereas others, being more susceptible, are already impaired and the plant must tolerate this particular deficiency. Moreover, there is no rule that "avoidance" should employ the same strategies for different processes. Growth, as an example, may be shifted to the low-stress night hours; this avoidance strategy is not available for photosynthetic CO_2 uptake, except in the highly specialized CAM plants.

It will therefore be useful to consider resistance strategies separately for each major process. In our case this means to aim at a classification applicable to photosynthesis. One of the major prerequisites for photosynthesis is the possibility for CO_2 uptake through open stomata. We may therefore without much difficulty interpret plant properties and reactions as avoidance strategies in regard to water stress on photosynthesis, whenever they help keep the stomata open despite lowered total water potentials. Where these strategies fail, the plant must tolerate a reduction in photosynthesis due to water stress. Some species are very successful as tolerators: *Viburnum lantana*, a shrub native in eastern Austria, did not open its stomata during a drought period for 19 successive days (HINCKLEY, DUHME, HINCKLEY and RICHTER, unpublished). It is obvious that this tolerance of reduced photosynthesis may. at the same time have constituted an example of stress avoidance for other processes affected only at still lower ψ_t values: the reduction in transpiration after stomatal closure saves water and helps in the avoidance of a further decrease in plant water potentials.

We should therefore classify osmotic adjustment in wheat as a stress avoidance reaction for the relief of water stress on photosynthesis. Likewise, stomatal closure at water potentials and water contents far below the turgor loss point is an avoidance strategy. In parenthesis we could mention that other features of wheat are also characteristic of an avoider in the sense defined here: an extended root system exploits soil water reserves, and old, unproductive leaves are continuously shed. Only where all these strategies do not provide for relief sufficient to keep the stomata of productive leaves open, the plant has to invoke tolerance as a resistance strategy. This will mean the endurance of more or less time without newly formed assimilates.

Let us now ask a final question: how important can a decrease in osmotic potential of a few bars actually become? TURNER and JONES (1980) seem to be quite skeptical, emphasizing the transience and limited degree of osmotic adjustment. We think that much depends on the way how stress develops. Low leaf water potentials will occur whenever there is a) a lack of precipitation and, therefore, a decrease in soil water potential, or b) high atmospheric demand which increases water transport and lowers potentials by internal friction. Osmotic adjustment in leaves is most probably not important as a strategy in cases of acute soil drought; only a minute quantity of water will become available for transpiration in this way. However, even plants rooted in soil at the field capacity will encounter periods of internal water stress during a daycourse under high evaporative demand. Osmotic adjustment can presumably provide in such cases for valuable extra time of productive photosynthesis, especially in the morning when stress develops and in the afternoon when it decreases. This will increase the competitive strength of the plant in the field.

5. ACKNOWLEDGEMENTS

We thank Dr. Peter RUCKENBAUER for seeds and expert advice in the wheat experiments, Dr. Heidrun KARLIC for supplying the data used in figure 1, and the Fonds zur Förderung der wissenschaftlichen Forschung for support from Projekt 3765.

REFERENCES
1. Hinckley, TM, Duhme, F, Hinckley, AR, Richter, H. 1980. Water relations of drought hardy shrubs: osmotic potential and stomatal reactivity. Plant, Cell and Environment 3, 131 - 140.
2. Hsiao, TC. 1973. Plant responses to water stress. Ann. Rev. Plant Physiol. 24, 519 - 570.
3. Karlic, H, Richter, H. 1982. Developmental effects on leaf water relations of two evergreen shrubs (*Prunus laurocerasus* L. and *Ilex aquifolium* L.). Flora, in press.
4. Levitt, J. 1972. Responses of plants to environmental stresses. New York and London, Academic Press.
5. Paleg, LG, and Aspinall, D. 1981. The physiology and biochemistry of drought resistance in plants. Sydney etc, Academic Press.
6. Ritchie, GA, and Hinckley, TM. 1975. The pressure chamber as an instrument for ecological research. Adv. Ecol. Res. 9, 165 - 254.
7. Turner, NC, Jones, MM. 1980. Turgor maintenance by osmotic adjustment: a review and evaluation. In: Turner and Kramer (1980), pp. 87 - 103.
8. Turner, NC, Kramer, PJ. 1980. Adaptation of plants to water and high temperature stress. New York etc, John Wiley & Sons.
9. Tyree, MT, Richter, H. 1981. Alternative methods of analyzing water potential isotherms: Some cautions and clarifications. I. The impact of non-ideality and of some experimental errors. J. Exp. Bot. 32, 643 - 653.
10. Tyree, MT, Richter, H. 1982. Alternative methods of analyzing water potential isotherms: Some cautions and clarifications. II. Curvilinearity in water potential isotherms. Can. J. Bot. 60, in press.

PHOTOSYNTHETIC ACTIVITY AND OSMOTIC VOLUMES OF ISOLATED INTACT CHLOROPLASTS AND OF CELLS IN LEAF TISSUE FROM VARIOUS PLANTS UNDER OSMOTIC STRESS

W.M.Kaiser

Botanisches Institut I der Universität, Mittlerer Dallenbergweg 64, 8700 Würzburg, FRG

ABSTRACT

   Photosynthesis of isolated intact spinach chloroplasts and of leaf slices from various hygro-, meso- and xerophytes and from plants adapted to salinity was measured under osmotic stress and compared with chloroplast and cell volumes. There was a direct relation between the reduction of the osmotic chloroplast or cell volumes and the inhibition of photosynthesis which was similar in the different plant species, irrespective of the highly different sensitivity of leaf tissue from various plants to hypertonic stress.

INTRODUCTION

   Moderate or severe water stress is known to inhibit photosynthesis both through stomatal control and at the level of the mesophyll cell or chloroplast (1). A main reason for the direct inhibition of the photosynthetic apparatus under hypertonic stress is the reduction in the volume of cells and organelles, which causes an increase in the concentration of internal solutes (2,3,4,5). This results in an inhibition of soluble enzymes (3). A transient leakage of chloroplasts was also observed upon hypertonic shock (3).

   As it was desirable to compare the response of different plants to osmotic stress on the basis of osmotic volumes of chloroplasts and cells, we have developed a method for measuring total protoplast volumes in leaf tissue from various plants at different states of osmotic dehydration. This permits a correlation between osmotic volume changes and rates of photosynthesis in different plant species.

MATERIALS AND METHODS

   The preparation of intact chloroplasts and of thin leaf slices was described earlier (4). Chloroplast volumes were measured by coulter counter distribution. Protoplast volumes in leaf tissue were obtained by infiltrating small leaf discs (5 mm Ø) with ^3H_2O and ^{14}C-sorbitol. After fast and careful washing of the leaf discs, protoplast volumes were obtained as the difference between ^3H_2O - space and ^{14}C - sorbitol space (2). Photosynthesis of chloroplasts was measured polarographically as CO_2-dependent O_2-evolution and photosynthesis of vacuum-infiltrated leaf slices as $^{14}CO_2$-fixation. All experiments were carried out under saturating white light (ca. 600 w m^{-2}) at 20° C. CO_2 was added as $KHCO_3$ (5 to 15 mmol l^{-1}). For further experimental details compare ref. 2.

RESULTS

Photosynthesis of isolated intact chloroplasts, isolated mesophyll protoplasts and of thin leaf slices from spinach was inhibited by hypertonic stress (sorbitol as osmoticum) to a simlar extent, with 50% inhibition occuring at external osmotic potentials (π^o) between 20 and 30 bar (fig.1). In spinach, this is about 2 to 4 times the internal osmotic potential of the cell sap (π^{cs}). Over a large range of osmotic potentials, intact chloroplasts responded as near-perfect osmometers, showing a linear relationship between volume and the reciprocal external osmotic potential ($1/\pi^o$) (fig.2, compare ref.6). The non-osmotic space at infinite osmotic potential can be extrapolated to be about 1/3 of the chloroplast volume at the isotonic point (π^o= 10 bar). In contrast to chloroplasts, protoplasts of leaf tissue from spinach or other plants responded to changes of π^o in a less ideal way. A plot of volume versus $1/\pi^o$ shows that in leaf tissue, protoplast extension is limited by the cell wall (fig.3). As expected, ideal osmometric behaviour was found only at π^o values higher then π^{cs}. In protoplasts, the non-osmotic space was usually only 5 to 10% of the maximal protoplast volume (fig.3). Since protoplast extension is limited by the cell wall, while isolated intact chloroplasts swell at low π^o, photosynthesis of both systems was compared only at $\pi^o \geqq \pi^{cs}$. Under isotonic conditions (10 bar in spinach), photosynthesis of leaf slices was usually not inhibited (fig.1).

In fig.4, the relative rate of photosynthesis of isolated intact chloroplasts and of leaf slices from spinach is plotted versus the relative osmotic volume of chloroplasts or of protoplasts in intact spinach leaf tissue, with 100% representing the values observed at the isotonic point.

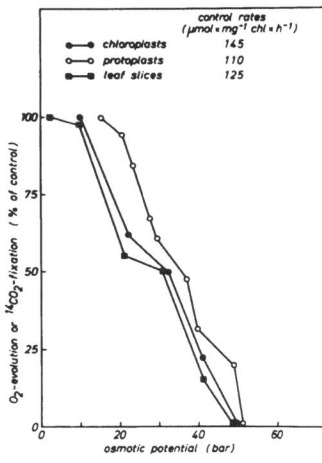

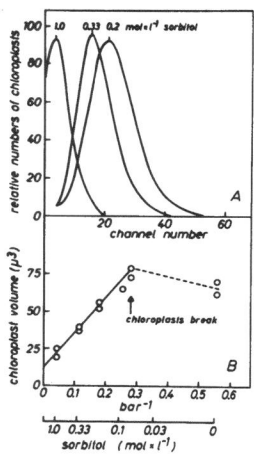

Fig.1: Rates of photosynthesis of isolated intact chloroplasts, isolated protoplasts and leaf slices (0.7 mm broad) from spinach, at increasing osmotic potentials in the external medium. Sorbitol was used as osmoticum. $KHCO_3$ was 5 mmol l^{-1} for chloroplasts and protoplasts, and 15 mmol l^{-1} for leaf slices. In all experiments the infiltration medium contained 25 mmol l^{-1} HEPES-KOH pH 7.6, 1 mmol l^{-1} $MgCl_2$, 1 mmol l^{-1} $MnCl_2$, 1 mmol l^{-1} EDTA.

Fig.2: Chloroplast volumes as measured by coulter counter distribution as a function of $1/\pi^o$. Upper part (A) shows the volume distribution of chloroplasts isolated from spinach at different π^o.

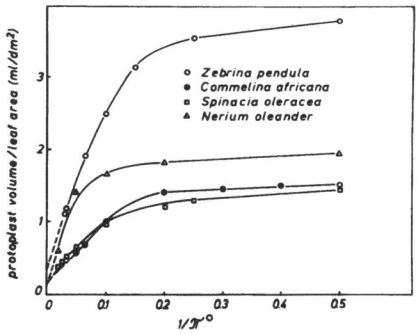

Fig.3: Total protoplast volumes in various leaf tissues plotted over $1/\pi^o$, with sorbitol as the osmoticum. Volumes were measured in vacuum-infiltrated leaf discs (10 to 20 discs /sample).

It can be seen that a given reduction of the chloroplast or protoplast volume causes a similar decrease of photosynthesis in both systems, irrespective of their different complexity. This was taken as an indication that chloroplast volumes are changed <u>in vitro</u> to a similar degree as <u>in situ</u>.

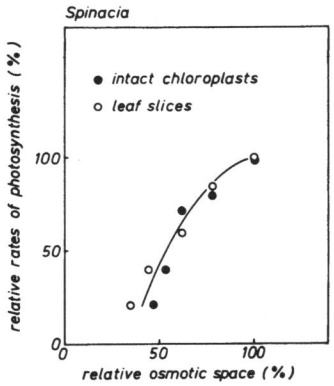

Fig.4: Relative rates of photosynthesis over the relative osmotic volume in isolated intact spinach chloroplasts or in spinach leaf slices. Data from fig. 1,2 and 3.

Rates of photosynthesis and total protoplast volumes were also measured in leaf tissue from several plant species. Photosynthesis of leaf slices from hygrophytes such as *Commelina* or *Zebrina pendula* proved to be more sensitive to osmotic stress than photosynthesis of the mesophyte *Spinacia deracea* or the xerophyte *Nerium oleandes* (fig.5). 50% inhibition of photosynthesis occured at 20 to 30 bar in spinach, and at 40 to 50 bar in Nerium. Surprisingly, similar differences were also found for the changes of the total osmotic protoplast volumes in the various leaf tissues (fig.3). The differential response of protoplast volumes to changes in π^o were due mainly to different internal osmotic potentials (π^{cs}) in various leaf tissues, which varied from about 4 bar in hygrophytes up to 25 bar in xerophytes (not shown). A plot of the relative rates of photosynthesis versus the relative osmotic protoplast volume showed again that a given decrease of the osmotic volume caused similar inhibition of photosynthesis in all plants tested (fig.6).

The relationship between volume changes and rates of photosynthesis was also examined in a plant species adapted to different salinities. *Mesembryanthemum cristallinum* is known as a plant in which CAM can be induced by high salinity or high osmotic potentials of the culture medium (7). Photosynthesis (measured as $^{14}CO_2$-fixation at the end of the de-

acidification phase) of plants grown under saline conditions was less
sensitive to high π^0 than that of control plants grown in NaCl-free culture
(fig.7). Similar differences were found for CO_2-fixation in the dark
(measured during the night-phase, not shown). Total protoplast volumes in
in leaf discs from both plant types were also measured. A plot of volume
versus $1/\pi^0$ revealed that in leaf tissue from plants grown under saline
conditions the non-osmotic space was dramatically increased compared to
control plants (about 45% compared to 3% of the maximal protoplast volume).
A plot of the relative rates of photosythesis versus relative osmotic
protoplast volumes showed that a similar reduction in protoplast volume
produced comparable inhibition of photosynthesis in both plant types (fig.9).
These data (fig.9) are similar to those shown in fig.4 and fig.6. Preliminary
experiments with spinach grown under saline and non-saline conditions in
hydroponic culture yielded results which were in all respects similar to
the results obtained with *Mesembryanthemum* .

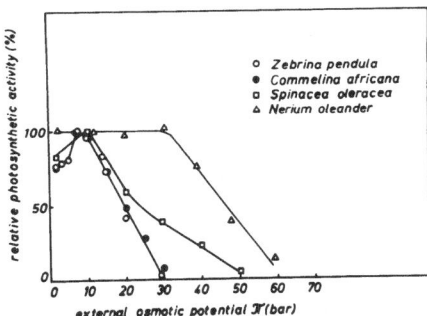

Fig.5: Photosynthesis of leaf slices
from two hygrophytes, a mesophyte
and a xerophyte under hypertonic
stress. Sorbitol was used as
osmoticum.

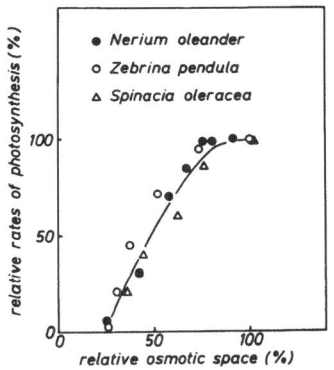

Fig.6: Relative rates of photosynthesis
versus the relative osmotic protoplast
volume in various leaf tissues.

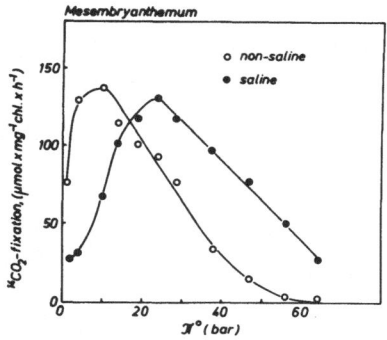

Fig.7: Rates of photosynthesis in leaf slices from *Mesembryanthemum cristallinum* grown hydroponically in modified Hoaglands solution with or without 400 mmol l^{-1} NaCl. Osmotic potentials in the incubation medium were adjusted with sorbitol.

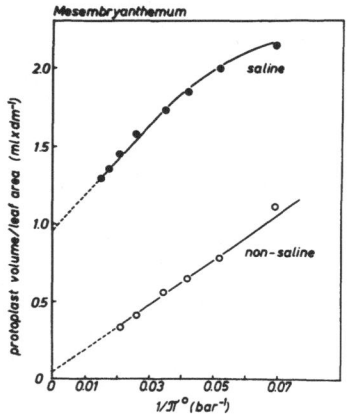

Fig.8: A plot of total osmotic protoplast volume versus $1/\pi^o$ in leaf discs from *Mesembryanthemum cristallinum* grown hydroponically in presence or absence of NaCl.

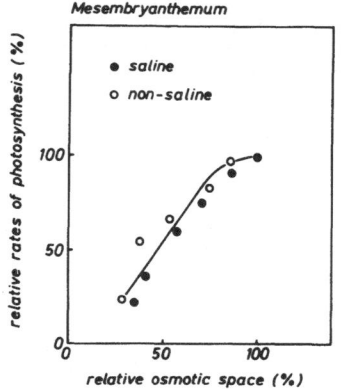

Fig.9: Relative rates of photosynthesis versus relative osmotic protoplast volumes in leaf tissue from

DISCUSSION

The data shown in figs. 4, 6 and 9 demonstrate that the same reduction
in osmotically active space brought about by osmotic dehydration of chloroplasts or cells causes comparable inhibition of photosynthesis in plants
which differ considerably in their sensitivity to drought. However, depending
on their drought resistance, different osmotic cencentrations in the sus-
pending media need to be applied to produce comparable dehydration of cells
from different plants. Internal osmotic potentials of xerophytes were higher
than those of sensitive species. The observation that inhibition of photosynthesis by dehydration is a function of the available osmotic volume
rather than of osmotic potentials (i.e. of solute concentrations which
are higher in xerophytes than in hygrophytes) appears at first sight to be
in contradiction to conclusions drawn earlier (5) that inhibition of photosynthesis by hypertonic stress is caused by increased solute concentrations.
However, different solutes differ in their effectiveness to inhibit photosynthetic reactions. Potassium ions and compensating anions are important
solutes of the metabolically active space of plant cells (2). Elevated
concentrations of potassium salts inhibited enzyme reactions in extracts
from different plants to a comparable extent indicating that species-
specific differences in drought sensitivity of plants cannot be explained
by species-specific differences in enzyme susceptibility to salt inhibition
(2). However, it was observed that the composition of solutes differed
considerably in plants differing in their sensitivity to drought.
Xerophytes such as *Nerium* which had higher osmotic potentials than hygrophytes or mesophytes contained a higher proportion of neutral solutes such
as soluble sugars (2). Enzyme reactions were usually less sensitive to
increased concentrations of sugars than to increased concentrations of
potassium salts, i.e. enzyme activity was less influenced by a reduction
in water potential than by the specific solute environment (3). Cellular
membranes such as thylakoids are known to maintain their properties even
in the presence of very high sugar concnetrations, while salts cause
membrane inactivation and protein dissociation from the membranes (8).
Thus, solutes such as sugars might be termed compatible solutes, while
salts, at high concentrations (exceeding 150 to 250 mmol l^{-1}) exert
damaging effects on membranes and inhibit enzyme reactions. The observation

62

reported in this work that a comparable reduction in mesophyll cell volume causes comparable inhibition of photosynthesis in different plant species irrespective of the osmotic potentials necessary to produce volume reduction strongly suggests that the content of potentially inhibitory solutes such as potassium salts in the metabolically active space of cells and organelles is comparable in different plant species while that of compatible solutes might differ widely depending on the necessity to maintain low cell water potentials. Alternatively, one might assume that it is not the concentration of osmotically relevant organic and inorganic solutes which limits enzyme activities in dehydrated cells and organelles, but increased protein concentrations. At the isotonic point, the concentration of soluble protein in spinach chloroplasts is about 0.25 g/ ml. One could speculate that any further increase of this high protein concentration might decrease enzyme activities due to protein-protein interactions. In that case, protein concentrations should be similar in chloroplasts (and cytoplasm) of all plants.

The observation that the non-osmotic space can be dramatically increased in plants like *Mesembryanthemum* or spinach under salinity, cannot be sufficiently explained at present. In control plants of *Mesembryanthemum* the ratio of fresh weight/ dry weight is usually about 60. In plants grown in 400 mmol l^{-1}NaCl, this ratio is changed to about 16, indicating considerable accumulation of dry matter (data not shown). This is paralleled by extremely increased concentrations of internal NaCl (also not shown). Still, further analysis of of changes in the solute composition of plants grown under salinity is needed for a better understanding of this phenomenon.

REFERENCES

1. Hsiao,T.C. 1973 Plant responses to water stress. Annu. Rev. Plant Physiol. 24, 519-570
2. Kaiser,W.M. 1982 Correlation between changes in photosynthetic activity and changes in total protoplast volumes in leaf tissue from hygro-, meso- and xerophytes under osmotic stress. Planta, in press
3. Kaiser,W.M., Heber,U. 1981 Photosynthesis under osmotic stress. Effect of high solute concentrations on the permeability properties of the chloroplast envelope and on activity of stroma enzymes. Planta 153, 423-429
4. Kaiser,W.M., Kaiser,G., Prachuab,P.K., Wildman,S., Heber,U. 1981 Photosynthesis under osmotic stress. Inhibition of photosynthesis of intact chloroplasts, protoplasts and leaf slices after exposure to high solute concentrations. Planta 153, 430-435

5. Kaiser,W.M., Kaiser,G., Schöner,S., Neimanis,S. 1981 Photosynthesis under osmotic stress. Differential recovery of photosynthetic activities of stroma enzymes, intact chloroplasts, protoplasts and leaf slices after exposure to high solute concentrations. Planta 153, 430-435
6. Nobel,P.S. 1969 Light-induced changes in the ionic content of chloroplasts in Pisum sativum. Biochim. Biophys. Acta 172, 134-143
7. Winter,K., Von Willert,D.J. 1972 NaCl-induzierter Crassulaceensäurestoffwechsel bei Mesembryanthemum cristallinum. Z. Pflanzenphysiol. 267, 166-170
8. Heber,U., Schmitt,J.M., Krause,G.H., Klosson,R.J., Santarius,K.A. 1981 Freezing damage to thylokoid membranes in vitro and in vivo, in:Effects of low temperatures on biological membranes. G.J. Morris, A.Clarke (eds.), Academic press, London, New York, Toronto, Sidney, San Francisco, pp. 263-283

THE INFLUENCE OF WATER STRESS ON PHOTOSYNTHESIS IN A BARLEY CROP

K.J. PARKINSON and W.DAY[*]

Rothamsted Experimental Station, Harpenden, Herts and[*] Long Ashton Research
Station, University of Bristol, Long Ashton, Bristol.

KEY WORDS: age, barley, photosynthesis, temperature, water stress.

ABSTRACT

Photosynthesis rates of individual leaves and ears were measured over
a range of light intensities and CO_2 concentrations in a study of drought
effects on the growth of a spring barley crop in the field. The effects
of stress treatment and other environmental factors on the parameter values
in fitted photosynthesis models are evaluated. The quantum efficiency
of leaves declined at temperatures above 25°C; mesophyll conductance
decreased with leaf age, and the decrease occurred earlier under water
stress. For ears, quantum efficiency was affected by both pre- and post-
anthesis stress.

INTRODUCTION

In our studies of the effects of water stress on the growth and yield
of barley growing in the field, we have sought to separate the different
physiological components of the response e.g. effects on leaf growth and
senescence, on stomatal resistance and on CO_2 exchange [9]. The present
paper deals with the effects of water stress treatments and other
environmental factors on components of the photosynthetic performance of
leaves and ears. We do not here consider stomatal effects, which will be
reported in a later paper, but rather allow for measured differences in
stomatal resistances in our analysis of the response of photosynthesis to
photon flux density and CO_2 concentration. To enable consideration of the
effects of water stress, leaf number and age, and temperature on these
responses, we have fitted simple models to the responses and the analysis
presented here concerns the values obtained for parameters in the models.

MATERIALS AND METHODS

The crop studied was spring-sown barley (cv. Julia) grown at
Rothamsted Experimental Station in 1979 on a site that could be protected
from rain by automatic shelters. The details of the site, shelters and
general agronomy are given elsewhere [3,5]: the principal features were

that the soil was a silty clay loam, with a large plant-available water
capacity (about 200 mm to 1·5 m), that the crop was sown on 18 April 1979,
reached anthesis around July 7 and was harvested at the end of August. The
fully irrigated plots gave a mean grain dry matter yield of 6·1 t ha^{-1}, and
those receiving no irrigation gave 3·7 t ha^{-1}. Two other treatments were
tested – early drought (irrigated each week after anthesis) and late drought
(irrigated each week until 20 June). Further details of yield responses
and water use are given elsewhere [5].

CO$_2$ exchange measurements were made on main stem leaves between 29 May
and 14 August and on main stem ears between 9 July and 19 August, using
single leaf and ear chambers [12]. Measurements of temperature and humidity
in the chambers were used to determine the stomatal resistance to water
vapour exchange. The depletion of CO$_2$ concentration of the gas that had
passed through the chamber was measured with an infra-red gas analyser
(ADC Ltd, Hoddesdon, U.K.) and the measurements were corrected for water
vapour effects [11] in order to determine the true CO$_2$ exchange rates.
Because of the fluctuating natural light levels in a summer with few clear
days, mercury vapour lamps (Type MBFR/U)were used to enhance natural
sunlight. Rates of CO$_2$ exchange at near ambient CO$_2$ concentration were
measured on each leaf or ear at a range of photon flux densities achieved
by using a series of metal gauzes. On the same organ, at a high photon
flux density, CO$_2$ exchange rates were also measured at a range of CO$_2$
concentrations from zero to 15 m mole m^{-3}, using a gas dilutor (ADC,
Hoddesdon [12]).

As a first step in data analysis, the CO$_2$ exchange rates of each
separate leaf or ear were used to determine values of parameters in simple
models of photosynthetic response to CO$_2$ concentration and photon flux
density. For leaves, a four parameter model based on that of Chartier and
Prioul [2] fitted the data well [4]. To allow for difference in stomatal
resistances, the substomatal CO$_2$ concentration, C_i, was calculated [1],
and the net photosynthesis rates, P_N, were fitted by the equation

$$P_N = \alpha Q(1 - D/C_f) - R_D \tag{1}$$

where C_f, the CO$_2$ concentration at the fixation site, is given by
$C_i - (P_N + R_D)r_m$, Q is the photon flux density (0·4 to 0·7 µm waveband)
r_m is the mesophyll resistance, D a photorespiration constant, α the
quantum yield, and R_D is dark respiration.

However, the fit of this model to ear data was not satisfactory. The best value for D was zero, and a better fit was generally obtained with a simple rectangular hyperbolic relation to C_i and Q.

$$P_N = 1/(1/\alpha Q + r_m/C_i) - R_D \qquad (2)$$

The models were fitted to the data using a maximum likelihood program [13] that determined the parameter values that minimised the residual variance.

In the field experiments, observations of leaf appearance were made; the mean date for appearance of the ligule on each leaf on the main stem has been used in the analysis of the effects of leaf age on the fitted photosynthetic parameters. Measurements were also made of leaf and soil water potentials, and soil water contents.

RESULTS

The fitted parameter values for individual leaves and ears were analysed in relation to drought treatment, age of organ and organ temperature. The treatments could be characterized by the organ water potentials, which were generally lower (by between 0·2 to 0·4 MPa) on the unirrigated treatments, or by soil water contents or potentials [5]. However, no direct relationship of parameter values to these variables was detected, and so water stress was classified solely by treatment. The measurements on leaves were made on successive leaves as they appeared, from leaf 3 to leaf 9 (flag). No significant differences between leaf numbers were apparent, and thus leaf age effects are considered for all leaves relative to the mean date of full lamina emergence (ligule appearance) for each leaf number.

Leaves

Quantum yield. No effects of leaf age or drought treatment were apparent, but there was a marked decline with increasing leaf temperature (Figure 1).

Photorespiration. The photorespiration constant increased with temperature (Figure 2) with a Q_{10} of about 1·15, but no other effects were apparent.

Low light response of net photosynthesis. The initial slope of the light response is related to both quantum yield and photorespiration as

$$\partial P_N/\partial Q\big|_{Q \to 0} \simeq \alpha(1 - D/C_i)$$

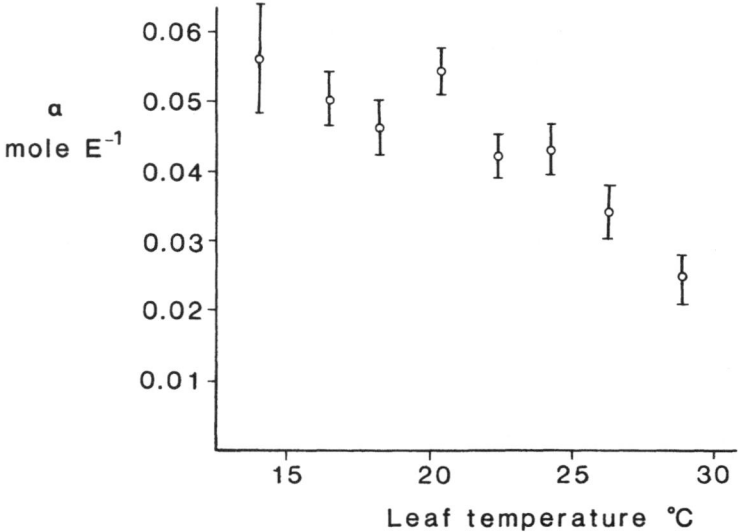

Figure 1. Quantum yield for leaves as a function of leaf temperature. The data are for all leaves and are grouped in 2°C categories.

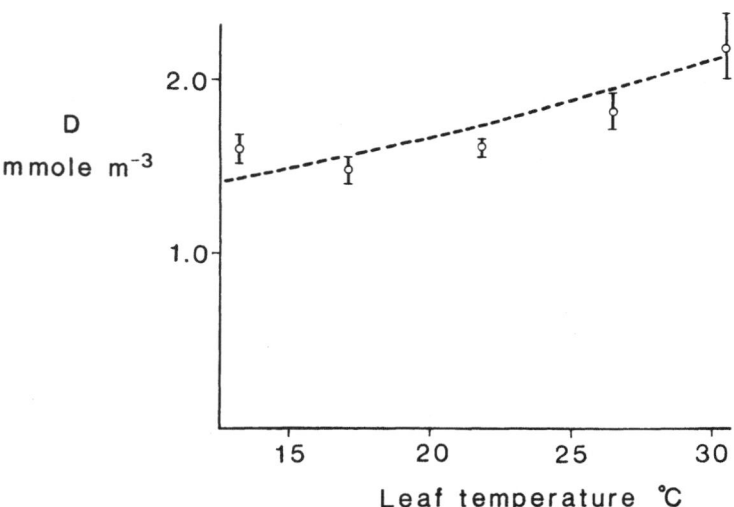

Figure 2. Photorespiration constant as a function of leaf temperature, with data grouped in 4°C categories.

Values of this net quantum yield are given in Table 1. The values are remarkably consistent, though the late drought treatment does have a lower efficiency in the short drought period prior to anthesis.

Table 1. Initial slope of light response for leaves (mole E^{-1}), adjusted to 20°C for effects on α and D shown in Figures 1 and 2. For the late drought treatment, pre-anthesis data is only for the 20 day period after irrigation stopped.

Treatment	Pre-anthesis	Post-anthesis
No drought	0·046 ± 0·003	0·041 ± 0·003
Late drought	0·032 ± 0·003	0·045 ± 0·003
Early drought)	0·042 ± 0·002	0·037 ± 0·002
Full drought)		0·039 ± 0·002

Mesophyll conductance. (= $1/r_m$). There was no significant temperature response, but conductances did decline with increasing leaf age (Figure 3). Excluding the full drought treatment, the results indicate that there was a period during which conductance was largely constant, after which conductance declined nearly linearly. Averaging over all leaves, the constant phase lasted till some 20 days after full lamina emergence, and the linear decline phase extrapolated to zero conductance around 50 days after full emergence. Data for the full drought treatment suggested an earlier decline - perhaps some 10 days earlier.

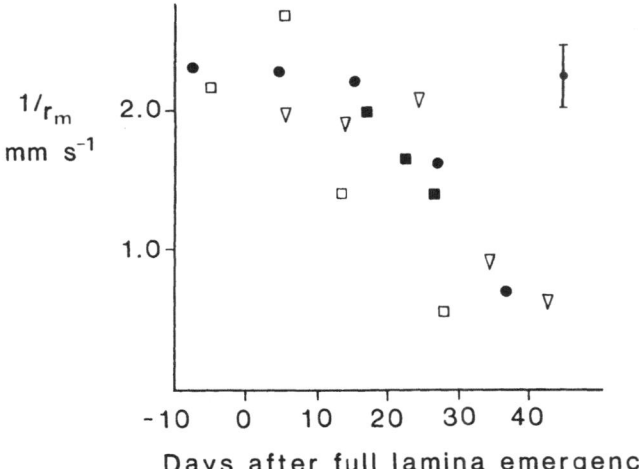

Figure 3. Mesophyll conductance as a function of leaf age after ligule appearance. The data are grouped in 10 day categories; no drought (●), full drought (□), early drought (■) and late drought (▽) ; a typical ± s.e. is shown.

Dark respiration. No treatment effects were apparent, but respiration rate increased with temperature (Figure 4) with a Q_{10} of about 1·74, and tended to decline with leaf age (Figure 5).

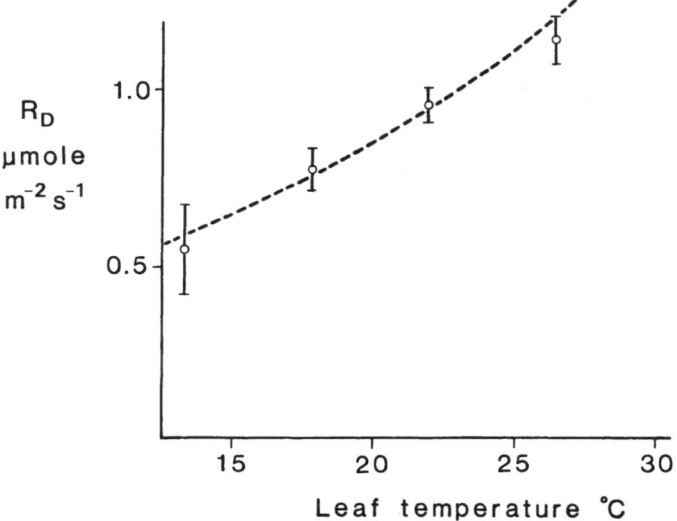

Figure 4. Dark respiration of leaves as a function of leaf temperature, with data grouped in 4°C categories, and adjusted for the ageing effect shown in Figure 5 to represent value at ligule appearance.

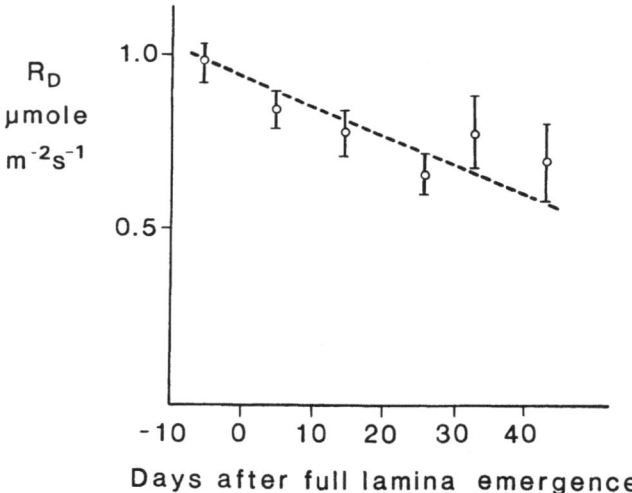

Figure 5. Dark respiration of leaves as a function of leaf age after ligule appearance, with data grouped in 10 day categories and adjusted to 20°C by the temperature effect shown in Figure 4.

Ears

Quantum yield. There was little evidence for any decrease in quantum yield with increasing temperature, except for the 'no drought' treatment (Figure 6). Other treatments showed no trend with temperature, but generally had lower quantum yields (Table 2).

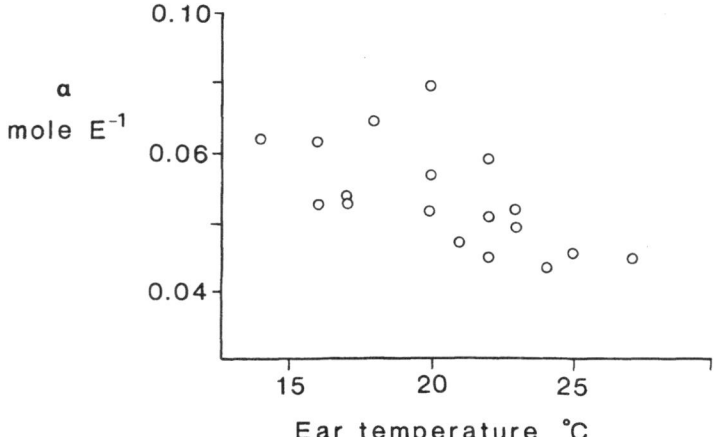

Figure 6. Quantum yield for ears as a function of ear temperature for the 'no drought' treatment only.

Table 2. Quantum yield (mole E^{-1}) for ears

No drought	$0 \cdot 052 \pm 0 \cdot 004$
Late drought	$0 \cdot 038 \pm 0 \cdot 004$
Early drought	$0 \cdot 033 \pm 0 \cdot 002$
Full drought	$0 \cdot 031 \pm 0 \cdot 003$

Mesophyll conductance. Results in Table 3 show a considerable decline of $1/r_m$ with age, and, early in grain filling, a higher conductance for the 'no drought' treatment.

Table 3. Mesophyll conductance for ears, early and late in the grain filling period.

Treatment	Early		Late	
	Mean date after anthesis	$1/r_m$ s mm^{-1}	Mean date after anthesis	$1/r_m$ s mm^{-1}
No drought	9	$1 \cdot 19 \pm 0 \cdot 07$	27	$0 \cdot 69 \pm 0 \cdot 07$
Late drought	11	$0 \cdot 88 \pm 0 \cdot 06$	21	$0 \cdot 72 \pm 0 \cdot 08$
Early drought	9	$0 \cdot 87 \pm 0 \cdot 05$	25	$0 \cdot 53 \pm 0 \cdot 09$
Full drought	13	$0 \cdot 78 \pm 0 \cdot 10$	19	$0 \cdot 73 \pm 0 \cdot 14$

Dark respiration. As observed for leaves, the only effects were an increase with temperature, Q_{10} about 1·38, and a decline with age (Figure 7).

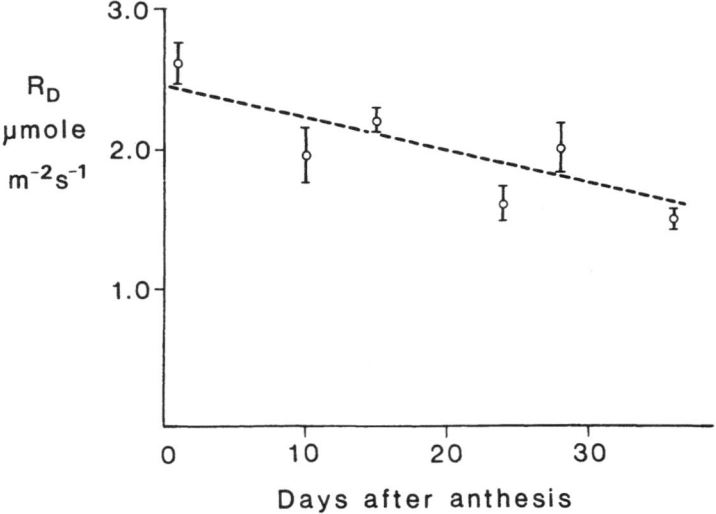

Figure 7. Dark respiration of ears, adjusted to 20°C, as a function of time after anthesis.

DISCUSSION

These results show the magnitude of environmental effects on some components of the photosynthesis of one particular crop of barley subjected to water stress treatments. Two major components of whole crop photosynthesis are not discussed at all – stomatal effects and differences in leaf area. The components considered here concern the potential photosynthesis of green tissue on the crop, and in general they show only small effects of water stress treatment.

For ears, the analysis has included some substantial simplifications: equation (2) treats all ear tissue as equivalent, though clearly the awns are quite distinct physically and probably photosynthetically from the bulk of the ear. Also, the expression of photosynthetic rates per unit area requires an estimate of area – we have used the product of total length (including awns) and the width half way along the ear to give a notional projected area. These simplifications should be noted when considering the results, particularly the absolute values of parameters.

The effects of treatment on quantum yield for ears include a decrease due to early drought that was apparent in the post anthesis period. The ear measurements were all made after anthesis, but the early drought

treatment, which received weekly irrigation after anthesis, had a quantum
yield as low as that of the full drought treatment. Slight differences in
ear conformation due to early drought cannot be ruled out as a possible
cause of this effect, though the estimates of projected ear area showed
no treatment differences.

The quantum yields are expressed as μmole CO_2 per μE incident.
Assuming that for leaves the absorption coefficient is 80%. this gives a
net quantum yield of about $0 \cdot 062$ mole E^{-1} at temperatures around 20°C.
This agrees well with other results for C_3 species [6] including wheat
[10]. However the decrease with temperature found here is considerably
greater than that reported for *Encelia californica* [6], for which the
decline by 30°C was only about 15%.

The quantum yield for leaves showed little direct water stress
effect, but the temperature response does suggest a possible indirect
effect. Water stress leads to stomatal closure, and this causes increases
in leaf temperature that can be quite substantial [7] Decreases in quantum
yield would accompany these temperature increases. This may explain some
of the differences in dry matter production observed in an earlier drought
experiment [9] where canopy air temperature often differed between
treatments by more than 4°C during the day. In that year, 1976, canopy air
temperatures were often near 25°C even in the irrigated treatment.

The magnitude and time course of mesophyll conductances for leaves
were similar to those reported for flag leaves of wheat [10]. The earlier
decline for the full drought treatment suggests premature senescence under
water stress - a response that is also observed for green leaf area [8].

The effects of temperature on respiration rates were much as expected.
The decline with time may be related to progressive changes in photosynthetic
capability of the organ - i.e. decreasing as the maximum photosynthesis
rate, which is proportional to mesophyll conductance, decreases [10] - but
for early leaves it may just reflect the changing light environment in
the canopy, as new leaves appear above them.

Acknowledgements

We are grateful to Mr J.Joyce, Mr B.K.French and Dr J.E.Leach for
their assistance in the field, to Dr Leach for the information on leaf
appearance, and to Mrs G.Tuck for the model fitting.

74

References

1. von Caemmerer S and Farquhar GD (1981) Some relationships between the biochemistry of **photosynthesis** and the gas exchange of leaves. Planta 153: 376-387.

2. Chartier P and Prioul JL (1976) The effects of irradiance, carbon dioxide and oxygen on the net photosynthetic rate of the leaf : a mechanistic model. Photosynthetica 10: 20-24.

3. Day W, Legg BJ, French BK, Johnston AE, Lawlor DW and Jeffers WdeC (1978) A drought experiment using mobile shelters : the effect of drought on barley yield, water use and nutrient uptake. J Agric Sci 91: 599-623.

4. Day W and Parkinson KJ (1982) Application to wheat and barley of two photosynthesis models for C_3 plants. Plant Cell Env (in press).

5. Day W, Parkinson KJ, Lawlor DW and Johnston AE (in preparation) The effect of drought on barley in two contrasting years : the response of yield, water use and nutrient uptake.

6. Ehleringer J and Björkman O (1977) Quantum yield for CO_2 uptake in C_3 and C_4 plants. Dependence on temperature, CO_2 and O_2 concentration. Plant Physiol 59: 86-90.

7. Idso SB, Reginato RJ, Reicosky DC and Hatfield JL (1981) Determining soil-induced plant water potential depressions in alfalfa by means of infrared thermometry. Agron J 73: 826-830.

8. Lawlor DW, Day W, Johnston AE, Legg BJ and Parkinson KJ (1981) Growth of spring barley under drought : crop development, photosynthesis, dry matter accumulation and nutrient content. J Agric Sci 96: 167-186.

9. Legg BJ, Day W, Lawlor DW and Parkinson KJ (1979) The effect of drought on barley growth : models and measurements showing the relative importance of leaf area and photosynthetic rate. J Agric Sci 92: 703-716.

10. Marshall B and Biscoe PV (1980) A model for C_3 leaves describing the dependence of net photosynthesis on irradiance. II Application to the analysis of flag leaf photosynthesis. J Exp Bot 31: 41-48.

11. Parkinson KJ (1981) Carbon dioxide infrared gas analysis. Effects of water vapour. J Exp Bot 22: 169-176.

12. Parkinson KJ, Day W and Leach JE (1980) A portable system for measuring the photosynthesis and transpiration of graminaceous leaves. J Exp Bot 31: 1441-1453.

13. Ross GJS (1975) Simple non-linear modelling for the general user. Proc 40th Session Intern Stat Inst Warsaw 2: 503-593.

EFFECTS OF WATER STRESS ON CO_2 EXCHANGE IN APPLE

H.G. JONES, L. FANJUL

Plant Physiology Division, East Malling Research Station, Maidstone, Kent, U.K.
(Present address of L. Fanjul: INIREB, Aptdo. Postal 63, Xalapa, Veracruz, Mexico)

ABSTRACT

The effects of mild or moderate water stress, maintained for a three week period on CO_2 exchange by leaves of small apple trees grown in pots were investigated. All components of the photosynthetic system were affected, with the maximum photosynthetic rate at non-limiting CO_2 declining to 67% of the control at -1.7 MPa and to 25% at -2.7 MPa. The corresponding values of stomatal conductance were, respectively, 48% and 12% of controls, and of mesophyll conductance 63% and 31%. Dark respiration was not much affected by stress, but the CO_2 compensation concentration increased with stress.

Methods for determining the stomatal limitation to photosynthesis are outlined. For the present data, the stomatal limitation increased from 39% of the total in controls to about 63% in severely stressed plants. The data also indicate that the stomatal response to humidity would have a relatively small effect on photosynthesis in well watered plants.

INTRODUCTION

Water stress is an important factor limiting plant productivity and crop yields, even in temperate climates such as the U.K. For example, one set of long-term trials with apple at East Malling (9) showed that irrigation increased yields over several years by an average of 17%. Both decreased photosynthesis per unit leaf area and decreased leaf area per unit ground area (leaf area index) can contribute to this reduced productivity.

In order to make most efficient use of irrigation and other crop management procedures such as pruning, it is necessary to determine how the photosynthetic system, including the stomata, responds to environmental stress. The most appropriate response for any plant depends on the particular climate in which it grows (12). For example, the probability of future rainfall is especially important in determining whether short-term drought avoidance responses such as stomatal closure or long-term relatively irreversible responses such as leaf abscission would prove to be more appropriate. Both types of response affect photosynthesis.

The effects of water deficits on photosynthesis probably include both 'unavoidable' effects of lowered leaf water potential (ψ) on metabolic processes, and adaptive responses such as stomatal closure, that act to conserve water and to limit the development of even lower water potentials. There is, however, increasing evidence that, at least with slowly developing stress, many components of the photosynthetic system decline together indicating close co-ordination. This acts to optimise the use of resources so that, for example, unnecessary production of photosynthetic machinery is avoided when stomatal closure is limiting photosynthetic rate (11).

There is also good evidence that leaves respond directly to the aerial environment, with stomata closing as air humidity falls (5, 16). This acts to conserve water in dry conditions, but it can also affect photosynthesis.

The experiments described in this paper investigate the role of stomata in apple in controlling photosynthesis, in response to relatively slowly developing soil water deficits and to different air humidities.

MATERIALS AND METHODS

Plant material. The studies were conducted on 3 year-old potted apple mini-trees (James Grieve on M.7 rootstock) grown and measured outdoors. Pots were covered with polyethylene bags to exclude rainfall. Control plants (C) were watered once or twice a day maintaining pot weights close to those at field capacity. Two drought treatments were imposed: S1 where the plants were watered daily to return each pot weight to 300 g less than it was at field capacity and S2 where the pot weights were returned to 500 g less than field capacity. These deficits were increased by 100 g each week over the three week stress period, at the end of which all plants were rewatered. The total weight of plant plus pot at field capacity was approximately 4.3 kg, with an average soil water content of approximately 1.2 kg. Treatments commenced on 10 August 1981. The leaf water potentials at midday in C and S1 treatments (Table 1) were comparable to those found in the field at East Malling for irrigated and unwatered trees, respectively (14).

<u>Gas exchange</u>. CO_2 and water vapour exchange of leaves near the middle of extension shoots were determined using a portable continuous-flow photosynthesis porometer as described by Fanjul *et al.* (6). All photosynthesis measurements were obtained at saturating photon irradiances of greater than 1500 μmol m^{-2} s^{-1} (400-700 nm). Where sunlight was inadequate to achieve this level, a 12V quartz-iodine lamp giving approximately 2000 μmol m^{-2} s^{-1} at the leaf, was used. Photosynthetic CO_2 response curves were obtained in normal air with 21% O_2 or with 1% O_2. An estimate of photorespiration rate (R_L) was obtained from the difference between photosynthetic rates in 1% O_2 and 21% O_2 with a cell wall CO_2 concentration (C_W) of 400 ng cm^{-3}, though it is recognised that this method overestimates the true photorespiration rate (see 7).

The stomatal conductance to CO_2 (g_l), was estimated from the rate of water loss measured with the porometer (5) and used to estimate C_W. The intracellular or mesophyll conductance (g_m) was estimated from the initial slope of the curve relating P to C_W. The photosynthetic rate at normal ambient CO_2 concentration (P_{amb}) was determined first for each leaf, before lowering the CO_2 concentration.

Further details of the experiments are given elsewhere (4).

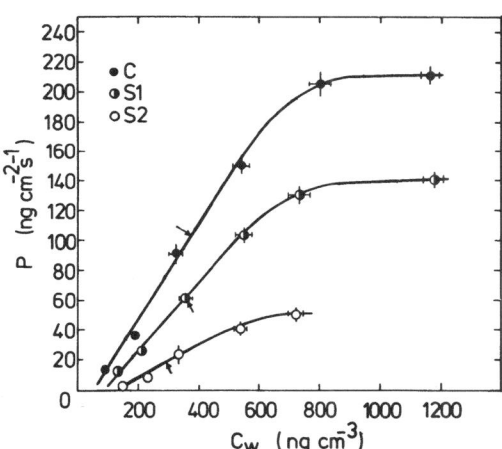

FIGURE 1. Relationship between P and C_W for apple leaves between 15 and 17 days after the start of stress. The points represent means of at least 4 leaves. (Bar = SEM; eye-fitted curves).

RESULTS

Figure 1 shows photosynthetic CO_2 response curves for mid-extension shoot leaves from each treatment, obtained between 15 and 17 days after the start of the drought treatments. The photosynthetic rate at normal CO_2 concentrations was decreased by stress as were the maximum rate (P_{max}), and mesophyll and leaf conductances, with the reduction being greater with more severe stress. The CO_2 compensation concentration (Γ) tended to increase with greater stress. These results and those after 24 days stress are summarised in Table 1, together with respiration data. The various photosynthetic parameters are also expressed as percentages of the corresponding controls.

Table 1. Effects of water stress on photosynthetic parameters of James Grieve mini-trees (± SEM).

		C	S1	S2	S1/C (%)	S2/C (%)
After 17-19 days stress:						
ψ	(MPa)	-1.1± 0.05	-1.7± 0.2	-2.7± 0.07	-	-
P_{amb}	(ng cm^{-2} s^{-1})	104	61	18	59	17
P_{max}	(ng cm^{-2} s^{-1})	214 ±28	143 ±13	54 ±19	67	25
g_l	(mm s^{-1})	5.0± 0.4	2.4± 0.2	0.6± 0.01	48	12
g_m	(mm s^{-1})	3.2± 0.2	2.0± 0.1	1.0± 0.2	63	31
Γ	(ng cm^{-3})	85 ±13	52 ±12	154 ±36	61	181
R_D	(ng cm^{-2} s^{-1})	7.2± 0.7	8.5± 1.1	6.3± 0.4	118	88
After 24 days stress:						
ψ	(MPa)	-1.3	-2.0	-2.8	-	-
P_{amb}	(ng cm^{-2} s^{-1})	120	76	42	63	35
g_l	(mm s^{-1})	5.2	2.5	1.0	48	19
g_m	(mm s^{-1})	3.9	2.3	1.7	59	44
Γ (21% O_2)	(ng cm^{-3})	89	57	129	64	145
Γ (1% O_2)	(ng cm^{-3})	13	2	50	15	385
R_D	(ng cm^{-2} s^{-1})	6.8	9.5	6.5	140	96
R_L	(ng cm^{-2} s^{-1})	49	31	17	63	35
$(R_D + R_L)/P_{amb}$		47	53	56	113	119

Figure 2 illustrates the relation between photosynthetic rate and stomatal conductance for measurements over the whole experimental period and includes some measurements taken during the recovery period after rewatering.

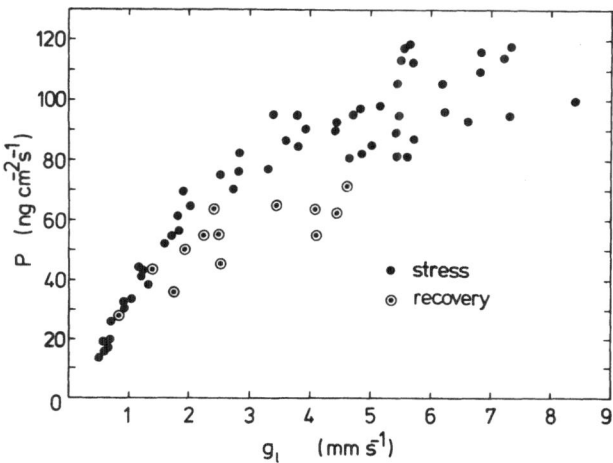

FIGURE 2. Relationship between P and g_l. Points represent measurements on different leaves over the stress period, with open points obtained after rewatering.

The relation between P and g_l approached linearity only at low values of g_l (below 2 mm s^{-1}). Increases of g_l beyond 5-6 mm s^{-1} had little effect on P.

Figure 3 shows the response to leaf-air vapour pressure difference (δe) of the stomatal conductance to water vapour (= $1.6 \times g_l$) for the control trees, where variation of δe arose by natural variation (it should be remembered that other factors such as radiation were also covarying). The correlation coefficient relating g_l and δe was -0.71 (P<0.05) for the control leaves. Also shown are humidity responses of seedling and rootstock leaves obtained in other experiments under more controlled conditions (5). Although vapour pressure difference can have large effects on stomatal conductance in well watered plants, the range of g_l (generally above 5 mm s^{-1}) was such that only small effects on P were likely to occur in this material. In contrast, no significant effect of δe on g_l was detectable in the stressed plants.

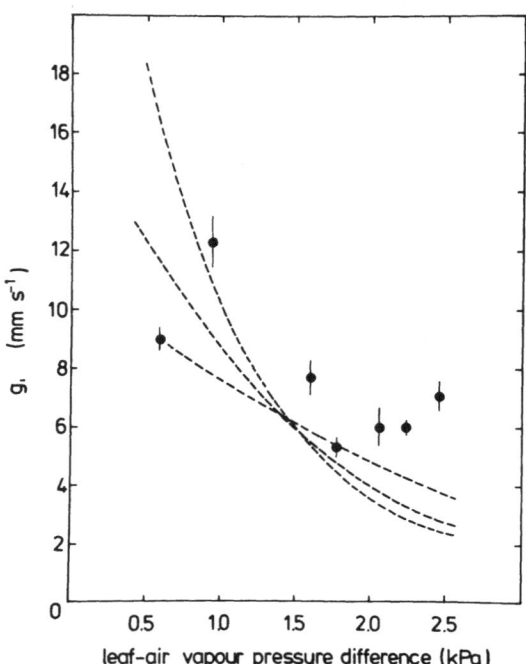

FIGURE 3. Response of leaf conductance for water vapour to leaf-air vapour pressure difference. Points are for leaves of control James Grieve. The dashed lines are for other apple material (data from ref. 5).

DISCUSSION

Marked effects of the drought treatment on almost all aspects of CO_2 exchange were demonstrated. Dark respiration (R_D), however, was least sensitive to water deficits, there being indications that dark respiration was promoted by mild stresses, and decreased only at severe stress. This type of response has been reported previously for other species (see e.g. 1). The decreases in estimated photorespiration rates almost exactly matched those in P_{amb}. This is probably a result of the close linkage between the photosynthetic carbon reduction cycle and the photorespiratory carbon oxidation cycles (see e.g. 15) and suggests that a common component (perhaps the amount of ribulose-bis phosphate carboxylase/oxygenase enzyme) is affected by stress.

Because of the small effect of water deficits on dark respiration, the ratio of total respiration $(R_D + R_L)$ to net photosynthesis increased with stress, as net photosynthesis declined. The tendency for Γ to increase is a further indication of CO_2 uptake decreasing relative to loss.

It is relevant to question the extent to which the changes observed with stress contribute directly to the reduction in P, and to what extent they may be adaptive responses preventing wasteful use of resources. For example large quantities of carboxylase are not required if stomatal closure is limiting photosynthesis. The close correlation between P and g_l (Fig. 2) could result from either of those possibilities. Jones (10, 13) and Farquhar & Sharkey (8) have discussed methods for determining the relative limitation to photosynthesis imposed by different component processes such as stomatal diffusion and mesophyll processes. As they point out, there is no absolute way of defining the relative importance of different processes.

The three most useful approaches that have been suggested for determining the stomatal limitation (l_g) as a fraction of the total photosynthetic limitation (i.e. a measure of the control exercised by the stomata) are

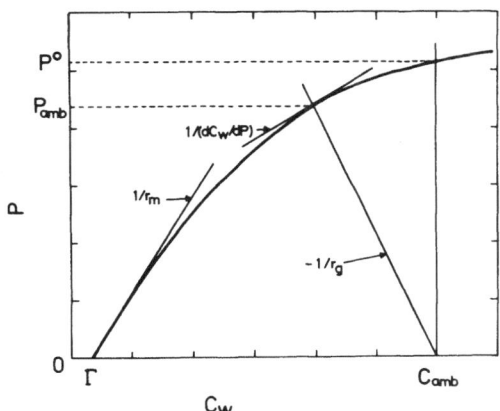

FIGURE 4. Illustration of calculation of the various parameters needed for determining stomatal limitations.

illustrated in Fig. 4 and are the following:

1) $l_g = r_g/(r_m + r_g)$

where the resistances r_g and r_m are the reciprocals of the corresponding conductances. This approach is only valid under CO_2 limiting conditions, where P increases linearly with increasing C_w.

2) $l_{g_o} = (P^o - P_{amb})/P^o$

where P^o is the photosynthetic rate at infinite stomatal conductance (estimated from a photosynthetic CO_2 response curve for $C_w = C_{amb}$, where C_{amb} is the CO_2 concentration in normal air). On this definition the stomatal limitation is simply the reduction in photosynthesis below what it would be if there was no stomatal limitation.

3) $l_g = r_l/(r_l + (d\, C_w/d\, P))$

where $d\, C_w/d\, P$ is the reciprocal of the slope of the P: C_w response curve at the normal operating point (i.e. normal air and the usual value of r_l). This approach, in contrast with 2), gives the marginal effect on photosynthesis of a small change in stomatal conductance at the normal operating point. This is particularly relevant when one considers the effect of small changes in stomatal aperture, for example in model studies of optimal stomatal behaviour (3).

All three methods give the same answer if the P: C_w response curve is linear up to values of C_w equal to normal ambient CO_2 concentrations, but this rarely holds (see e.g. 2, 11).

However, for the current data set the photosynthetic response was nearly linear over this range, so the differences between the methods were small.

Table 2. Effect of stress on the stomatal limitation to photosynthesis, calculated using three methods.

		C	S1	S2
l_g	(method 1)	0.39	0.46	0.63
l_g	(method 2)	0.38	0.46	0.62
l_g	(method 3)	0.39	0.46	0.61

Interestingly, however, in a similar experiment carried out in a glasshouse (4) the operating point was on the curved part of the CO_2 response curve. The reasons for this difference are not known.

Not only was the stomatal conductance more sensitive to stress than any other photosynthetic parameter (Table 1), but all the limitation analyses demonstrated an increase of l_g from about 0.39 in C to about 0.63 in S2. Although these results imply that stomatal closure was a major factor in the reduction in photosynthetic rate in the stressed plants, the alterations in intracellular photosynthetic activity are clearly also significant. In particular, it should be noted that the limitation analyses do not adequately take account of altered respiration, yet this can affect P, even without any change in g_m or P_{max}. Figure 2 shows that stomatal conductance recovered more rapidly after relief of stress than did the intracellular limitation to photosynthesis.

The stomatal response to humidity apparently provides a useful mechanism limiting water use in conditions of high evaporative demand, that has only a minimal cost in terms of decreased photosynthesis, at least when soil water is adequate.

REFERENCES
1. Bunce JA, Miller LN. 1976. Differential effects of water stress on respiration in the light in woody plants from wet and dry habitats. Can J Bot 54: 2457-2464.
2. Caemmerer Svon, Farquhar GD. 1981. Some relationships between biochemistry of photosynthesis and the gas exchange of leaves. Planta 153: 376-387.
3. Cowan IR. 1977. Stomatal behaviour and environment. Adv Bot Res 4: 117-228.
4. Fanjul L. 1982. Effects of water stress on photosynthesis in apple leaves. Ph.D thesis submitted at Wye College, University of London.
5. Fanjul L, Jones HG. 1982. Rapid stomatal responses to humidity. Planta 154: 135-138.
6. Fanjul L, Jones HG, Treharne KJ. 1980. A portable system for simultaneous measurements of transpiration and CO_2 exchange. Photosynth Res 1: 83-92.
7. Farquhar GD, Caemmerer Svon, Berry JA. 1980. A biochemical model of photosynthetic CO_2 assimilation in leaves of C_3 species. Planta 149: 78-90.
8. Farquhar GD, Sharkey TD. 1982. Stomatal conductance and photosynthesis. Annu Rev Pl Physiol 33: 317-345.
9. Goode JE, Higgs KH, Hyrycz KJ. 1978. Nitrogen and water effects on the nutrition, growth, crop yield and fruit quality of orchard-grown Cox's Orange Pippin apple trees. J. hort Sci 53: 295-306.
10. Jones HG. 1973. Limiting factors in photosynthesis. New Phytol 72: 1089-1094.
11. Jones HG. 1973. Moderate-term water stresses and associated changes in some photosynthetic parameters in cotton. New Phytol 72: 1095-1105.

84

12. Jones HG. 1981. The use of stochastic modelling to study the influence of stomatal behaviour on yield-climate relationships. In Charles-Edwards DA, Rose DA, eds. Mathematics and Plant Physiology, pp. 231-244. London: Academic Press.
13. Jones HG. 1983. Plants and the aerial environment. Cambridge: Cambridge University Press (in press).
14. Jones HG, Higgs KH. 1979. Water potential-water content relationships in apple leaves. J exp Bot 30: 965-970.
15. Lorimer GH, Andrews TJ. 1981. The C_2 chemo- and photorespiratory carbon oxidation cycle. In Hatch MD, Boardman NK, eds. The biochemistry of plants. Vol 8, pp. 329-374. New York: Academic Press.
16. Ludlow MM. 1980. Adaptive significance of stomatal responses to water stress. In Turner NC, Kramer PJ, eds. Adaptation of plants to water and high temperature stress, pp. 123-138. New York: Wiley.

MORPHOLOGICAL AND PHYSIOLOGICAL ADAPTATIONS FOR MAINTAINING
PHOTOSYNTHESIS UNDER WATER STRESS IN APPLE TREES

A.N. LAKSO
NEW YORK STATE AGRICULTURAL EXPERIMENT STATION
GENEVA, N.Y. 14456   USA

ABSTRACT

The apple tree has evolved several mechanisms, both morphological and
physiological, to ameliorate the effects of water stress on photosynthesis.
Drought escape or avoidance occurs by rapid leaf area development in the
spring. Drought tolerance at high water potentials is accomplished by good
water use efficiency through tight coupling of photosynthesis and stomatal
conductance. Stomatal responses to humidity in field trees appears to be
modified by the stomatal sensitivity to photosynthetic activity. In addition
to stomatal behavior, the reduction in leaf area and the folding of exposed
leaves reduce radiation interception and increase water use efficiency.
Increases in the root hydraulic conductivity in response to evaporative demand
also helps maintain high water potentials. Drought tolerance at low water
potentials is mediated by marked osmotic adjustment in mature leaves although
shoot tips show little osmotic adjustment. These mechanisms are effective
primarily due to the slow physiological aging of apple leaves.

INTRODUCTION

Maintenance of photosynthetic productivity under water stress can be
accomplished by many different mechanisms. In evaluating the resistance of a
crop species to water stress, the range of resistance mechanisms should be
considered rather than just one or a few. Landsberg and Jones [15] have
recently reviewed the water relations of the apple, thus this paper will
evaluate the morphological and physiological adaptations in relation to
maintenance of photosynthesis under water stresses that occur at different
times and intensities.

Turner [19] classified these resistance mechanisms into three major
categories: drought escape, drought tolerance at high water potential, and
drought tolerance at low water potentials. These categories provide a useful
framework for analysis.

DROUGHT ESCAPE

Because of their perennial nature apple trees display a growth habit that is effective in drought escape. The canopy develops on basically two types of shoots: short shoots that bear the fruit (spurs) and extension shoots. Due to stored reserves, existing tree framework and rosette-like short stems, the spurs develop quickly in the spring to provide rapid light interception and a substantial amount of photosynthesis before summer droughts occur. The remaining leaf area develops on longer extension shoots that respond more to drought since these leaves grow well into the summer.

This developmental mechanism for avoiding drought requires the leaf area that develops early in the season to have a very slow physiological aging rate to maintain the photosynthetic productivity for the whole season. Apple leaves have slow photosynthetic aging curves [1]. Essentially no change in photosynthetic rates for almost 4 months in exposed spur leaves in the field has been reported [17]. Autumnal photosynthetic senescence and leaf fall in apple appear to be primarily controlled by temperature [9,12] explaining the need for mechanical leaf removal in apple in the tropics [7].

DROUGHT TOLERANCE AT HIGH WATER POTENTIALS

The two primary mechanisms involved are (1) reduction of water loss and (2) maintenance of water uptake, which both serve to maintain high water potentials as stress occurs.

In apple the reduction of water loss is accomplished in several ways. One very important mechanism is the tight coupling of stomatal conductance (g_s) to photosynthesis (Fig. 1). Over a wide range of experimental treatments on outside-grown apple trees essentially the same relationship was found between g_s and photosynthesis regardless of whether the treatments induced water stress or inhibited translocation [R.S. Johnson and A. Lakso, unpublished). This allows for very good water use efficiency (WUE) since the stomates respond to the changes in photosynthetic activity.

The important response to humidity in apple leaves is quite variable (Fig. 2), although the reasons are not entirely clear [13,21]. Thorpe et al [15,18] have modeled a linear response of apple stomates to humidity,;however, this author has observed cases of no stomatal response in actively growing field trees over a day even though the VPD changed 15-20 mbars (Fig. 2). Later in the season stomates on leaves on the same trees showed both the expected inverse response of g_s to VPD as well as a hysteresis on another day.

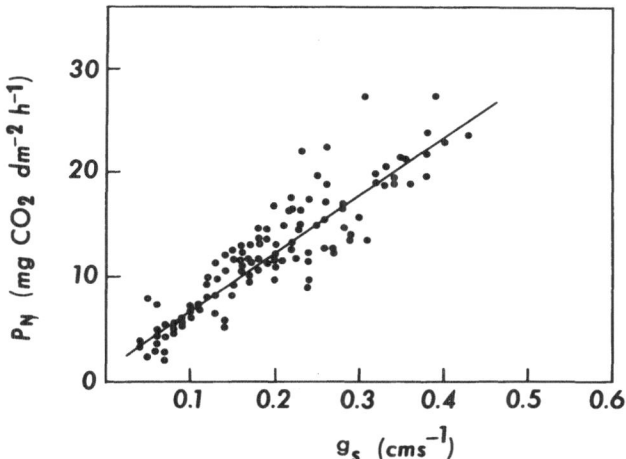

FIGURE 1. The relationship of net photosynthesis (Pn) to stomatal conductance (g_s) in field (●) and potted (o) apple tree leaves (from [10]).

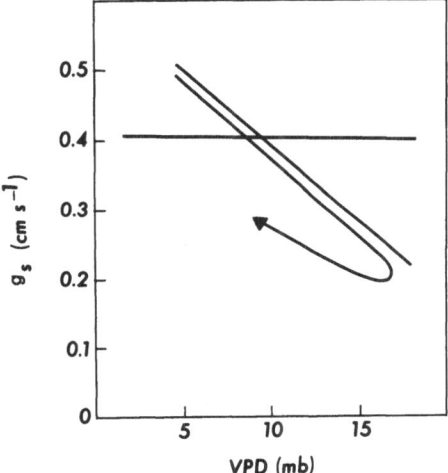

FIGURE 2. Relationships of apple leaf stomatal conductance and leaf-air vapor pressure deficit on three different sunny days in the field during a season.

These responses are believed to be due to the tight coupling of g_s to photosynthesis (presumably on internal CO_2 response). The apparent poor humidity response seems to be a lack of stomatal opening in low VPD's due to the over-riding response to photosynthesis. Data calculated from West and Gaff [22] shows no stomatal response up to 15 mb VPD then a decrease in g_s with higher VPD when measured at 300 μl l^{-1} CO_2 (Fig. 3). When measured in CO_2-free air, g_s values decreased with increasing VPD in a consistent manner, supporting the conclusion that apple stomatal responses to VPD can be over-ridden by the CO_2 response.

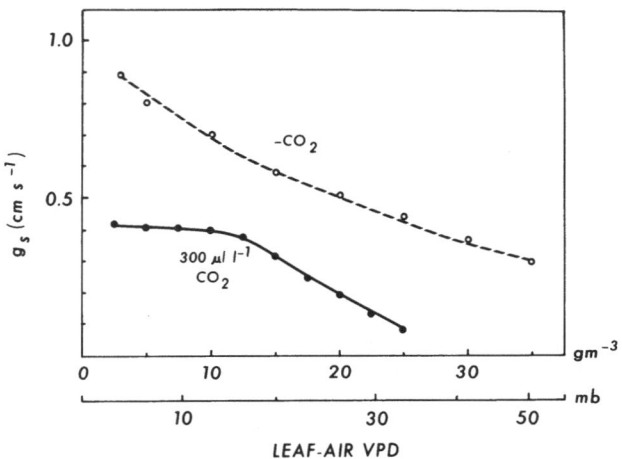

FIGURE 3. Response of apple leaf stomatal conductance to VPD in 300 ul l^{-1} CO_2 and CO_2-free air (calculated from the data of West and Gaff 22).

Several contradictory responses of apple stomates to VPD in the field are also consistent with the photosynthetic activities of the plants. Outside-grown potted trees that showed typical decreases in g_s as the VPD increased during the day also showed the same diurnal g_s pattern the following day while in a growth chamber with constant VPD. In another study on field apple trees radiation, temperature and g_s were monitored diurnally. The g_s correlated poorly to VPD but well to the time after exposure to continuous full sunlight (Fig. 4). Apple leaf photosynthesis is unaffected by low to moderate VPD since the g_s is sufficient to be non-limiting. Under higher VPD's (more than about 25 mb) the stomatal closure should begin to limit the photosynthesis.

Greenhouse or outside potted trees under mild evaporative demand can show marked humidity responses typical of many other plants [4; A. Lakso, unpublished]. Greenhouse trees in the autumn that showed a strong humidity response demonstrated a much reduced VPD response (primarily to low VPD) after only a few days of moderate evaporative stress while concurrently showing an increase in response to CO_2 (A. Lakso and F. Lenz, unpublished).

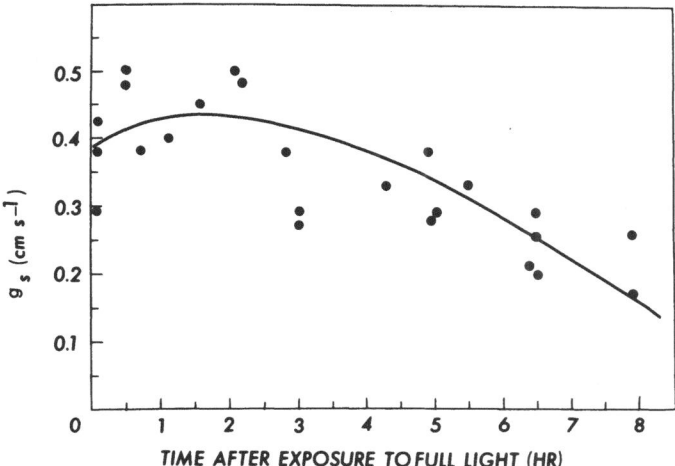

FIGURE 4. Patterns of stomatal conductance of several apple leaves on a clear day in the field. Time 0 varied for the leaves from 0800 to 1315 while the VPD varied from 4 to 23 mbars.

Thus, in field apple trees the reduction in water loss by stomatal behavior appears to be one of maintaining a high water use efficiency as drought develops. This ameliorates the onset of stress and helps to maintain photosynthesis. Under severe stress apple stomates close to conserve water as typical of most plants.

In addition to the stomatal behavior, reduction of water loss is also aided by the reduction of leaf area. As discussed earlier, spur leaf area develops quickly in the spring, but substantial leaf area on extension shoots develop longer into the season. This later leaf area development is affected by water stress in several ways. First, it occurs later in the season, thus the probability of drought is greater. Second, leaf size is reduced by water stress. Third, leaf numbers are reduced since a determinate terminal bud sets on stressed extension shoots. This bud can revert to a new flush of growth if the stress is relieved within 1-4 weeks after bud set in mid-season; however, the interactions of environmental and internal factors controlling this bud is complex.

The leaf area that does develop displays another morphological adaptation, folding about the midrib, that reduces radiation absorption by the

exterior exposed leaves and raises leaf water potentials. The leaf folding is
positively correlated to radiation exposure [11]. This has a relatively small
effect on total canopy water loss since total radiation interception is
affected little, but it provides a more uniform distribution of radiation into
the canopy for higher total photosynthesis and thus better water use
efficiency.

Maintenance of water uptake is an alternate way to maintain high leaf
water potentials. Besides changes in the extent and growth of root systems
that obviously can obtain additional water for high water potentials, apple
trees have also been shown to markedly change hydraulic conductances in
response to the growing conditions. Apple trees growing outside for several
weeks developed three-fold greater root conductances than greenhouse-grown
trees giving much less response to a water stress imposed later [3]. The
photosynthesis was similarly less affected by the stress on the outside-grown
trees [2].

It should be noted that exposed field apple leaves normally operate at -
12 to -20 bars leaf water potential on sunny days even with high soil moisture
due to low plant hydraulic conductances typical of tree species. Leaf water
potentials can reach -30 bars with good soil moisture if the evaporative
demand is high, so the terms "high" and "low" water potentials have different
magnitudes relative to annual plants.

DROUGHT TOLERANCE AT LOW WATER POTENTIALS

For crop production plants drought tolerance at low water potentials is
primarily that of turgor maintenance. The most effective mechanism for turgor
maintenance is osmotic adjustment [20]. Osmotic adjustment of several bars
has been shown in apple [5,6,8]. In response to somewhat drier conditions our
studies have shown that stomatal closure in response to excision drying can
occur at leaf total water potentials that vary over 20 bars during a season
[10]. This variation can be explained almost completely by active osmotic
adjustment (determined at 100% RWC) that changes both downward and upward in
response to the stress periods (Fig. 5). Besides the turgor maintenance by
osmotic adjustment, adapted field apple leaves have been found to have
measurable photosynthesis below -50 bars total potential [10] indicating that
the photosynthetic mechanism can withstand low water potentials if sufficient
turgor is available for stomatal opening.

The apple leat osmotic adjustment observed can occur with good soil moisture if the evaporative demand is sufficient to induce the leaf stress levels necessary to induce osmotic accumulation. Thus, soil moisture itself is not necessarily a good indicator of stress or stress adaptation in apple.

Diurnally the exposed apple leaf may lower its osmotic potential 5 to 10 bars on a sunny day to help maintain mid-day turgor [6]. About 2 to 4 bars of this adjsutment is due to the apparent accumulation of photosynthates not yet translocated and the remainder by passive concentration as the RWC decreases.

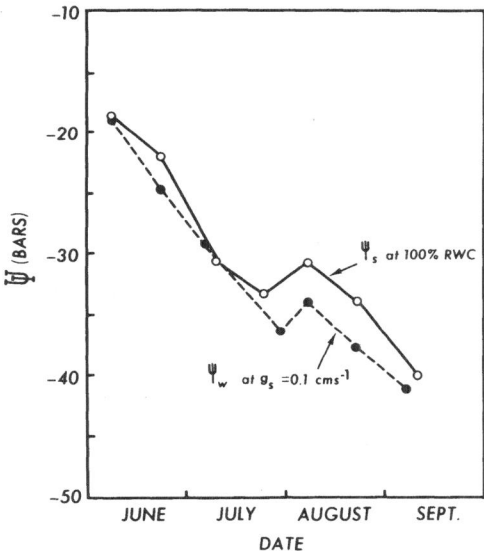

FIGURE 5. Seasonal changes in mature apple leaf osmotic potential at 100% RWC (determined by pressure-volume curves) and leaf water potential giving a g_s = 0.1 cm s^{-1} on excised leaves [14].

The combination of long term osmotic adjustment to periodic stresses and diurnal adjustment provide an effective mechanism to maintain sufficient turgor for stomatal opening and photosynthetic productivity. To this end we have observed several cases of open stomates and active photosynthesis under drought conditions that casued obvious wilting and even leaf drop in nearby plantings of annual crops and other perennial fruit crops (cherries and grapes).

Interestingly, the expanding shoot tip and young leaves of apple do not show significant osmotic adjustment while the mature leaves on the same shoots show osmotic adjustment [14]. Under severe drying stress the shoot tip and youngest leaves show 3-5 bars of adjustment, although leaf expansion was almost stopped by that point. These results are opposite those found in wheat [16] where the apex and young leaf showed marked osmotic adjustment. In the context of the leaf area development and the slow physiological aging of matured leaves, the lack of osmotic adjustment in the apple shoot tip is effective in reducing leaf area in response to stress while the mature leaves adjust to the stress to maintain photosynthetic production.

In summary, the apple shows several morphological and physiological mechanisms to maintain photosynthesis under water stress. Patterns of leaf area development allow for rapid canopy development in the spring to escape stress. Once the canopy is established, tight stomatal coupling to photosynthesis provides efficient water use. As stress develops further, osmotic adjustment in mature leaves maintains photosynthesis at low water potentials while leaf area development is reduced. A key physiological aspect that allows these stress adaptations to be effective is the slow physiological aging of apple leaves. Taken together these mechanisms confer a high degree of stress adaptability to the apple tree.

LITERATURE CITED

1. Barden JA (1978) Apple leaves, their morphology and photosynthetic potential. HortScience 13:644-646.
2. Davies FS and Lakso AN (1979) Water stress responses of apple trees. I. Effects of light and soil preconditioning treatments on tree physiology. J Amer Soc Hort Sci 104:392-395.
3. Davies FS and Lakso AN (1979) Water stress responses of apple trees. II. Resistance and capacitance as affected by greenhouse and field conditions. J Amer Soc Hort Sci 104:395-397.
4. Fanjul L and Jones HG (1982) Rapid stomatal responses to humidity. Planta (in press).
5. Goode JE (1970) The cumulative effects of irrigation on temperate fruit crops and some recent studies on the water relationships of apple trees. Proc 18th Int Hort Cong 4:187-197.
6. Goode JE and Higgs KH (1973) Water, osmotic and pressure potential relationships in apple leaves. J Hort Sci 48:203-215.
7. Janick J (1974) The apple in Java. HortScience 9:13-15.
8. Jones HG and Higgs KH (1979) Water potential-water content relationships in apple leaves. J Exp Bot 30:965-970.
9. Jonkers H (1980) Autumnal leaf abscission in apple and pear. Fruit Sci Rpts 7:25-29.
10. Lakso AN (1979) Seasonal changes in stomatal response to leaf water potential in apple. J Amer Soc Hort Sci 104:58-60.

11. Lakso AN (1985) Apple leaf morphology in relation to light exposure and its physiological significance. Proc 21st Int Hort Cong (Hamburg) (in press).
12. Lakso AN and Lenz F (1983) Regulation of apple tree photosynthesis in the autumn by temperature. In Lakso AN and Lenz F, eds. Regulation of Photosynthesis in Fruit Trees, NY State Agr Exp Sta Spec Bul (in press).
13. Lakso AN and Seeley EJ (1978) Environmentally induced responses of apple tree photosynthesis. HortScience 13:646-650.
14. Lakso AN, Frackelton AS and Carpenter SG (1983) Seasonal osmotic relations in mature and immature apple leaves. J Amer Soc Hort Sci (submitted).
15. Landsberg JJ and Jones HG (1981) Apple orchards. In Kozlowski TT, ed. Water Deficits and Plant Growth, vol VI, pp 419-469.
16. Munns R, Brady CJ and Barlow EWR (1979) Solute accumulation in the apex and leaves of wheat during water stress. Aust J Plant Physiol 6:379-389.
17. Porpiglia PJ and Barden JA (1980) Seasonal trends in net photosynthetic potential, dark respiration, and specific leaf weight of apple leaves as affected by canopy position. J Amer Soc Hort Sci 105:920-923.
18. Thorpe MR, Warrit B and Landsberg JJ (1980) Responses of apple leaf stomata: a model for single leaves and a whole tree. Plant Cell Environ 3:23-27.
19. Turner NC (1979) Drought resistance and adaptation to water deficits incrop plants. In Mussell H and Staples RC, eds. Stress Physiology in Crop Plants, pp 343-372. Wiley-Interscience, New York.
20. Turner NC and Jones MM (1980) Turgor maintenance by osmotic adjustment: a review and evaluation. In Turner NC and Kramer PJ, eds. Adaptation of Plants to Water and High Temperature Stress, pp 87-103. Wiley-Interscience, New York.
21. Warrit B, Landsberg JJ and Thorpe MR (1980) Responses of apple leaf stomata to environmental factors. Plant Cell Environ 3:13-22.
22. West DW and Gaff DF (1976) The effect of leaf water potential, leaf temperature and light intensity on leaf diffusion resistance and the transpiration of leaves of Malus sylvestris. Physiol Plant 38:98-104.

WATER RELATIONS OF PLANT CELLS

E. STEUDLE, S.D. TYERMAN and S. WENDLER
Arbeitsgruppe Membranforschung am Institut für Medizin,
Kernforschungsanlage Jülich, D-5170 Jülich, FRG.

1. INTRODUCTION

An adequate and complete description of water relations of higher
plants at the tissue and organ level requires reliable data of the water
relations at the cellular level. An understanding of cell water relations
is a necessary prerequisite for characterizing the hydraulic resistances
to water transport in the different pathways in tissues, i.e., in the
apoplasmic, symplasmic and transcellular (cell-to-cell) pathways.
Furthermore, the driving forces for the transport across the tissue have
to be known. It is commonly agreed that there is no active water transport
in plants (for energetic reasons) and that water in cells and tissues
is moved by the gradient of the chemical potential of water or water po-
tential(ψ), which incorporates turgor pressure (P), osmotic pressure (π)
and a matric component (τ), i.e.:

$$\psi \;=\; P - \pi - \tau \tag{1}$$

Thus, we have to measure the different components of the water potential
to get more information about the driving forces and, furthermore, we
have to quantify the hydraulic resistances to water movement in the
different tissue pathways.

2. METHODS

Conventional methods such as the pressure chamber and psychrometric
methods yield average water potentials of entire tissues. The methods
"average" turgor and osmotic pressure of all cells in a tissue or organ
regardless of the fact that there are different types of cells present. The
methods, furthermore, require water flow equilibrium within the tissue and
assume that the cell membranes are ideally semipermeable to the solutes.

96

In many cases these assumptions have not been really checked and this may result in errors in both the measurement and in the interpretration of the results. Furthermore, the pressure bomb and the psychrometric methods are not suited to measure the kinetics of water movement and to evaluate the amounts of water transported in the different paths of a complex tissue.

2.1. Pressure probe technique

In contrast to these methods is the pressure probe technique (Hüsken et al. 1978 ; cf. Zimmermann and Steudle 1978) which allows determinations of water relations parameters at the cellular level. Both, components of the water potential and hydraulic conductivities can be measured. The pressure probe (Fig. 1) consists of a microcapillary (tip diameter: 2 to 7 μm) filled with silicone oil and connected to a small pressure chamber containing a pressure transducer.

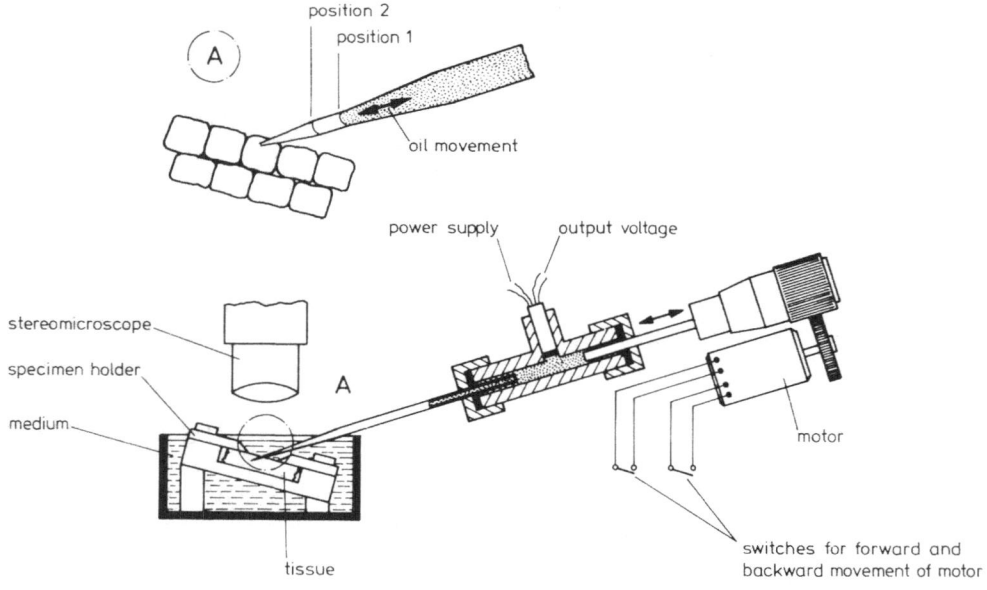

Figure 1. Schematic presentation of the pressure probe for measuring cell turgor, water flows, hydraulic conductivity and cell wall elasticity of higher plant cells. For further explanation, see text.

When the microcapillary tip is introduced into a cell, the cell turgor (P)
pushes back the oil into the capillary, forming a meniscus at the cell sap/
oil boundary in the tip. The meniscus can be positioned at a certain point by
an electronic feedback mechanism (Hüsken et al. 1978; not shown in Fig.1)
or manually, and is then moved by a defined amount to change the cell volume
(V) and cell turgor. Changes in cell volume (ΔV) are obtained from the capillary diameter and the distance between two positions of the meniscus (Fig.1),
whereas the corresponding ΔPs are recorded by the pressure transducer. Since
the elastic modulus of the cell (ε) is given by:

$$\varepsilon = V \frac{\Delta P}{\Delta V} , \qquad (2)$$

this parameter can be evaluated when the cell volume is known.

In order to determine water exchange rates (half-times of water exchange
of cells with their surroundings, $T_{1/2}$) the meniscus is adjusted to a certain
position and water flow is allowed to equilibrate. Then the meniscus is moved
"nearly instantaneously" to a second position and is fixed there again. The
subsequent relaxation of turgor to a new stationary value is recorded and is
used to evaluate the hydraulic conductivity (Lp) of the cell membrane. Since

$$T_{1/2} = \frac{V}{A} \frac{\ln 2}{Lp(\varepsilon + \sigma \cdot \pi_0^i)} , \qquad (3)$$

Lp can be calculated if V/A and ε as well as the osmotic pressure of the
cell, π_0^i, are known. $\sigma \cdot \pi_0^i$ can be estimated from the cell turgor when the
tissue is equilibrated in distilled water or from separate measurements of
the osmotic pressure of the cell sap. σ is the reflection coefficient of the
solutes in the cell.

The pressure probe has the advantage that turgor pressure can be measured
with high accuracy (ca. 30 mbar in the pressure range from 0 to 15 bar).
Averaging of turgor and cell wall elasticity as with the conventional methods is avoided. Since P can be recorded in single cells for long
periods of time (sometimes for a whole day) it is possible to measure the
effect of osmotic stress on tissue cells. Furthermore, not only hydrostatic
experiments (as described above) can be performed but also osmotic experiments in which the cell (or tissue) is exposed to a change in the osmotic
pressure of the medium. In these experiments similar results for the water
relation parameters (Lp, $T_{1/2}$ etc.) have been obtained as for hydrostatic

experiments and this is a proof for the validity of the technique (Tyerman and Steudle 1982).

The half-time of water exchange ($T_{1/2}$) is perhaps the most important cellular water relations parameter for tissues under water stress, since $T_{1/2}$ affects the overall rate of shrinking or swelling of a tissue. If we neglect the apoplastic water transport, the propagation of a change in external water potential (or of cell turgor and volume) will proceed through a tissue following a diffusion type of kinetics and water will be propagated only in the cell-to-cell pathway. The diffusion coefficient, D, of the process (the so-called "diffusivity") will be given by (Philip 1958a):

$$D = \frac{\alpha}{2} Lp \cdot \ell \cdot (\epsilon + \pi_0^i) \sim \frac{1}{T_{1/2}} \qquad (4)$$

where α is a shape factor which can be taken as unity to a first approximation and ℓ is the thickness of the cells in the direction of the propagation of the change. It is easily verified that the diffusivity of the cell-to-cell path is inversely proportional to $T_{1/2}$ and that the factor of proportionality incorporates only cell shape and geometry factors.

On the other hand, if water propagates in both the apoplast and the cell-to-cell path it can be shown that the diffusivity is given by (Molz and Ikenberry 1974):

$$D = \frac{\ell(Lp_w \cdot a + Lp \cdot \ell \cdot A/2)}{C_w + C_c} \qquad (5)$$

where a,A = cross-sectional areas of the apoplasmic or cell-to-cell pathway, respectively, Lp_w = hydraulic conductivity of the cell wall (in $cm^2 \cdot s^{-1} \cdot bar^{-1}$), C_w = water storage capacity of the cell wall and $C_c = V/\epsilon + \pi_0^i$ = storage capacity of the cell.

Eq. (5) has been derived assuming that there is local water flow equilibrium between individual cells and its surrounding wall space. This assumption seems to be justified considering the short half-times which have been measured in higher plant cells using the pressure probe (see below).

In principle, the pressure probe technique provides a simple tool to estimate the relative amounts of water transported in the apoplasmic and cell-to-cell pathways by comparing the diffusivities obtained for the cell-to-cell path (according to Eq. (4)) with those obtained for entire tissues (e.g. from shrinking or swelling the tissue) which should incorporate both pathways (Eq. (5)). However, this comparison is only justified provided that the assumption of local equilibrium holds.

2.2 New developments in the pressure probe technique

2.2.1.   Pressure clamp technique. The pressure clamp technique introduced by Wendler and Zimmermann (1982) provides a method for measuring water relations parameters and in addition to this cell dimensions (cell volumes). The latter is very important, since the estimation of cell surface areas and cell volumes up to now causes the biggest error in the determination of Lp and ε, respectively.

In the technique the stationary cell turgor, P, is changed instantaneously with the aid of the pressure probe and is then clamped at a value of P+ΔP by moving the meniscus in the tip of the capillary to compensate for water flow across the cell membrane. When the movement of the meniscus in the tip is plotted against time, a "volume relaxation" curve is obtained from which a half-time can be evaluated. It can be shown (Wendler and Zimmermann 1982) that this half-time will be given by:

$$T_{1/2} = \frac{V}{A} \cdot \frac{\ln 2}{Lp \cdot \sigma \cdot \pi_o^i} \tag{6}$$

The half-time is by a factor of $(\varepsilon + \sigma \cdot \pi_o^i)/\sigma \cdot \pi_o^i$ larger than that which would have been obtained from a pressure relaxation experiment (see Eq.(3)) using the same cell. This is an advantage, because for higher plant cells the $T_{1/2}$-values from pressure relaxations are sometimes rather short (see below) and this causes experimental problems. Lp can be evaluated either from $T_{1/2}$ using Eq.(6) or from the initial slope, s_v, of the volume relaxation. It is valid that (Wendler and Zimmermann 1982):

$$Lp = - \frac{s_v}{A \cdot \Delta P} \tag{7}$$

Since the cell does not shrink or swell during the experiment, ε is not involved and need not be known.

When a new water flow equilibrium is reached after a volume relaxation, it can be shown that ΔV^∞ (the volume change in the capillary at $t \to \infty$) is a direct measure of the cell volume (Wendler and Zimmermann 1982):

$$V = - \frac{\Delta V^\infty}{\Delta P} \cdot (\sigma \cdot \pi_o^i + \Delta P) \tag{8}$$

Again $\sigma \cdot \pi_0^i$ can be obtained from the cell turgor at water flow equilibrium when the cell (or tissue) is bathed in distilled water. Up to now, the pressure clamp technique has been applied to giant cylindrical internodes of Chara, but should, in principle, be also applicable to higher plant cells to get better values of V and (if the shape of the cells is known) perhaps also of A to evaluate ε and Lp with higher accuracy. For cells with a short half-time, the method should enhance the resolution of the relaxation measurements and, thus, also increase the accuracy of the determination of Lp.

  2.2.2. Measurement of solute permeability and reflection coefficients. The pressure probe technique cannot only be used to measure typical water relation parameters such as $T_{1/2}$, ε and Lp, but also solute properties such as the reflection coefficients, σ, and the permeability coefficient, P_s. In the presence of a permeable osmoticum the pressure/time courses recorded with the probe are biphasic, i.e. after a (usually short) "water phase" and a pressure minimum (P_{min}), turgor is again increasing and returning to the original value (P_0; see Fig. 2). This second "solute phase" is due to an equilibration of the osmoticum across the cell membrane and is strictly exponential. It can be shown (Tyerman and Steudle 1982; Steudle and Tyerman 1982) that in most of the cases the "solvent drag" effect for the solute flow can be neglected and that for the pressure/time curves an analytical expression can be given which has already been derived   (Philip 1958b; Dainty 1963). The reflection coefficient of a permeable solute (σ_s) can be calculated from the change in the osmotic pressure of the medium ($\Delta\pi_s^0$) and the measured change in turgor at the minimum ($P_0 - P_{min}$), i.e.:

$$\sigma_s = \frac{P_0 - P_{min}}{\Delta\pi_s^0} \; \frac{\varepsilon + \pi_0^i}{\varepsilon} \; \exp\left(P_s \frac{A}{V} \, t_{min}\right) \tag{9}$$

whereas the permeability coefficient (P_s) is evaluated from the rate constant of the solute phase ($k_s = P_s \frac{A}{V}$). The biphasic pressure/time curves are reversible, i.e. when the osmoticum is removed from the medium the curve goes through a maximum (not shown in Fig. 2).

  In principle, the three important coefficients (Lp, P_s, σ_s) that govern the shrinking and swelling properties of a cell following an osmotic shock can be obtained with the pressure probe. The method has been developed using Chara internodes (Steudle and Tyerman 1982) and cells from the isolated epidermis of Tradescantia virginiana (Tyerman and Steudle 1982). Some of the data of σ and P_s for these species are summarized in tables 1 and 2 and are compared

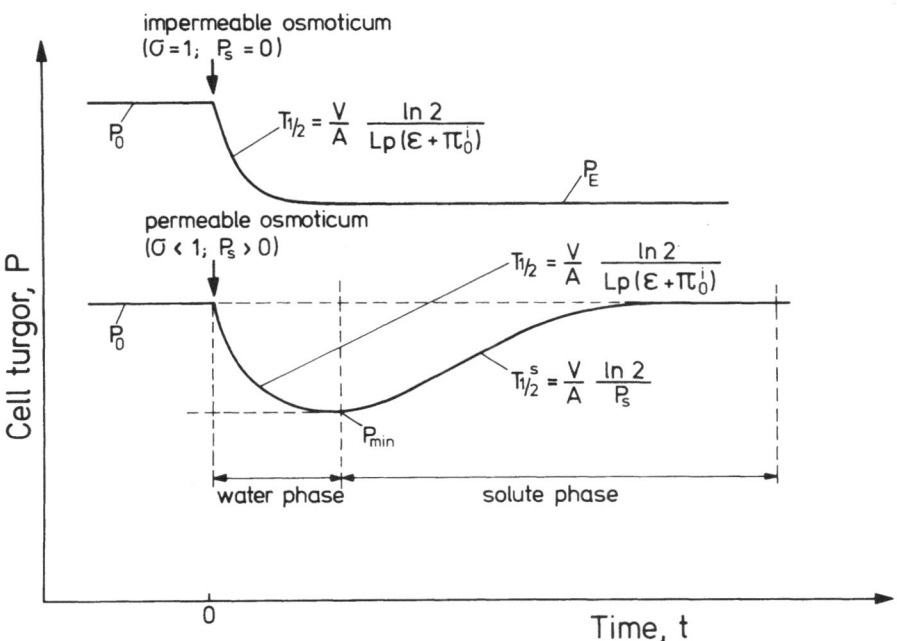

FIGURE 2. Pressure/time curves of an isolated plant cell after changing the osmotic pressure of the medium (diagrammatic presnetation). In the upper trace an impermeable osmoticum is added at $t=0$ which causes a monophasic change in turgor. In the lower trace a permeable solute leads to a transient change in turgor to a minimum value (P_{min}) which is followed by an increase in turgor due to the equilibration of solute across the membrane. From the first (water) phase Lp can be calculated (Eq. (3)) and from the second (solute) phase the solute permeability (P_s) is obtained. The reflection coefficient of the solute (σ_s) is determined from the change in turgor at the minimum ($P_0 - P_{min}$; Eq. (9)).

with literature data. In the determination of P_s of rapidly permeating substances (e.g. monohydroxy alcohols; see Table 2) unstirred layers have to be taken into account. For <u>Chara</u> external unstirred layers could be reduced by stirring and some corrections for internal layers could be applied. For this alga the P_s-values were independent of external concentration over a large concentration range but σ as well as Lp decreased with increasing external concentration (Steudle and Tyerman 1982). The decrease of σ seemed to be related to the decrease in Lp with increasing concentration (which is well

Table 1. Reflection coefficients of Chara corallina (a) and epidermis cells of Tradescantia virginiana (b) as obtained by the pressure probe according to Eq.(9). For comparison also data for Chara corallina (c) and Nitella translucens (c) are given which have been determined by the transcellular osmosis technique. The data for Nitella flexilis (d) have been measured with the pressure probe without correcting for solute flow.

Osmoticum	Chara corallina[a]	Tradescantia virginiana[b]	Chara corallina[c]	Nitella translucens[c]	Nitella flexilis[d]
Sucrose	0.95	1.04	-	-	0.97
Formamide	0.99	0.99	1	-	0.79
Mannitol	1.02	1.06	-	-	-
Dimethyl-formamide	0.76	-	-	-	-
Methanol	0.38	0.15	0.30	0.50	0.31
Ethanol	0.40	0.25	0.27	0.44	0.34
n-Propanol	0.24	-0.58	0.22	0.40	0.17
iso-Propanol	0.45	0.26	-	-	0.35
n-Butanol	0.14	-	-	-	-
iso-Butanol (2-Methylpropanol-1)	0.21	-	-	-	-
Aceton	0.17	-	-	-	-

(a) Steudle and Tyerman 1982; (b) Tyerman and Steudle 1982;
(c) Dainty and Ginzburg 1964a;   (d) Steudle and Zimmermann 1974;

known for Charceen spp; cf. Zimmermann and Steudle 1978) and such a relationship is expected from the frictional model of a lipid-pore membrane (Kedem and Katchalsky 1958; Dainty and Ginzburg 1963). From this model σ_s is given by:

$$\sigma_s = 1 - \frac{\overline{V}_s \cdot P_s}{Lp \cdot RT} - \frac{K_s^C \cdot f_{sw}}{f_{sw} + f_{sm}} \qquad (10)$$

where $\overline{V}_s$ = molar volume of the osmoticum; K_s^C = partition coefficient of the solute between water-filled pores and the external solution and f_{sw}, f_{sm} = frictional coefficients which express frictional forces between water and solute and solute and membrane as they travel across the membrane. From the model

Table 2. Permeability coefficients, P_s, for some osmotica for Chara corallina internodes (a) which have been obtained from the rate constant of the solute phase (Fig.2). For comparison some literature data (b) are given which have been determined by tracer experiments using C-14 labelled alcohols.

| Osmoticum | Permeability coefficients, P_s x 10^4 (cm·s^{-1}) of | | |
	Chara corallina[a)	Chara corallina[b)	Nitella translucens[b)
Dimethylformamide	0.81	-	-
Methanol	3.29	4.0	4.8
Ethanol	2.36	2.8	4.3
n-Propanol	2.63	-	-
iso-Propanol	1.86	2.0	2.1
n-Butanol	2.51	-	-
iso-Butanol (2-Methylpropanol-1)	2.30	-	-
Aceton	3.36	-	-

(a) Steudle and Tyerman, 1982; (b) Dainty and Ginzburg, 1964b;

it is expected that σ_s depends linearly on 1/Lp and this has been found for Chara corallina (Steudle and Tyerman 1982). The relationship between σ and 1/Lp could be used to evaluate P_s on an independent way and the results were similar to those obtained from the rate constant of the solute phase (see Fig. 2).

Although the P_s-values of the solutes determined up to now may incorporate to some extent unstirred layers, they are nevertheless of practical importance because they would determine the osmotic properties of the cell. The P_s-values of tissue cells for solutes determine the rate at which a tissue regains turgidity following an osmotic shock (i.e. by high salinity) and are thus crucial for our understanding of osmotic processes in plants under stress conditions. Up to now there are only a few data on P_s for higher plant tissue cells in the literature.

Values of σ for osmotic solutes and especially for solutes which are also present in the cell sap are also very rare in the literature but are as important as the P_s-values for describing the effects of osmotic stresses in tissues. In the usual techniques for measuring water relations (pressure chamber; psychrometry; see above) it is assumed that σ of the solutes is very close to unity. This assumption may not be justified in all cases and may cause significant errors in using these techniques, especially if estimates of cell turgor are made from the difference between water potential and osmotic pressure (see Eq. (1)).

3. WATER RELATION PARAMETERS OF HIGHER PLANT CELLS

Table 3 shows some results obtained by the pressure probe, i.e. $T_{1/2}$-values, diffusivities and Lp-values. The results collected for quite different types of plants and tissues indicate that for higher plant cells the half-times of water exchange are short (order: seconds in most of the cases). The short half-times are due to high Lp-values rather than to high ε-values (not shown in the table; cf. Eq. (3)) and lead to rather high diffusivities of the tissue ranging between 10^{-7} and 10^{-5} $cm^2 \cdot s^{-1}$ in most of the cases. For comparison the diffusion coefficient for mannitol in water is $6 \cdot 10^{-6}$ $cm^2 \cdot s^{-1}$ at $20^{\circ}C$. It should be noted that the diffusivities given in Table 3 are those for the cell-to-cell pathway only (Eq. (3)) and are thus underestimates for the diffusivities of the tissue which also incorporate the parallel apoplasmic pathway. Most of the Lp-values obtained with the pressure probe on higher plant cells range between 10^{-7} and 10^{-5} $cm \cdot s^{-1} \cdot bar^{-1}$ and are thus in the upper part of the range given in the literature which is between 10^{-10} and 10^{-5} $cm \cdot s^{-1} \cdot bar^{-1}$ (cf. Dainty 1976; Zimmermann and Steudle 1978). The high hydraulic conductivity and diffusivities have led to the conclusion that, at least for some of the tissues given in Table 3, the cell-to-cell pathway for water may be substantial as compared with the apoplasmic pathway.

For the leaves of Tradescantia virginiana it has been found (Zimmermann et al. 1980; Tomos et al. 1981) that for the mesophyll Lp and D were low, whereas for the epidermis Lp and D were high. Therefore, it has been concluded that for this plant the cell-to-cell transport of water in the epidermis could be a preferred path for both the stationary water transport from the xylem vessels to the stomata and for signalling changes of the water status of the plant to the stomata. The first conclusion can be drawn from the high Lp-value and high cross-sectional area of the cell-to-cell path (as compared with the apoplast). The second conclusion is directly derived from the short $T_{1/2}$-value: which should result in a quick propagation of a change in water potential along the epidermis provided that the hydraulic resistance between mesophyll

Table 3. Half-times of water equilibration of individual tissue cells, diffusivities and hydraulic conductivities of cell membranes as obtained by the pressure probe for different plants and tissues. The diffusivities refer to the cell-to-cell path only (see Eq. (4)) and are, thus, underestimates of the tissue diffusivity. For some cells (Elodea and Chenopodium), D is given for a propagation in the direction of the longitudinal as well as in the direction of the transverse cell axes.

Species	Plant type	Tissue/cell type	Half time of water exchange $T_{\frac{1}{2}}$ [s]	Diffusion coefficient for propagation of pressure (volume) in tissue [cm² · s⁻¹]	Hydraulic conductivity, Lp[cm·s⁻¹·bar⁻¹]	Remarks & references
Capsicum annuum	cultivated plant	fruit tissue; mesophyll cells	65 – 250	3 – 6 · 10⁻⁷	4 – 6 · 10⁻⁷	Hüsken, Steudle, Zimmermann (1978)
Tradescantia virginiana	mesophyte	leaf tissue { epidermis	1 – 35	0.2 – 6 · 10⁻⁶	0.2 – 11 · 10⁻⁶	Zimmermann, Hüsken, Schulze (1980) Tomos, Steudle, Zimmermann, Schulze (1981) Tyerman, Steudle 1982
		subsidiary	3 – 34	1·10⁻⁷ – 2·10⁻⁶	2 – 35 · 10⁻⁷	
		mesophyll	55 – 95	1 · 10⁻⁸	4 – 6 · 10⁻⁷	
		isolated epidermis	9 – 54	0.5 – 3 · 10⁻⁷	(6.4±4.5)10⁻⁷(15)	
Chenopodium rubrum	salt tolerant plant	vasc. parenchyma	11 – 35	2–8·10⁻⁷(long); 2–5·10⁻⁸(transv)	0.3–1.2·10⁻⁶	cells immobilized in 6% alginate; Büchner, Bentrup, Zimmermann (1980)
		suspension culture cells	12 – 40	—	0.2–2.2·10⁻⁶	
		immobilized suspended cells	15 – 20	—	1.0–1.8·10⁻⁶	
Kalanchoë daigremontiana	CAM plant	leaf mesophyll cells	2 – 9	6 · 10⁻⁶	0.2 –1.6 · 10⁻⁵	from leaf slices; Steudle, Smith, Lüttge (1980)
Pisum sativum	cultivated plant	growing epidermis	1 – 27	—	0.2–2 · 10⁻⁶	Cosgrove, Steudle (1981)
		epicotyl cortex	0.3 – 1	3.2·10⁻⁶	0.4–9 · 10⁻⁵	
Elodea densa	aquatic plant	epidermis { upper	3 – 35	2.3–27 · 10⁻⁷(long) 1 – 12 · 10⁻⁷(transv)	(7.8 ± 5.5)·10⁻⁷ (22); P>4bar	Lp pressure dependent; Steudle, Zimmermann, Zillikens, 1982
		lower	55 – 21	0.6–24 · 10⁻⁷(long) 2 – 6 · 10⁻⁸(transv)	(5.6 ± 2.0)·10⁻⁷ (10); P>4bar	

and epidermis is large as compared with the resistance between adjacent epidermis cells. The possibility that the epidermis could be a preferred path for leaf water and that cell-to-cell transport can be substantial differs from the usual concept that in the leaf water travels mainly in the mesophyll apoplast to reach stomatal pores. A final decision about the different concepts requires more data about the hydraulic resistances and diffusivities in the mesophyll and epidermis in more than only one species. It should be noted that a preferred transport of water in the epidermis of T. virginiana and Hedera helix has been also proposed by Sheriff and Meidner (1974; 1975), who used epidermal strips to measure lateral water movement in the epidermis. Sheriff and Meidner concluded that apoplasmic water transport was predominant in the epidermis. However, their data have been questioned by Tyree and Yianoulis (1980) on the grounds that the values are far too high to be explained in terms of either a cell-to-cell or apoplasmic transport and are most likely due to an artifact.

There is even more evidence in favour of a substantial cell-to-cell transport from the data obtained for the mesophyll tissue of the leaves of the CAM plant Kalanchoë daigremontiana (Steudle et al. 1980) and for the growing tissue of the pea epicotyl (Cosgrove and Steudle 1981). For Kalanchoë the diffusivity of the cell-to-cell path can be calculated from $T_{1/2}$ and the cell dimensions to be $6 \cdot 10^{-6}$ $cm^2 \cdot s^{-1}$ (Eq. (4); $\alpha = 1$), which would result in an average half-time ($\bar{t}_{1/2}$) for the shrinking or swelling of a tissue slice of say 2 mm in thickness of $\bar{t}_{1/2}$ = 6 min. This value is in agreement with the $\bar{t}_{1/2}$-values found in uptake measurements by weighing tissue slices of the same thickness ($\bar{t}_{1/2}$ = 8 min; Lüttge et al. 1977). If there were a substantial water transport in the apoplast (parallel to the cell-to-cell path) the latter figure would have been substantially smaller.

For growing tissue of the pea up to now the highest Lp- and D-values have been measured (Cosgrove and Steudle 1981; Table 3) and the half-times of water exchange of individual cells were as short as 0.3 to 1.2 s (Table 3). The diffusivities derived from the $T_{1/2}$ corresponded well with those for the osmotic shrinking or swelling of entire peeled segments of the epicotyl and, therefore, we may conclude that in this tissue the cell-to-cell path is dominating the water relations. It could be shown in osmotic experiments that the responses in cell turgor of cortex cells of the epicotyl to changes in the osmotic pressure of the medium depended strongly on the position of the cell, i.e. a cell deep in the tissue had a much slower response than one right at the surface. Since a rate limitation by solute diffusion could be excluded, this means that the cell-to-cell path was not by-passed by an apoplasmic pathway with a much higher hydraulic conductance.

The results summarized in Table 3 suggest that higher plant cells in general have short $T_{1/2}$ and high Lp- and D-values. If this is true, then there would be some consequences for the osmotic relations of plants: (1) The half-times of water flow equilibration of tissues would be of the order of minutes and water flow equilibrium would be maintained in most of the tissues to a fairly good approximation under "normal" conditions. At least over short distances (i.e. a few cell layers), the differences in water potential which can be built up within a tissue should be rather small.

(2) The hydraulic resistance of the cell-to-cell path for both stationary and transient water movement seems to be more important than it has been thought in the past. The old models for water transport in tissues favouring the cell wall path (cf. Weatherley 1963; Briggs 1967; Tanton and Crowdy 1972) should be re-considered.

(3) It has been postulated (cf. Dainty 1972) that plants may adapt to short-term (e.g. diurnal) water stress by a variation of cellular water parameters (ε, Lp, $T_{1/2}$). It has been proposed that the equilibration time of tissues may become as long as the half-time of the environmental changes so that no negative effects on the turgidity could result. For tissue cells with values of $T_{1/2}$ of the order of seconds, this possibility seems to be unlikely because ε and/or Lp would have to decrease substantially to get reasonably long values of $T_{1/2}$. However, the possibility that stationary water flow across tissues may be affected under stress conditions by changes in Lp cannot be excluded. At present, we are lacking data on Lp in higher plants under different stress conditions.

(4) On the other hand, if it turns out that the cellular water relations parameters are not really affected by water stress then plants may need to have efficient adaptive mechanisms at the solute transport level to survive under these conditions. As shown, the pressure probe can be used to determine changes in solute transport.

Future work has to quantify the amounts of water transported in the diff-erent tissue pathways and especially to work out methods for determining the hydraulic resistance of the apoplast. The pressure probe can be used for modelling tissue water transport because it allows measurements of water flow parameters of cells (as the tissue components) as well as the speed by which changes in water potential (turgor, volume) are propagated across a complex tissue. Comparing probe measurements of single cells with the osmotic properties of tissues as a whole should allow a complete and quantitative description of tissue water relations.

ACKNOWLEDGEMENTS

For discussing the manuscript we thank Prof. U. Zimmermann,KFA Jülich. This work was supported by a grant from the Deutsche Forschungsgemeinschaft, Zi 99/8.

108

REFERENCES

1. BRIGGS GE. 1967. Movement of water in plants. Botanical Monographs. Ed. J.H. Burnett, Blackwell, Oxford.

2. BÜCHNER K-H, ZIMMERMANN U, BENTRUP F-W. 1980. Turgor pressure and water transport properties of suspension-cultured cells of Chenopodium rubrum L. Planta, 151, 95-102.

3. COSGROVE D, STEUDLE E. 1981. Water relations of growing pea epicotyl segments. Planta, 153, 343-350.

4. DAINTY J. 1963. Water relations of plant cells. Adv. Bot. Res., 1 , 279-326.

5. DAINTY J. 1972. Plant cell-water relations: The elasticity of the cell wall. Proc. Roy. Soc. Edinburgh A,70, 89-93.

6. DAINTY J. 1976. Water relations of plant cells. In: Transport in Plants II. Part A: Cells (Eds. U. Lüttge and M. G. Pitman), Encycl. Plant Physiol. New Ser. Vol. 2A, Springer-Verlag, Berlin, pp. 12-35.

7. DAINTY J, GINZBURG BZ. 1963. Irreversible thermodynamics and frictional models of membrane processes, with particular reference to the cell membrane. J. Theoret. Biol., 5, 256-265.

8. DAINTY J, GINZBURG BZ. 1964a. The reflection coefficient of plant cell membranes to certain solutes. Biochim. Biophys. Acta, 79, 129-137.

9. DAINTY J, GINZBURG BZ. 1964b. The permeability of the protoplasts of Chara australis and Nitella translucens to methanol, ethanol and isopropanol. Biochim. Biophys. Acta, 79, 122-128.

10. HÜSKEN D, STEUDLE E, ZIMMERMANN U. 1978. Pressure probe technique for measuring water relations of cells in higher plants. Plant Physiol., 61, 158-163.

11. KEDEM O, KATCHALSKY A. 1958. Thermodynamic analysis of the permeability of biological membranes to non-electrolytes. Biochim. Acta 27, 229-246.

12. LÜTTGE U, BALL E, GREENWAY H. 1977. Effects of water and turgor potential on malate efflux from leaf slices of Kalanchoë daigremontiana. Plant Physiol., 60, 521-523.

13. MOLZ FJ, IKENBERRY E. 1974. Water transport through plant cells and cell walls: Theoretical development. Soil Sci. Soc. Am. Proc., 38, 699-704.

14. PHILIP JR. 1958a. Propagation of turgor and other properties through cell aggregations. Plant Physiol., 33, 271-274.

15. PHILIP JR, 1958b. The osmotic cell, solute diffusibility, and the plant water economy. Plant Physiol., 33, 264-271.

16. SHERIFF DW, MEIDNER H. 1974. Water pathways in leaves of Hedera helix L

and <u>Tradescantia virginiana</u> L. J.Exp. Bot., <u>25</u>, 1147-1156.

17. SHERIFF DW and MEIDNER H. 1975. Water movement into and through <u>Trades-cantia virginiana</u> L leaves. J. Exp. Bot., <u>26</u>, 897-902.

18. STEUDLE E, ZIMMERMANN U. 1974. Determination of the hydraulic conductivity and of reflection coefficients in <u>Nitella flexilis</u> by means of direct cell-turgor pressure measurements. Biochim. Biophys. Acta, <u>332</u>, 399-412.

19. STEUDLE E, SMITH JAC, LÜTTGE U. 1980. Water relation parameters of individual mesophyll cells of the CAM plant <u>Kalanchoë daigremontiana</u>. Plant Physiol., <u>66</u>, 1155-1163.

20. STEUDLE E, ZIMMERMANN U, ZILLIKENS J. 1982. Effect of cell turgor on hydraulic conductivity and elastic modulus of <u>Elodea</u> leaf cells. Planta, <u>154</u>, 371-380.

21. STEUDLE E, TYERMAN SD. 1982. Determination of permeability coefficients, reflection coefficients and hydraulic conductivity of <u>Chara corallina</u> using the pressure probe: effects of concentrations. J. Membrane Biol., submitted.

22. TANTON TW, CROWDY SH. 1972. Water pathways in higher plants. J. Exp. Bot., <u>23</u>, 600-625.

23. TOMOS AD, STEUDLE E, ZIMMERMANN U, SCHULZE E-D. 1981. Water relations of leaf epidermal cells of <u>Tradescantia virginiana</u>. Plant Physiol., <u>68</u>, 1135-1143.

24. TYERMAN SD, STEUDLE E. 1982. Comparison between osmotic and hydrostatic water flows in a higher plant cell: determination of hydraulic conductivities and reflection coefficients in isolated epidermis of <u>Tradescantia virginiana</u>. Aust. J. Plant Physiol., in press.

25. TYREE MT, YIANOULIS P. 1980. The site of water evaporation from sub-stomatal cavaties, liquid path resistances and hydroactive stomatal closure. Ann. Bot., <u>46</u>, 175-193.

26. WEATHERLEY PE. 1963. The pathways of water movement across the root cortex and the leaf mesophyll of transpiring plants. In: The water relations of plants. Ed. A.J. Rutter and F.M. Whitehead, Blackwell, London, pp. 85-100.

27. WENDLER S, ZIMMERMANN U. 1982. A new method for the determination of hydraulic conductivity and cell volume of plant cells by pressure clamp. Plant Physiol., <u>69</u>, 998-1003.

28. ZIMMERMANN U, STEUDLE E. 1978. Physical aspects of water relations of plant cells. Adv. Bot. Res., <u>6</u>, 45-117.

29. ZIMMERMANN U, HÜSKEN D, SCHULZE E-D. 1980. Direct turgor pressure measurements in individual leaf cells of <u>Tradescantia virginiana</u>. Planta, 149, 445-53

PHOTOSYNTHESIS, STOMATAL CONDUCTANCE AND LEAF WATER POTENTIAL DURING WATER STRESS SITUATIONS IN YOUNG RUBBERTREES (*Hevea brasiliensis*) UNDER TROPICAL CONDITIONS

R. CEULEMANS, I. IMPENS and A.P. NG

Department of Biology, University of Antwerpen, U.I.A., Universiteitsplein 1, B-2610 WILRIJK, Belgium   and
Plant Science Division, Rubber Research Institute of Malaysia, KUALA LUMPUR 01-02, West-Malaysia

ABSTRACT

Net photosynthesis, stomatal conductance and water potential of leaves of three *Hevea brasiliensis* clones subjected to water stress, were measured with a portable gas exchange unit under tropical conditions in West-Malaysia. Both net photosynthesis and stomatal conductance showed a sigmoid shaped, declining curve as a function of increasing water stress situations. Although net photosynthesis reached zero after 9 days of drought stress (at leaf water potentials of around 1.3 MPa), almost complete recovery was observed one day after rewatering. It was concluded from slower photosynthetic decrease, higher stomatal conductances at the end of the drying cycle and higher net photosynthesis after rewatering, that rubber clone FX 25 possessed a better water stress resistance than clones PR 107 and RRIM 701. The interesting potentialities of a mobile gas exchange unit for studying plant resistance to water stress are also discussed.

INTRODUCTION

In several regions of the world water stress is considered as the most important factor of yield decrements. Extensive laboratory studies have already been made on the effects of water stress situations and decreasing soil moisture content on gas exchange processes of different plant species as sunflower and soybean (2, 3, 10), sugarbeet (7), bean and tomato (12), eggplant (1),poplar (4) and other woody plants (6), azalea (5), sorghum (11) and lettuce (13).

   It is generally accepted that water stress induces a progressive reduction in photosynthesis and transpiration (3, 12), but only few authors report results on the effects of water stress under tropcial conditions in the field. Wien et al. (14) studied the effects of drought stress on the growth and gas exchange processes of cowpea and soybean under tropical conditions in Nigeria. Although regular and frequent showers occur commonly both during the dry and the wet season, short or longer periods of drought can occur frequently also in the tropical regions of Southeast Asia. Hence, the need to understand the effects of drought on plant gas exchange and growth, and to develop new cultivars or clones that can withstand moisture stress will increase.

Since we are not aware of any such study concerning the effects of water stress on gas exchange properties of rubbertrees, it was the aim of this communication to report some quantitative results concerning the effects of drought stress situations on net photosynthetic rate and stomatal conductance to water vapour of different *Hevea* clones under tropical field conditions, measured by means of a portable and simple gas exchange unit.

MATERIAL AND METHODS

Plant material

Twenty young and uniform cuttings of three different clones (i.e. clones FX 25, PR 107 and RRIM 701) of the rubbertree (*Hevea brasiliensis* Muell. Agr.) were followed during a drying cycle experiment in November 1980. Plants were grown from budded stumps in black plastic bags (volume ca. 15 dm^3) and were put under an open glasshouse at the Experiment Station of the Rubber Research Institute of Malaysia at Sungei Buloh near Kuala Lumpur (3.09 °N, 101.43 °E) in West-Malaysia.

Air temperatures in the glasshouse during the measurements were between 30 and 35 °C, relative air humidity around 80 % and photosynthetic photon flux density - measured with a Lambda (LiCor, USA) quantum sensor - between 300 and 800µ E $m^{-2}s^{-1}$.

Prior to the drought experiment rubberplants were watered daily and up to field capacity just before the experiment started. Different water stress situations were applied by subjecting the plants to a drying cycle of fourteen days. After the *Hevea* plants have been denied water for two weeks, rewatering has started and measurements continued for another two days.

The glasshouse facilities, the young experimental *Hevea* plants and the gas exchange equipment are shown in Figure 1.

Figure 1. The open glasshouse, young experimental *Hevea* plants and gas exchange equipment (left) at the Experiment Station of the RRIM in West-Malaysia during the drought experiment.

Gas exchange equipment

Net photosynthesis of four leaves per plant was measured with a mobile gas exchange unit (Figure 1) consisting of a portable, battery operated infrared gas analyzer, a simple clamp-on cuvette and a twin set of air pumps arranged in an open circuit. The clamp-on cuvette (Figure 2) was a flat, half-open and ventilated plexiglass leaf chamber with an inner area of 20 cm^2 and was constructed by the workshop of the University of Antwerpen. Only the lower side of the leaves was measured in this cuvette type. A quantum sensor (Lambda, LiCor, USA) measuring incident photon flux density on the leaf was attached on top of the cuvette (Figure 2), while the leaf chamber itself was mounted on an extensible tripod in order to reach leaves at different levels easily.

Reference air from outside the cuvette and sample air from the cuvette were sucked into the differential CO_2 infrared gas analyzer (Leybold-Heraeus, type Binos 1, Germany) through two membrane pumps (Hartmann-Braun, type Gf 30, Germany). The flow rate of the air through the system was between 60 and 100 l h^{-1} and was monitored just before entering the infra-red gas analyzer by means of two calibrated flow meters (Brooks, type R 215/B, USA). A zero check of the infrared gas analyzer was carried out

Figure 2. The flat, half-open and ventilated plexi gas exchange cuvette (inner dimensions 5x4x1 cm) used for measuring net photosynthesis of the *Hevea* leaves. The quantum sensor, inner fan and in and out sucking tubes are also shown.

each time before switching to another clone. Differences in CO_2 concentrations were measured on a 50 ppm scale of the Binos 1 with an estimated error of 2 %. Only net photosynthesis values at photon flux densities of around 500 $\mu E \ m^{-2} s^{-1}$ were used for calculations.

Stomatal conductances to water vapour were measured simultaneously with an automatic, non-ventilated Lambda diffusion porometer (LiCor, model LI-60, USA) on the same *Hevea* plants. Calibration of the porometer with the aid of a calibration plate with known conductances was made before and after each measuring period.

Five cuttings per clone were sampled for the photosynthetic and the stomatal conductance measurements. On the remaining plants leaf water potentials were frequently determined by means of a leaf pressure chamber (PMS-Instrument, model 1000, USA) as described originally by Scholander et al. (9). At least four replications per clone were made each time.

RESULTS AND DISCUSSION

Mean values of net photosynthetic rates, stomatal conductances to water vapour and leaf water potentials during the fourteen day drying cycle experiment have been represented on Figure 3 for the three *Hevea* clones. The net photosynthesis on this figure has been expressed per unit of leaf area.

The first day after the plants had been watered to field capacity, the initial net photosynthesis (at about 500 $\mu E \ m^{-2} s^{-1}$) increased for all clones with more than 40 %. This phenomenon has also been observed by

Hansen (7) and by Lemeur (personal communication, 1982), although the
reason for this sudden rise is not well understood. Maybe it has something
to do with an increased root activity after the plants had been watered
to field capacity. The net photosynthesis under these conditions was comparable to the values obtained by Samsuddin and Impens (8) for different
Hevea clones under laboratory conditions, although the stomatal conductances found here were slightly higher.

From the second day on, net photosynthesis declined sharply - together
with stomatal conductances - and felt to zero after 9 days of water stress.
Leaf water potentials were 1.2 MPa (for clones FX 25 and PR 107) and 1.5
MPa (for clone RRIM 701) at that moment, while stomatal conductances to
water vapour reached minimal values of circa 0.1 cm s^{-1}.

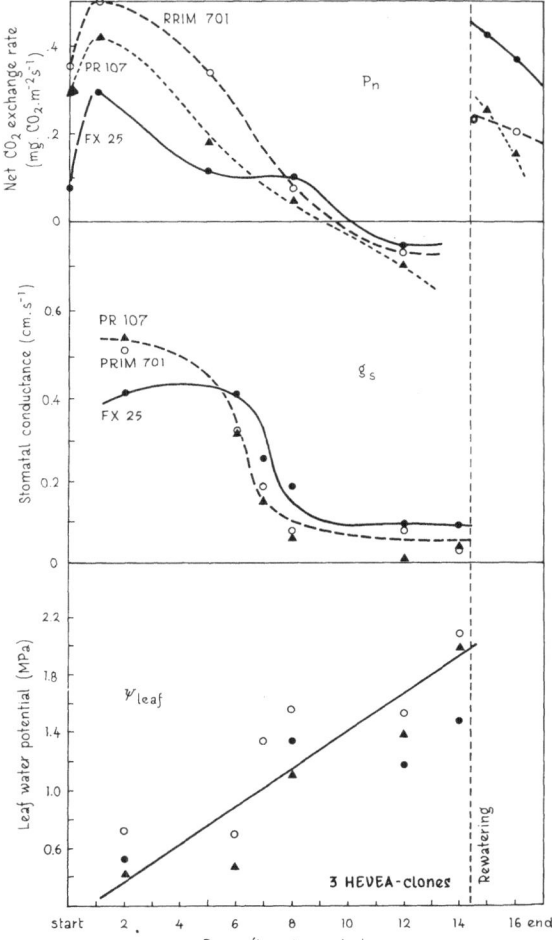

Figure 3.
Net photosynthesis, stomatal
conductance and leaf water
potential of three *Hevea*
clones during a two week drying
cycle experiment under tropical
conditions. The linear regression line shown at the bottom
of this figure is significant
at the 1 % level. Values represent means of 20 replicates per
clone.

From day 10 on until the end of the drought experiment (at day 14)all experimental plants only were respiring, while stomatal conductances remained fairly constant at about 0.1 cm s^{-1} (clone FX 25) or less (clones PR 107 and RRIM 701). The shape of the curve of net photosynthesis versus days of stress is somewhat sigmoid as found earlier in sunflower (3), sugarbeet (7) and in different woody plants(6). This sigmoid shape of the curve is typical for plants subjected to dehydration.

Although leaf water potentials give a better and more general idea about the physiological water status of the plant in comparison to e.g. duration of stress, we found a significant linear relationship between leaf water potential of the *Hevea* plants and the number of days after the start of the drying cycle (bottom of Figure 3). Leaf water potentials increased from 0.6 MPa at the start of the drought experiment until 2.1 MPa at the end.

Differences in net photosynthetic rates between clones RRIM 701 and PR 107 - both initially having higher net photosynthesis - were not very pronounced. Nevertheless, clone FX 25 initially showed the lowest net photosynthesis (i.e. 0.3 mg $CO_2 m^{-2} s^{-1}$) but maintained better performance and lower respiration rates at the end of the drying cycle beside a much slower decline of its net photosynthesis during the increasing water stress. This *Hevea* clone also showed highest net photosynthesis after rewatering, indicating a better recovery. All these determinations suggest a better water stress resistance of clone FX 25 and this tendency is also confirmed by the stomatal conductances during the drought experiment. At the beginning of the experiment, *Hevea* clone FX 25 had lower stomatal conductances to water vapour (0.4 cm s^{-1}) in comparison with the two other clones (0.5 cm s^{-1}), but at the end of the experiment conductances of clones RRIM 701 and PR 107 remained much lower than those of clone FX 25 which possessed a still reasonable conductance, hence transpiration rate.

The shape of the stomatal response curve as a function of increasing water stress situations is similar to those reported earlier by Hansen (7) for sugarbeet and by Sionit and Kramer (10) for sunflower and soybean, although Davies and Kozlowski (6) found a slightly different series of curves for a number of woody plants.

Immediately (i.e. one day) after rewatering net photosynthesis increased again and attained almost initial values for clones RRIM 701 and PR 107, but a much higher rate for *Hevea* clone FX 25. Such an increased net photosynthesis above the level of the photosynthesis prior to the drying cycle

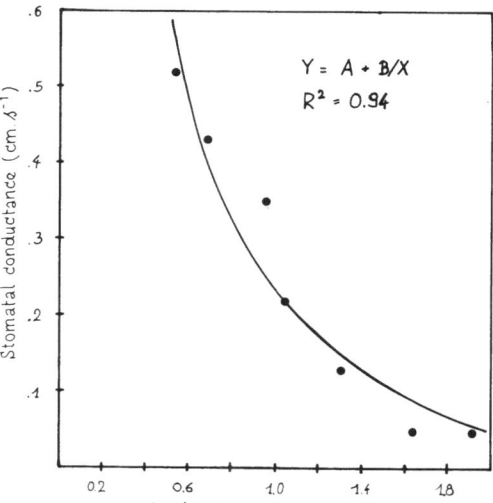

Figure 4.
Relation between stomatal conductance to water vapour and
leaf water potential during
the drying cycle experiment
under tropical field conditions.
Mean values for the three *Hevea*
clones have been shown.

experiment has also been reported by Hansen (7),while Davies and Kozlowski
(6) found a poor recovery of the leaf gas exchange in woody plants following rewatering after drought. In lettuce (13) and eggplant (1) net photosynthesis did not recover completely due to aftereffects of stress on the
stomates. Most likely severity of water stress and different drought resistance of species play an important role in recovering capacity. The
second day after rewatering has started, we however observed a certain
decrease in the net photosynthesis of all three rubber clones (Figure 3).
A similar pattern - a complete recovery of photosynthetic rate in one day
followed by a slight decrease the second day after rewatering - has also
been observed for *Beta vulgaris* plants after a drying cycle of six days (7).

The recovery of the *Hevea* plants was almost complete, even after having been stressed up to leaf water potentials of 2.0 MPa. Woody plants
can even tolerate easily leaf water potentials of 2.5 MPa as observed by
Davies and Kozlowski (6). Sunflower leaves stressed to 1.6 MPa recovered
a few days after rewatering (10), but died when stressed to 2.0 MPa or more
(3, 10).

Finally, a tight relationship between stomatal conductances to water
vapour and leaf water potentials was found during the drying experiment
(Figure 4). In this figure the relation between both water related parameters for the three rubber clones has been significantly (at the 1 %

level) described by a hyperbolic function. Similar observations have already been made earlier by several authors (2, 6, 7, 10).

CONCLUSIONS

It is possible to study differences in water stress resistance among different *Hevea* clones under tropical field conditions by means of a simple gas exchange unit which can be easily transported and consists primarily of an infrared gas analyzer, a set of pumps and flow meters as well as a portable diffusion porometer and a leaf pressure chamber.

Although differences among the three *Hevea* clones were not very pronounced, clone FX 25 seems to possess a better resistance against water stress situations, based on a serious of preliminary measurements of net photosynthesis, stomatal conductances and leaf water potentials during a two week drying cycle experiment. Although Tariq Al-Ani and Bierhuizen (12) stated that stomatal conductances alone could be used as a tool by which the effects of soil moisture stress could be predicted, we suggest that a combination of measurements of net photosynthesis, stomatal conductances and leaf water potentials should be used.

On the basis of these experiments it can however not be concluded that the reduction in the net photosynthesis during a drying period will always be fully reversible; on the contrary, one might expect that drought over a much longer period will incur an irreversible reduction of the net photosynthesis.

If the photosynthesis by rewatering is temporarily increased in some clones (e.g. FX 25) above the level of the net photosynthesis prior to the drying period, the loss in production will probably be limited when the drought period is short. More elaborate field gas exchange experiments could be helpful when trying to predict water stress resistance of new clones or cultivars in the tropical regions.

ACKNOWLEDGEMENTS

As this was a joint Belgian - Malaysian scientific effort, we gratefully acknowledge the cooperation shown by Datuk Dr. Haji Ani bin Arope, Director of the RRIM and by Dr. P.K. Yoon, Head of the Plant Science Division as well as the material and financial support of the University of Antwerpen (UIA). The help of Dr.ir. R. Gabriëls, Research Scientist at the Research Station for Ornamental Plant Growing (Melle) is also highly appreciated

REFERENCES

1. Behboudian MH (1977) Responses of eggplant to drought : II. Gas exchange parameters. Scientia Hort 7:311-317.
2. Berger A (1970) Le potentiel hydrique et la résistance à la diffusion dans les stomates indicateurs de l'état hydrique de la plante. In Unesco. Réponse des plantes aux facteurs climatiques, pp. 201-212. Uppsala : Actes Coll.
3. Boyer JS (1976) Water deficits and photosynthesis. In Kozlowski TT, ed. Water deficits and plant growth, vol.IV, pp. 153-190. New York : Academic.
4. Ceulemans R, Impens I, Lemeur R, Moermans R and Samsuddin Z (1978) Water movement in the soil - poplar - atmosphere system. II. Comparative study of the transpiration regulation during water stress situations in four different poplar clones. Oecol Plant 13:139-146.
5. Ceulemans R, Impens I and Gabriëls R (1979) Comparative study of leaf water potential, diffusion resistance and transpiration of azalea cultivars subjected to water stress. Hort Science 14:507-509.
6. Davies WJ and Kozlowski TT (1977) Variations among woody plants in sto-matal conductance and photosynthesis during and after drought. Plant and Soil 46:435-444.
7. Hansen GK (1971) Photosynthesis, transpiration and diffusion resistance in relation to water potential in leaves during water stress. Acta Agric Scand 21:163-171.
8. Samsuddin Z and Impens I (1979) Photosynthesis and diffusion resistances to carbon dioxide in *Hevea brasiliensis* Muell.Agr. clones. Oecologia 37: 361-363.
9. Scholander PF, Hammel HT, Bradstreet D and Hemmingsen EA (1965) Sap pressure in vascular plants. Science 148:339-346.
10. Sionit N and Kramer PJ (1976) Water potential and stomatal resistance of sunflower and soybean subjected to water stress during various growth stages. Plant Physiol 58:537-540.
11. Sullivan CY and Ross WM (1979) Selecting for drought and heat resistance in grain sorghum. In Mussell H and Staples RC, eds. Stress physiology in crop plants, pp. 263-281. New York : J.Wiley.
12. Tariq Al-Ani A and Bierhuizen JF (1971) Stomatal resistance, transpiration and relative water content as influenced by soil moisture stress. Acta Bot Neerl 20:318-326.
13. Van Holsteijn HMC, Behboudian MH and Bongers HCML (1977) Water relations of lettuce : II. Effects of drought on gas exchange properties of two culti-vars. Scientia Hort 7:19-26.
14. Wien HC, Littleton EJ and Ayanaba A (1979) Drought stress of cowpea and soybean under tropical conditions. In Mussell H and Staples RC, eds. Stress physiology in crop plants, pp. 283-301. New York : J.Wiley.

ENHANCEMENT OF MAINTENANCE RESPIRATION UNDER WATER STRESS

H. MOLDAU, M. RAHI
Institute of Astrophysics and Atmospheric Physics, Estonian Academy of
Sciences, 202444 Toravere, Tartu, Estonia, U.S.S.R.

ABSTRACT

The daily courses of the CO_2 and water vapour exchange of single intact vegetative Phaseolus vulgaris plants were monitored in plant chamber under an optimal water supply and during the depletion of moisture in the root zone. The coefficients of the maintenance and growth respiration were calculated from CO_2 exchange in a steady-state regime under optimum conditions. Using the same coefficients, the dynamics of the rate of synthesis of structural dry matter and of the amount of reserve materials were calculated during the depletion of moisture. Comparing the calculated increments with the measured increase of sugars and starch, it is concluded that under a rapidly developing water stress the maintenance coefficient of dark respiration increases, while the growth coefficient remains unchanged. It is suggested that under an increasing water stress appears an adaptational component of dark respiration associated with resynthesis and redistribution of biomass between different parts of the plant.

1. INTRODUCTION

Plant production has often been discussed within the concept of growth and maintenance components of dark respiration [7, 20]. This concept has quantitatively been evaluated on biochemical level [13, 14] and some sophisticated mechanistic models of the plant dry matter accumulation have been developed [2, 17]. These papers have made a sustantial contribution to the understanding of the plant production process and have given rise to experimental studies in which the efficiency of the growth and maintenance processes on a variety of species over a range of irradiance [16, 18], temperature [16, 7] and plant age [3] has been studied. These experiments have revealed a relatively high stability of growth efficiency and large variations in maintenance losses, even within the same species [19].

The influence of water stress on the growth (R_G) and maintenance (R_M) components of dark respiration (R) is poorly understood due to the difficulties in separating R_M and R_G. In both experimental papers hitherto published [11, 21] R_G per unit dry matter has been found to be independent of water deficit. For water-stressed sorghum plants a slight decrease in R_M per unit dry matter has been established [21]. In [11] there was found no change in R_M per unit dry matter of bean plants when water deficit developed slowly but under a rapidly progressing water stress its value tended to increase.

In both papers the usual methods of experimental separation of R_G and R_M [6, 7, 15] were used. But in [11] model predictions were also employed. The model, first described in [9], enables us to calculate the increase in the amount of the free assimilates in a plant from the continuous recording of its CO_2 exchange provided that the initial amount of free assimilates and the growth and mainte- nance losses are known. Since the calculated increase in free assimilates is very sensitive to R_M, the model can be used to detect changes in R_M, if the calculated content of free assimilates is compared with the measured content. In [11] no chemical analyses were made. This paper presents the results of expe- riments where conclusions drawn from model predictions are supported by the results of the chemical analyses of the content of free assimilates.

2. MATERIALS AND METHODS
2.1. Experimental procedure

Dwarf bean plants (<u>Phaseolus vulgaris</u> L. var. "Flitz") were grown in a growth chamber in 1 dm^3 stainless steel pots containing 1200 g (dry weight) of sand and Knop nutrient solution. The irradiance at the plant level from a mercury fluo- rescent lamp was 160 W m^{-2} PAR) during the 14 h light period, the air tempera- ture 25-28°C and the relative humidity 50-55% throughout the day. The CO_2 con- centration was natural and the sand moisture content was kept within an optimum range (40-50% of the field capacity).

Sixteen days after germination, a plants was transferred into the air- and temperature-controlled two-compartment (for shoots and roots) plant chamber [5] for the continuous measurement of transpiration and of CO_2 exchange. Conditions in the plant chamber were : irradiance 100 W m^{-2} PAR, relative humidity around the shoots in the light 40-50% and in the dark 10-15%. The air temperature was 27±0.3°C, the CO_2 concentration in the inlet air was 340 μl l^{-1} and in the plant chamber its value was held within a a range of 310-360 μl l^{-1} by the re- adjustment of air flow rates. Under these standard conditions the plant was allowed to adapt itself in the cours of about 48 h. As a rule, the CO_2 and water vapour exchange characteristics of the plant became stabilized by the start of the second day of the adaptation period. This was an indication that a steady state of the plant had been reached. The first day was discarded in data processing. Accordingly, the beginning of the light period of the second day was regarded as the start of the experiment (time t = 0) and the second day itself (the time interval t = 0 ÷ 24 h) ¬ as a steady-state day. At t = 0 the sugar and starch content of the experimental plants was estimated by chemical analyses of similarly treated plants. Watering was withheld on the following day and registration continued until the plant wilted. In the control experiments watering was continued. At the end of the experiment the plant dry weight and sugar + starch content were measured. Sugars were extracted with ethanol and starch was hydrolysed with 0.5% H_2SO_4. Sugar content was determined colorimetrically.

2.2. Model

The gas exchange data were treated according to a model [9] which assumes the plant dry matter to consist of two fractions : structural dry matter (SDM) and the reserve materials (RM) (Fig. 1). The latter is the immediate product of gross photosynthesis, P_G, and the substrate for the SDM synthesis. During the synthesis of SDM some carbon is lost in growth respiration, R_G. Some more carbon is lost in the maintenance respiration, R_M, originating from the process of breakdown and resynthesis of SDM. It is assumed that this process presents recycling of substances inside the SDM (presented in Fig. 1 as a dotted contour).

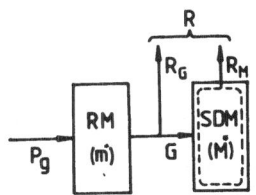

FIGURE 1. Scheme of the CO_2 balance of a plant, adopted for calculations of the reserve materials (RM) and of the structural dry matter (SDM). For explanations see text.

Denoting the amounts of RM and SDM (in mg dry weight) by m^* and M^*, respectively, the change rates of these fractions can be described by the following equations :

$$E_{RM} \frac{dm^*}{dt} = P_G - R_G - G, \tag{1}$$

$$E_{SDM} \frac{dm^*}{dt} = G - R_M, \tag{2}$$

where G is the rate of synthesis of SDM in CO_2 equivalents and E_{RM}, E_{SDM} are the conversion factors of RM and SDM to CO_2 equivalents.

We also assume that R_M is proportional to the amount of SDM and that G is proportional to R_G :

$$R_M = CE_{SDM}M*,$$ (3)

$$G = gR_G,$$ (4)

where C is the maintenance coefficient (dimension : $time^{-1}$) and g is the growth coefficient (dimensionless). The latter can be converted to the more widely used growth conversion efficiency Y_G by the formula $Y_G = g/(1 + g)$, since Y_G is defined as $R_G/(R_g + G)$ [17].

2.3. Determination of E_{SDM}, g and C.

Additional experiments (each in two replications) were performed to establish the coefficients of the model.

Since RM was assumed to comprise mainly sugar and starch, for E_{RM} the theoretical value $E_{RM} = 1.47$ mg CO_2 $(mgRM)^{-1}$ was adopted. As SDM is the dominating fraction in the dry matter, E_{SDM} was calculated from experiments, where the amount of CO_2 incorporated under the standard, steady-state conditions during 3 days was compared with the increase of SDM. The increase of SDM was calculated by substracting from dry matter increase, $(\Delta m* + \Delta M*)$, the increase of the amount of sugar + starch, $(\Delta m*)$. Dry matter increase itself, in turn, was estimated from the increment of the plant leaf area, as from a simultaneously grown group of plants it was established that under standard, steady-state conditions the relative dry weight increase was 1.25 times the relative leaf area increment. The content of sugar + starch in the experimental plants was taken to be the same as it was determined for a similarly exposed group of plants. From two experiments $E_{SDM} = 2.17$ mg CO_2 $(mgSDM)^{-1}$ was calculated; the same value was also used when the plants were subjected to water stress.

To evaluate g, the content of RM at the end of the experiment (t = n), c(n), was defined as :

$$c(n) = \frac{m(n)}{M(n)} = \frac{E_{RM}m*(n)}{E_{SDM}M*(n)},$$ (5)

where m and M are the amounts of RM and SDM in CO_2 equivalents.

From (5) :

$$m(n) + M(n) = E[m(n) + M(n)],$$ (6)

where $E = E_{RM}E_{SDM}[1 + c(n)]/[E_{RM} + c(n)E_{SDM}]$ is the conversion factor for the SDM + RM mixture.

The amount of dry matter in CO_2 equivalents at the start of the experiment,

$m(0) + M(0)$, is obtained from :

$$m(0) + M(0) = m(n) + M(n) - \int_{o}^{n} (Pg - R)dt, \tag{7}$$

where the integral is the amount of CO_2 incorporated during the experiment.

If the length of a steady-state experiment is one (or several) full light-dark cycle(s), the content of RM at t = 0, $c(0)$, is equal to $c(n)$ [10]. In this case the change in the amount of RM during the experiment, $\Delta m = m(n) - m(0)$, is equal to $c(n)\Delta M$, with $\Delta M = M(n) - M(0)$ as the change in the amount of SDM, and from (7) we obtain :

$$\Delta m = \frac{c(n)}{1 + c(n)} \int_{o}^{n} (Pg - R)dt \tag{8}$$

On the other hand, substituting G in (1) from (2) and assuming C and g to be constant, the integration of (1) yields

$$\Delta m = \int_{o}^{n} P_g dt - (1 + g) \int_{o}^{n} Rdt - C\overline{M}\Delta t, \tag{9}$$

where $\overline{M} = \int_{o}^{n} Mdt/\Delta t$ is the mean amount of SDM during interval t = 0÷n.

Considering (8) formula (9) gives for g

$$g = \frac{\int_{o}^{n} (P_g - R)dt + \left[1 + c(n)\right] C\overline{M}\Delta t}{\left[1 + c(n)\right] \left[\int_{o}^{n} Rdt - C\overline{M}\Delta t\right]} . \tag{10}$$

To get the integrals in this formula, a plant was exposed to standard conditions; the CO_2 exchange was monitored in a steady-state during one full light-dark cycle. The value of $c(n)$ was obtained from sugar + starch determination.

The maintenance coefficient C in (10) was determined from an experiment where a plant was exposed to continuous darkness to suppress the growth component of respiration. It was achieved on the 3rd day when measurements indicated a slight decrease in the leaf area. All respiration under these conditions was assumed to evolve from maintenance requirements. Substracting 3% of sugars (still present mainly in the roots) from the final dry weight from the two replications of experiment C was calculated to be 9.53×10^{-4} and 1.14×10^{-4} $mgCO_2(mgCO_2SMD\ h)^{-1}$. The mean value $C = 9.73 \times 10^{-4}$ was inserted into (10) and from the two similar one-day steady-state experiments the values of g = 4.05 (Y_G = 0.80) and g = 3.48 (Y_G = 0.78) were obtained. The mean value of g = 3.74 was used throughout the experiments.

2.4. Data processing.

The CO_2 exchange data were processed as described in [10] except that the content of RM at the start of the experiments (always the start of a light pe-riod), c(0), was assumed to be the sugar + starch content of similarly exposed plants. Similar to [10], the gas exchange data during the steady-state day were used to calculate an individual maintenance coefficient C for any plant accor-ding to the formula, derived in [10].

$$C = \frac{2 \left| g \int_0^{24} R dt - \Delta M \right|}{(1 + g) \left[2M(0) + M \right]}, \tag{11}$$

where ΔM is the increase of SDM during the steady-state day and g = 3.74.

Thereafter the change of SDM and RM during the steady-state day was obtained, using in (1) and (2) time step Δt = 2 h and the value of C calculated from (11). With Δt = 2 h the dynamics of SDM and RM was calculated also for the remaining part of the experiments, using again g = 3.74 but assuming the constant or changing value of C. The calculated RM content at the end of the experiment was compared with the measured sugar + starch content of the plant.

3. RESULTS

Fig. 2a represents the time course of the primary data of CO_2 and water vapour exchange in a control, steady-state experiment, where a plant was well watered. Changes in the net CO_2 uptake and transpiration of the shoots during the light periods are caused by an endogeneous rythm in stomatal movements, as reported previously [8].

The b and c parts of Fig. 2 were calculated from the CO_2 exchange data using constant values of the growth and the maintenance coefficients throughout the experiment. Due to an increase in SDM, the maintenance respiration R_M increased by 10% during the experiment, while the growth respiration R_G decreased by 50% As a result the specific growth rate of SDM, $\mu_S = (G - R_M)/M$, decreased by 60%. But in spite of these significant ontogenetic trends, the content of RM, c, remained relatively stable, showing only a slight day-to-day increase superimpo-sed on the regular daily amplitude. A decrease in μ_S during the last hours of dark periods was probably caused by a decrease in the RM content [10].

Fig. 3 represents the results of an experiment where during the first hours of the second light period CO_2 uptake and transpiration rates were reduced by

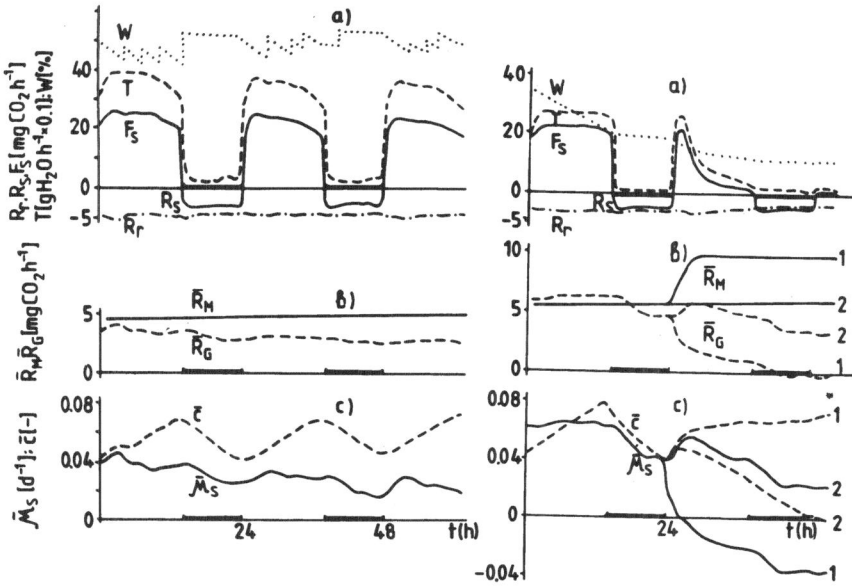

FIGURE 2. Time course of the CO_2 exchange, transpiration and dry matter accumulation in a well-watered plant. Black strips on the abscissa denote dark periods.
a) Net CO_2 uptake of shoots (F_S), respiration of shoots (R_S), respiration of roots
(R_r), transpiration (T) and sand moisture content (W). Note the doubled scale
below zero ordinate. b) Maintenance respiration ($\bar{R}_M$), growth respiration ($\bar{R}_G$).
c) The specific growth rate of SDM ($\bar{\mu}_S$) the content of RM($\bar{c}$). $\bar{R}_m$ and $\bar{R}_G$ are
halved 2-hour totals, $\bar{\mu}_S$ is a 2-hour mean reduced to 24 h and $\bar{c}$ is a 2 hour mean.
Parameters : C = 9.14 x $10^{-1}h^{-1}$, c(0) = 0.04, g = 3.74.

FIGURE 3. As in Fig. 2, but plant under progressive water deficit. Notation as
in Fig. 2. Branching of the curves on parts b) and c) is caused by the use of
two version : 1 - C = const = 8.26 x $10^{-4}h^{-1}$ throughout the experiment, 2 - C
increased from 8.26 x 10^{-4} to 14.09 x 10^{-4} between t = 24 h and 28 h.
c(0) = 0.04, g) 3.74. The measured sugar + starch content at the end of the
experiment (t = 52 h) is denoted by an asterisk (*) in part c).

the onset of water deficit. The sharp decrease was accompanied by a temporary
increase in root respiration.

The gas exchange data were at first treated as in the control experiment,
assuming the values of g and C during the experiment to be constant. This led
to zero c and to the positive μ_S at the end of the experiment (curves 2 in Fig.
3c). Such low value of c is not in line with measured sugar + starch content
and the positive value of μ_S does not agree with the decrease of the leaf area
in these conditions.

However, more probable values were obtained, assuming that during the first hours of the second light period C increased linearly but thereafter remained constant. In this case the dark respiration at the end of the experiment arises wholly from the maintenance requirements. The value of g was previously assumed to be constant. This version yielded the final content of RM close to the measured sugar-starch content and negative specific growth rates of SDM during the development of water stress (curves 1 in Fig. 3c).

The same final result can be achieved, alternatively assuming that C remains constant and g undergoes a change. However, as was already shown [11], in this case the values of g = 0.5-1.0 (Y_G = 0.3-0.5) must be introduced. No one has so far measured such low values, these are also theorethically improbable [14].

Therefore, the most probable version seems to be an increase in the maintenance requirements under water stress. Simultaneous decrease in the synthesis of SDM causes an increase in the RM content.

To confirm the increase of RM content under water deficit, further experiments were undertaken, where plants were subjected to a moderate water deficit (20% of field capacity) for some days before transfer into the plant chamber, where, as before, a severe water stress was allowed to develop. The CO_2 exchange data of these pre-adapted plants will be analysed in a further paper, here only the results of the chemical analyses at the end of the experiment are given together with the mean data obtained from the above analysed experiments. From the Table it follows that in pre-adapted plants the sugar and starch content was even higher than in the plants exposed to a severe deficit without preadaptation. It can also be seen that the increase in the starch content under water deficit is more pronounced than that of the sugar content.

Table. Sugar (c_{SR}) and starch (c_{SH}) content in the control and stressed bean plants (all measured 4 h after the start of the light period)

Treatment	c_{SR}	c_{SH}	$c_{SR} + c_{SH}$
Control	0.040	0.010	0.050
Stressed	0.053	0.039	0.092
Pre-stressed for 4 days	0.073	0.064	0.137
for 7 days	0.086	0.073	0.159

4. DISCUSSION

Estimation of the growth and maintenance coefficients of dark respiration for dwarf bean plants in this series of experiments yielded about 10% higher values for g than in our previous estimate [9] whereas the maintenance coefficient C is about 70% higher. As C is known to decrease with age [4], the higher value of C can be explained by the use of younger plants in this series. Some difference may also arise from a different variety. Nevertheless, in both cases the values of g and C fall into the range reported for agricultural plants in the vegetative phase [15].

The values of E_{SDM} lower than 1.67 were calculated from the chemical composition of dry matter [12]; from the more recent data [2] the values of E_{SDM} between 1.8-2.0 can be derived. Our new estimate $E_{SDM} = 2.17$ $mgCO_2(mgSDM)^{-1}$ is even higher but some sugars exudated by roots were found in the sand and this probably accounts for the difference.

Using the constant values of g and C throughout the experiment, reliable time courses of the RM content c and of specific growth rate of SDM μ_S were derived from the CO_2 exchange data of well-watered plants (Fig. 2). When the same assumption was extended to the plants undergoing water deficit, unreliable values of c and μ_S were calculated (Fig. 3). The discrepancy between the calculated and the measured values was small when an enhancement of C under rapidly developing water deficit was assumed. The same conclusion was drawn in a previous report [11] where an increase in C was also measured directly exposing water-stressed non-growing plants to continuous darkness.

Therefore under a developing water stress an additional component of dark respiration seems to be generated. It includes some additional expenditure of energy to maintain ionic balance on membranes and to support changes in the protein turnover and some other processes of readjusment on organelle level [13] But probably it also includes the losses associated with the redistribution of some structural substances between the plant organs [11]. This process is widely known to be accelerated under rapidly progressing stress conditions to stabilize the functional equilibria of plant organs. The transient increase in root respiration (Fig. 3) seems to be a typical example of such processes which reflect an increase in root metabolic activity. In the above calculations, this transient part of respiration was included in growth respiration because it probably comprised some losses for growth at the expense of the primary reserve sugars available in the roots. But it could also be included in maintenance respiration as a manifestation of adaptive responses to maintain plant functions; in this case agreement of calculated and measured RM contents in Fig. 3 would be even better.

Probably an enhancement of maintenance requirements under a developing water stress can be classified as a category of extra losses which arise due to the

generation of processes of maintaining the basic functions of plants under unfavourable environmental conditions. Such processes can be qualified as adaptive, and for this reason it was proposed to separate a third, adaptational
component [11] in dark respiration. A proposal has been made to introduce a
"regrowth respiration" component [6]. This term seems to be capable of including also the losses associated with normal ontogenetic redistribution of materials occurring mainly during the reproductive phase [1] and it therefore can
include a broader group of processes.

We hope that attempts to introduce new functional components of dark respiration will contribute to further understanding of particular processes involved
in respiration and to the quantitative determination of their relative role in
the whole production process. A more rapid progress will be achieved if measurements of the gas exchange were accompanied by detailed chemical analyses of
plant dry matter.

REFERENCES

1. Austin RB, Morgan CL, Ford MA and Blackwell RD (1980 Contributions to grain
   yield from pre-anthesis assimilation in tall and dwarf barley phenotypes in
   two contrasting seasons. Ann Bot 45, 309-319.
2. Greenwood DJ and Barnes A (1978) A theoretical model for the decline in the
   protein content in plants during growth. J Agric Sci Camb 91, 461-466.
3. Hansen GK (1978) Utilization of photosynthates for growth, respiration, and
   storage in tops and roots of Lolium multiflorum L. Physiol Plant 42, 5-13.
4. Jones MB, Leafe EL, Stiles W and Collett B (1978) Pattern of respiration of
   a perennial reygrass crop in the field. Ann Bot 42, 693-703.
5. Karolin A and Moldau H (1976) A controlled environment chamber for recording of transpiration and CO_2-exchange of plant shoots and roots. Plant
   Physiol, Moscow 23, 630-634 (in Russian).
6. Kuperman IA, Khitrovo EV and Semikhatova OA (1981) Comparison of methods of
   respiration separation into components. Physiology and Biochemistry of Cultivated Plants, Kiev 13, 563-576 (in Russian).
7. McCree KJ (1974) Equations for the rate of dark respiration of white clover
   and grain sorghum, as functions of dry weight, photosynthetic rate, and temperature. Crop Sci 14, 509-514.
8. Moldau H (1979) Versatility of stomatal control. In : Structure, function
   and ecology of stomata. Sen DN (ed.), Dehra Dun, India, 175-188.
9. Moldau H and Karolin A (1977) Effect of the reserve pool on the relationship between respiration and photosynthesis. Photosynthetica 11, 38-47.
10. Moldau H and Sober J (1981) Growth rate - reserve content relationship as
    influenced by irradiance, CO_2 concentration and temperature. Photosynthesis
    Research 1, 217-231.
11. Moldau H, Sober J and Rahi M (1980) Components of dark respiration of bean
    plants under water stress. Plant Physiol, Moscow 27, 5-10 (in Russian).
12. Penning de Vries FWT (1972) Respiration and growth. In : Crop processes in
    controlled environment. Rees EA et al. (eds.), Acad Press, London and New-
    York, 327-347.
13. Penning de Vries FWT (1975) The cost of maintenance processes in plant
    cells. Ann Bot 39, 77-92.

14. Penning de Vries FWT, Brunstig AHM and van Laar HH (1974) Products, requirements and efficiency of biosynthesis : a quantative apporach. J Theor Biol 45, 339-377.
15. Ruget F (1981) Respiration de croissance et respiration d'entretien : méthodes de mesure, comparaison des résultats. Agronomie 1, 601-610.
16. Ryle GJA, Cobby JM and Powell CE (1976) Synthetic and maintenance respiratory losses of $^{14}CO_2$ in Uniculm barley and maize. Ann Bot 40, 571-586.
17. Thornley JHM (1977) Mathematical models in plant physiology. New York : Academic Press.
18. Wahua TAT and Miller DA (1978) Effects of shading on the N_2-fixation, yield and plant composition of field-grown soybeans. Agric J 70, 387-392.
19. Wilson D (1975) Variation in leaf respiration in relation to growth and photosynthesis of Lolium. Ann Appl Biol 80, 323-328.
20. Wilson D (1982) Response to selection for dark respiration rate of mature leaves in Lolium perenne and its effects on growth of young plants and simulated swards. Ann Bot 49, 303-312.
21. Wilson DR, van Bavel CHM and McCree KJ (1980) Carbon balance of water-deficient grain sorghum plants. Crop Sci 20, 153-159.

NET PHOTOSYNTHESIS AND WATER STATUS OF APPLE SEEDLINGS UNDER DEVELOPING
DROUGHT CONDITIONS : EFFECTS OF FUNGICIDE PRE-TREATMENT.

P. SIMON and R. MARCELLE, Research Station of Gorsem, Laboratory of Plant
Physiology, B-3800 Sint-Truiden, Belgium.

1. ABSTRACT

The resistance to water-stress of plants which were pre-treated with
five different fungicides was investigated in two experiments. In both
experiments net photosynthetic CO_2-uptake and total leaf water potential
were monitored during a period of developing water stress. The second
experiment also included measurements of leaf osmotic and turgor potential,
relative leaf water content, and leaf surface area.

The plants pre-treated with Fenarimol maintained the highest photosynthetic activity in both experiments demonstrating the possibility to change
the resistance to water stress by fungicide pre-treatment. An analysis of
the leaf water relations showed that only in the case of Fenarimol the
higher photosynthetic activity could be ascribed to a more favourable plant
water status which was the result of a lower evaporating leaf area. Osomotic
adjustment as a mechanism to maintain turgor at low leaf water potentials
was not enhanced in any of the treatments.

2. INTRODUCTION

The physiology and morphology of fruit trees can be modified by pesticides.
Insecticides, acaricides, and fungicides have been shown to influence photosynthesis and/or respiration (2,6). Fungicides may also suppress shoot
elongation and reduce leaf surface area (1).

The present paper reports results showing that the resistance to a developing water-stress as measured by the net photosynthetic CO_2 uptake (Pn) could
be influenced by pre-treatments with fungicides. These fungicides were chosen
according to their ability to accelerate or to prevent the appearance of leaf
spots on apple leaves; a description of this disorder has been given earlier
(7).

3. MATERIALS AND METHODS

3.1. Plant material

Seedlings of *Malus domestica* Borkh., cv Bittenfelder, were grown indivi-
dually in plastic pots containing compost under greenhouse conditions. Maximum
temperature in the greenhouse did not exceed 26°C while minimum night temperature rarely fell below 12°C. In a first experiment, the plants received the
pretreatment with the fungicides when they were about 7 months old (mean
number of leaves per plant : 29). In a second experiment, the plants were about
4 months old (mean number of leaves per plant : 16).

3.2. Fungicide applications

In both experiments the seedlings were sprayed to run-off with one of the
following fungicides applied at commercial rates (only the concentrations in
gl^{-1} of the active ingredient are given between brackets) :

Captan :N-(trichloromethylthio)-4-cyclohexene-1,2-dicarboxymide (1.04);

Fenarimol :α-(2-chlorophenyl)-α-(4-chlorophenyl)-5-pyrimidinemethanol (0.04);

Carbendazim : (2-methoxycarbomyl)-benzimidazole (0.25);

Biloxazol : β-[(1,1'-biphenyl)-4-yloxy]-α-(1,1-dimethylethyl)-1H-1,2,4-tria-
            zole-1-ethanol (0.13);

Tolylfluanide : N,N-dimethyl-N'-tolyl-N'-(dichlorofluor-methylthio)-sulfamide
               (0.67).

In the first experiment, the plants received a total of 4 applications at
7-day intervals; in the second experiment, they received a total of 6 applications at 5-day intervals.

Compared to Captan, which was taken as the control in the first experiment,
Fenarimol and Biloxazol are known to prevent or to retard the appearance of
leaf spots while Carbendazim and Tolylfluanide accelerate the appearance of
the symptoms. Captan itself however cannot be considered as an absolute crite-
rion as it is also known to predispose apple trees to leaf spot. In the second
experiment, plants which were only sprayed with water were also included.

3.3. Photosynthesis measurements

Pn was measured with a CO_2-IRGA (ADC, series 225) in an open system as
described earlier (8). The quantum flux density during the measurements was
250 μ mole m^{-2} s^{-1} (Lambda quantum sensor); the temperature and the vapour
pressure deficit inside the cuvette were 22° ± 1°C and about 1 kPa respectively. The leaves used for measuring Pn were adult leaves having received all
the fungicide pre-treatments (4 or 6 according to the experiment). These were
washed with distilled water one day (first experiment) or six days (second

experiment) before the photosynthetic activity was measured. The same leaves (1 per plant) were measured during the whole experiemnt. There were two replicates (first experiemnt) or six replicates (second experiment) per treatment.

3.4. Water relations measurements

In the first experiment, only total leaf water potential (ψ_w) was measured with a pressure bomb using leaves of the same age as for the Pn measurements, but taken on comparable and similarly treated plants.

In the second experiment, the following parameters of plant water relations were measured or calculated : ψ_w, leaf osmotic potential (ψ_s), relative leaf water content (RWC), and leaf turgor potential (ψ_p). Osmotic adjustment, as a mechanism to maintain ψ_p at low ψ_w was also evaluated. Two opposite but comparable leaves were used for those measurements. ψ_w was measured with a pressure bomb on one leaf while RWC and ψ_s were measured on the other leaf. RWC was determined on two leaf discs of 1 cm^2 according to Barrs (3). The rest of the same leaf tissue with the exception of the central vein was enclosed and frozen in pieces of small section PVC tubing sealed with parafilm. ψ_s was determined with a vapour pressure osmometer (Wescor, model 5100C) on a 8 µl subsample of the sap squeezed from the thawed tissue. No correction was made for the dilution of the symplastic water by apoplastic water during the thawing and subsequent squeezing process. ψ_p was calculated as $\psi_w - \psi_s$. Differences in osmotic adjustment were evaluated by comparing the calculated ψ_s for RWC = 100% (ψ_{s100}). Assuming that most of the tissue water is part of the osmotic system, then :

$$\psi_{s100} \simeq \psi_s \times \frac{RWC}{100} .$$

Water-stress was induced by withholding irrigation. Day 0 was the last day the plants received water; in the first experiment, this was 23 days after the last fungicide pre-treatment; in the second experiment, this was only 7 days after the last pre-treatment.

Leaf surface areas were measured with an area meter (Li-3000 from Li-Cor).

3.5. Statistics

The means of the two or six replicate measurements of each parameter were subjected to a two-way analysis of variance (treatments x days). In such an analysis the effects due to the treatments are averaged over the days and the effects due to the days are averaged over the treatments. Least significant differences were calculated for $p = 0.05$.

4. RESULTS AND DISCUSSION

4.1. First experiment

During the whole period of developing water-stress, the photosynthetic activity Pn of the leaves pre-treated with the fungicides Fenarimol, Carbendazim and Biloxazol remained higher than that of the leaves pre-treated with Captan and Tolylfluanide (Table 1). Furthermore, Pn did not change during the first seven days after water was withheld indicating that stress had not yet occurred during that period.

Total leaf water potential ψ_w was not influenced by the fungicide pre-treatments but during the development of water-stress, ψ_w became more and more negative (Table 1).

At the end of this first experiment, it was obvious that pre-treatments with fungicides could alter the resistance of apple seedlings to water-stress but this could not be related to changes of ψ_w.

Table 1. Net photosynthesis (Pn) and total leaf water potential (ψ_w) for the different fungicide treatments and for the different days after withholding water in the first experiment. Underlined data are statistically different from the control.

		Pn^1	ψ_w^2
Treatments	Captan (Control)	2.46	-1.68
	Fenarimol	5.81	-1.69
	Carbendazim	2.90	-1.59
	Biloxazol	3.58	-1.61
	Tolylfluanide	2.84	-1.76
	F-test	**	NS
	LSD	0.41	-
Days	0 (control)	3.97	-1.03
	2	3.87	-0.98
	5	4.22	-1.27
	7	4.01	-1.44
	9	3.11	-2.31
	12	1.24	-2.98
	13	0.76	-
	F-test	***	***
	LSD	0.49	0.46

1 μ mole CO_2 m^{-2} s^{-1}
2 MPa

Table 2. Net photosynthesis (Pn), total leaf water potential (ψ_w), relative leaf water content (RWC), leaf osmotic potential (ψ_s), leaf osmotic potential at RWC = 100 % (ψ_{s100}), and leaf turgor potential (ψ_p) for the different treatments and for the different days after withholding water in the second experiment. Underlined data are statistically different from the control.

		Pn^1	ψ_w^2	RWC^3	ψ_s^2	ψ_{s100}^2	ψ_p^2
TREATMENTS	H$_2$O (control)	3.91	-1.64	67	-1.74	-1.15	0.09
	Captan	3.69	-1.67	74	-1.71	-1.26	0.04
	Fenarimol	_4.42_	_-1.36_	75	-1.60	-1.23	_0.24_
	Carbendazim	_3.98_	-1.56	69	-1.72	-1.18	_0.17_
	Biloxazol	4.08	-1.66	68	-1.75	-1.17	0.10
	Tolylfluanide	_3.41_	-1.63	70	-1.71	-1.18	0.08
	F-test	6.93**	8.74**	2.78 ns	2.19 ns	1.04 ns	5.71**
	LSD	0.41	-0.12	–	–	–	0.10
DAYS	3 (control)	4.76	-1.17	76	-1.56	-1.18	0.39
	5	_4.24_	_-1.52_	_69_	_-1.66_	-1.16	_0.15_
	7	_2.29_	_-2.08_	_66_	_-1.90_	-1.25	_-0.18_
	F-test	128.6***	264***	14.14**	45.23***	2.99 ns	166.62***
	LSD	0.29	-0.09	4	-0.08	–	0.07

1 μ mole CO_2 m^{-2} s^{-1}
2 MPa
3 %

4.2. Second experiment

In this experiment, the seedlings were grown in smaller pots and development of water-stress was therefore faster. As seen in Table 2, the photosynthetic activity during the water-stress of the Fenarimol-treated leaves remained higher than that of the leaves pre-treated with the other fungicides or with water. This confirmed the results obtained in the first experiment. In the leaves pre-treated with Tolyfluanide, the photosynthetic activity was significantly lower than in the leaves pre-treated with water but not significantly different from the activity in the leaves pre-treated with Captan. The latter result also confirmed the observations of the first experiment.

Compared to Captan, the Carbendazim and Biloxazol pre-treatment increased somewhat the photosynthetic activity but this increase was not significant whereas it was in the first experiment. This discrepancy could be due to the fact that the water-stress experienced by the plants in the second experiment was not as severe and not as long as in the first experiment so that the treated plants could not express maximally their response to the imposed stress.

An examination of ψ_w and its component potentials shows that the better photosynthetic activity of the plants pre-treated with Fenarimol was probably the result of the general better water status of these plants, expressed by a less negative ψ_w, a higher turgor potential ψ_p, and a tendency for a higher relative water content RWC (Table 2). In all the other treatments, no significant differences could be found for any of the plant water status para- . meters. The lower photosynthetic activity measured in the leaves pre-treated with Tolylfluanide could therefore not readily be ascribed to a more severe water-stress as measured by ψ_w and its component potentials.

Maintenance of ψ_p at low ψ_w by the mechanism of osmotic adjustment did not prevail in any of the treatments as compared to the control (H_2O) and also no significant differences between the dates were found (Table 2). It is possible that the drought period was too short and developed too rapidly for osmotic adjustment to occur such as it was observed in unirrigated apple trees over a 3-months period (5). However, in another study on stressed trees in pots and in the orchard, no osmotic adjustment occurred over a period of several weeks (4).

The data of the second experiment suggest that the better plant water status after pre-treatment with Fenarimol was probably mainly the result of a reduction of total leaf area in these plants (Table 3) and so allowing a slower depletion of the available water in the pots.

Table 3. Mean leaf surface area per plant (10^{-4} m^2) in the second experiment

Treatments	
H_2O (control)	419
Captan	474
Fenarimol	357
Carbendazim	440
Biloxazol	441
Tolylfluanide	450
F	3.93 **
LSD	57

In conclusion, fungicide pre-treatment of plants can change the resistance of these plants to water-stress as indicated by the differential response of photosynthesis of the plants during developing drought conditions. For one fungicide (Fenarimol), the increased resistance to water stress can be ascribed to a better water status of the plants probably due to a smaller evaporative leaf surface area. For the other fungicides, the differential response of photosynthetic activity to water stress cannot be explained by changes in the plant water status. Further work on this subject including measurements of stomatal and mesophyll conductances are needed.

5. ACKNOWLEDGEMENTS

The technical assistance of Mr. W. Brugmans and Mrs. C. Missotten and the financial support of I.W.O.N.L., Brussels are gratefully acknowledged.

6. REFERENCES

1. Abdel-Rahman M (1977) Morphological effects of fungicides on apple trees. Abstract. Proc Amer Phytophathol Soc 4:213 (only).
2. Ayers Jr JC and Barden JA (1975) Net photosynthesis and dark respiration of apple leaves as affected by pesticides. J Amer Soc Hort Sci 100:24-28.
3. Barrs HD (1968) Determination of water deficits in plant tissues. In Kozlowski TT, ed. Water deficits and plant growth. pp 235-368. New-York : Academic Press.
4. Davies FS and Lakso AN (1979) Diurnal and seasonal changes in leaf water potential components and elastic properties in response to water-stress in apple trees. Physiol Plant 46:109-114.
5. Goode JE and Higgs KH (1973) Water, osmotic and pressure potential relation- ships in apple leaves. J Hort Sci 48:203-215.
6. Heinicke DR and Foott JW (1966) The effect of several phosphate insecticides on photosynthesis of Red Delicious apple leaves. Can J Plant Sci 46:589- 591.
7. Jonkers H (1973) Review on leaf spot and leaf drop : a physiological disorde of the "Golden Delicious" apple. Sci Hort Amsterdam 1:231-237.
8. Marcelle R (1975) In Marcelle R, ed. Environmental and Biological control of Photosynthesis, pp 349-359. The Hague : Junk.

THE EFFECT OF WATER STRESS ON GREENING OF PRIMARY BARLEY (*HORDEUM VULGARE* L. CV. MENUET) LEAVES

R. VALCKE & M. VAN POUCKE, Dept. SBM, Limburgs Universitair Centrum, Universitaire Campus, B 3610 Diepenbeek, Belgium

1. ABSTRACT

Atmospheric water stress applied to greening barley seedlings by low relative humidity of the ambient air, impairs chloroplast development. This is first illustrated in the kinetics of protein and chlorophyll synthesis. There are also significant effects on the photochemical activities. An explanation for these effects is proposed in terms of the efficiency of molecular assembly and energy transfer in the photosystems. Changes are also observed in the ultrastructure of plastids greening in plants under atmospheric water stress. These changes are discussed in relation to similar effects, known to occur in other plants in normal or in stress conditions.

2. INTRODUCTION

Reduction of leaf water potential inhibits photosynthesis in green plants. Photosynthetic electron transport (30), photophosphorylation (19) as well as chloroplast enzyme activities (13,23) are affected.

Plastid differentiation during greening is also responsive to leaf water stress. Effects on chlorophyll accumulation in etiolated or greening tissue are well documented (1,8). In this paper, data are presented especially on possible changes in polypeptide pattern of the internal plastid membranes, on the development of photochemical activity and on ultrastructural changes during greening of etiolated seedlings, submitted to low and high leaf water stress, induced by different levels of relative humidity of the ambient air.

3. MATERIALS AND METHODS

3.1. Plant material

Seeds of barley were imbibed in darkness at 28°C in washed seasand. After 24 h the germinated seeds were transferred to vermiculite and grown in continuous darkness at 25°C for 5 days in either 80 % or 40 % relative humidit

(RH). The seedlings are further referred to as normal and water stressed
seedlings respectively. For greening experiments, the plants were then transferred to continuous light (240 µE m^{-2} h^{-1}) under the same conditions of
temperature and RH.

3.2. Photosynthetic electron transport

Plastid fragments were isolated according to the method of Plesnicar and
Bendall (25). Reduction of ferricyanide was followed by measuring O_2 production
at light saturation in a Rank oxygen electrode at 25°C. The reaction medium
used was as described by Van Assche et al (31). Reduction of methylviologen
with DCPIP/HAsc as electrondonor was performed according to Oben and Marcelle
(20). Total chlorophyll analyses were performed according to Bruinsma (6). The
method of Bradford (4) was used for protein analyses.

3.3. Isolation of internal plastid membranes

Primary leaves were homogenised in 0.1 M Tricine buffer pH 7.9 containing
0.6 M glycerol and 3 mM $Ca(NO_3)_2$ according to Høyer Hansen & Simpson (12).
A Braun MX 32 mixer as adapted by Kannangara (15) was used. The isolated plastids were schocked osmotically in 25 mM HEPES buffer pH 7.5 containing 10 mM
EDTA. The internal plastid membranes were further purified by flotation on a
discontinuous sucrose gradient as described by Chua and Bennoun (7). The
purified membranes were collected from the sucrose gradient and washed in
HEPES buffer. For gel electrophoresis the final pellet was solubilized in
50 mM Na_2CO_3, 50 mM DTT, 12 % sucrose and 2 % SDS.

3.4. Gel electrophoresis

SDS-polyacrylamide gel electrophoresis (SDS-PAGE) on continuous gels (7,5-
15 % in acrylamide) was performed as described by Chua & Bennoun (7) using the
discontinuous buffer system of Neville (18). After electrophoresis, gels were
stained for protein in 0.25 % Coomassie Brillant Blue R 250 and destained
according to Fairbanks (9).

3.5. Electron microscopy

Pieces of the primary leaves were fixed for electron microscopy in 2 %
glutaraldehyde in 0.1 M Na-cacodylatebuffer pH 7.0, postfixed in 1 % osmium
tetroxide in 0.1 M Ca-Mg-Veronal buffer pH 7.0. After several washings in
distilled water, the material was stained overnight in 2 % uranylacetate,
dehydrated through an ethanol series and embedded in Spurr's resin (29).
Sections were poststained with Reynolds lead citrate (26). Rotary photographic enhancement was performed as described by Markham et al (17).

4. RESULTS

Total chlorophyll synthesis in continuous light is shown in fig. 1.
Under both conditions of RH, there is a lag phase of approximately 2 h after
onset of illumination. Then follows a period of rapid accumulation of chlorophyll at a constant rate during 16 h after which period accumulation rate
decreases. Accumulation as well as total content of chlorophyll are significantly lower in conditions of low RH. After 24 h of continuous light under
low RH, the total amount of chlorophyll on a fresh weight basis is about 70 %
of the value obtained under high RH.

In contrast to the pigments, protein content of the plastid fraction
increases very slowly during greening (fig. 2). In stressed seedlings it
reaches approximatively 80 % of the control value, irrespective of the
greening time.

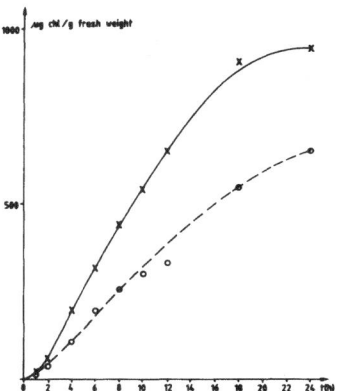
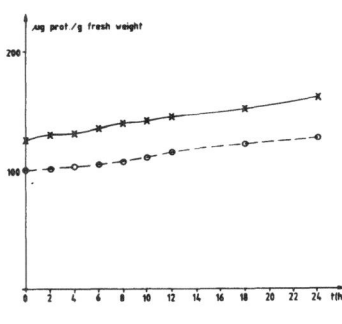

Fig. 1 Accumulation of chlorophyll (a+b) Fig. 2 Accumulation of proteins
     RH conditions : x-x 80 %, 0--40 %   RH conditions : cf. fig 1

Fig. 3 shows the polypeptide composition of purified membranes at different
stages of greening in both high and low RH. The 97, 45, 34, 31, 29 and 26 kD
bands appear during this process; they correspond to the chlorophyll-protein
complexes chl a-P_1(P700 chlorophyll-protein complex), chl a-P_2 (chl a - AP_2,
reaction center of photosystem II), chl a/b-P_1,P_2 and AP_2 (light-harvesting
chlorophyll protein complex) respectively (16). It is clear that water stress
does not affect the polypeptide pattern (e.g. at 9 h of illumination), but

retards the appearance of the bands, corresponding to the chlorophyll-
protein complexes (compare the pattern at 40 % and 80 % RH after 3 h of
illumination).

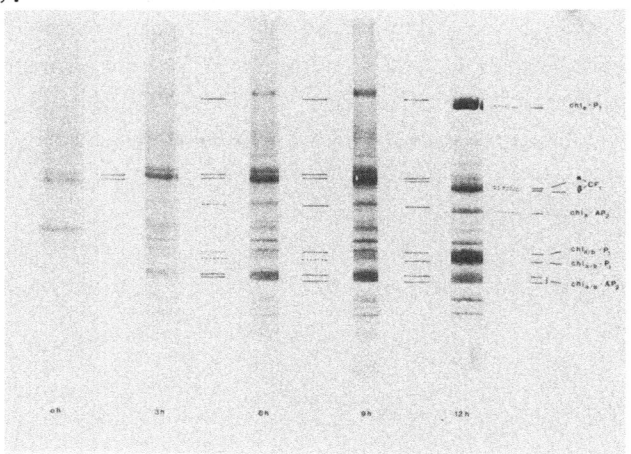

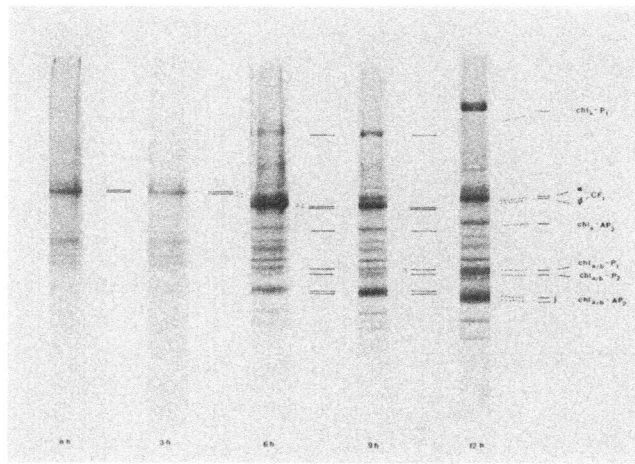

Fig. 3. Polypeptide composition of internal membranes.
       Upper row : "normal" seedlings (80 % RH)
       Lower row : "stressed" seedlings (40 % RH)

Expressed on a protein basis, PS1 is already detected after 2 h of
illumination for both conditions of RH (fig. 4). In control plants, this
activity constantly rises but declines after about 16 h when chlorophyll
accumulation also decreases. Under water stress however, PS1 is significantly less active; its activity rate already diminishes after approximately

8 h with chlorophyll accumulation still increasing.

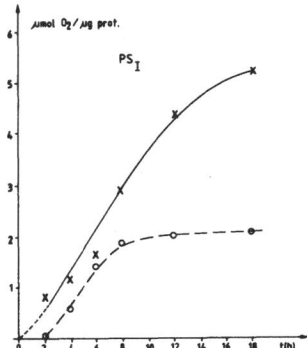

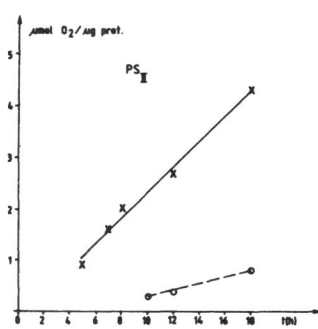

Fig. 4 Photochemical activity of PS I   Fig. 5 Photochemical activity of PS II
     RH conditions : x-x 80 %; O--O 40 %   RH conditions cf. fig. 4

The structural changes in the developing plastids have been extensively described (3, 11, 34). A delayed sequence of the same events is observed in stressed seedlings. Their plastids are characterised by swollen thylakoids and a large number of plastoglobuli and by the presence of aggregates in the stroma (fig. 6). In longitudinal view, these aggregates appear to be fibrils of variable length; in transverse section, they show an inconstant number of elements (fig. 7). Rotary photographic enhancement of these aggregates with n=6 suggests a hexagonal arrangement in which each element is surrounded by six others. However, other values of n indicate different cristalline arrangements as can be seen in fig. 7. These results indicate that the elements of the aggregates are not arranged in a simple cristalline lattice, but rather in a very complex superstructure.

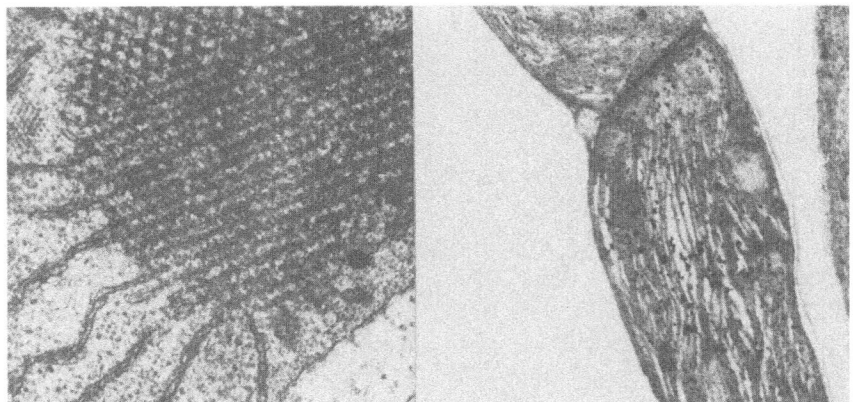

Fig. 6 Electron micrographs of an etioplast with aggregates and of a chloroplast (after 24 h of continuous greening) with swollen thylakoids.

5. DISCUSSION

The results presented show that atmospheric water stress as a result of
low RH generally retards the greening process as a whole. Rate of synthesis
of pigments and proteins, photochemical activities as well as the dynamics of
ultra-structural development in the plastids are alle quantitatively delayed,
but in each of these processes the events are not significantly changed
neither in character nor in sequence.

Concerning the effect of RH on the general pattern of chlorophyll synthesis
its accumulation rate and final content, our results with barley are confirmed
by those obtained in wheat by Duysen and Freeman (8). However, they are not in
agreement with the conclusions of Alberte et al on black bean. These authors
showed that a lag phase in chlorophyll accumulation only occurs at low
(35 %) RH (1), whereas in our material a lag-phase is a common phenomenon in
regimes of both low and high RH.

Comparison of fig. 1 and 4 show that, after the lag-phase, chlorophyll
accumulation and PS I activity increase simultaneously and at a constant rate
in both regimes of RH. After 8 h of illumination in low RH, however, a
discrepancy arises between chlorophyll accumulation, proceeding uninterruptedly
and PS I activity, which levels off. From then onwards, the activity pigment
ratio is very low, as compared to high RH. These results suggest that in a lea-
developing under atmospheric water stress, the functional development of the
energy transfer mechanism in the chlorophyll/PS I protein complex is impaired.

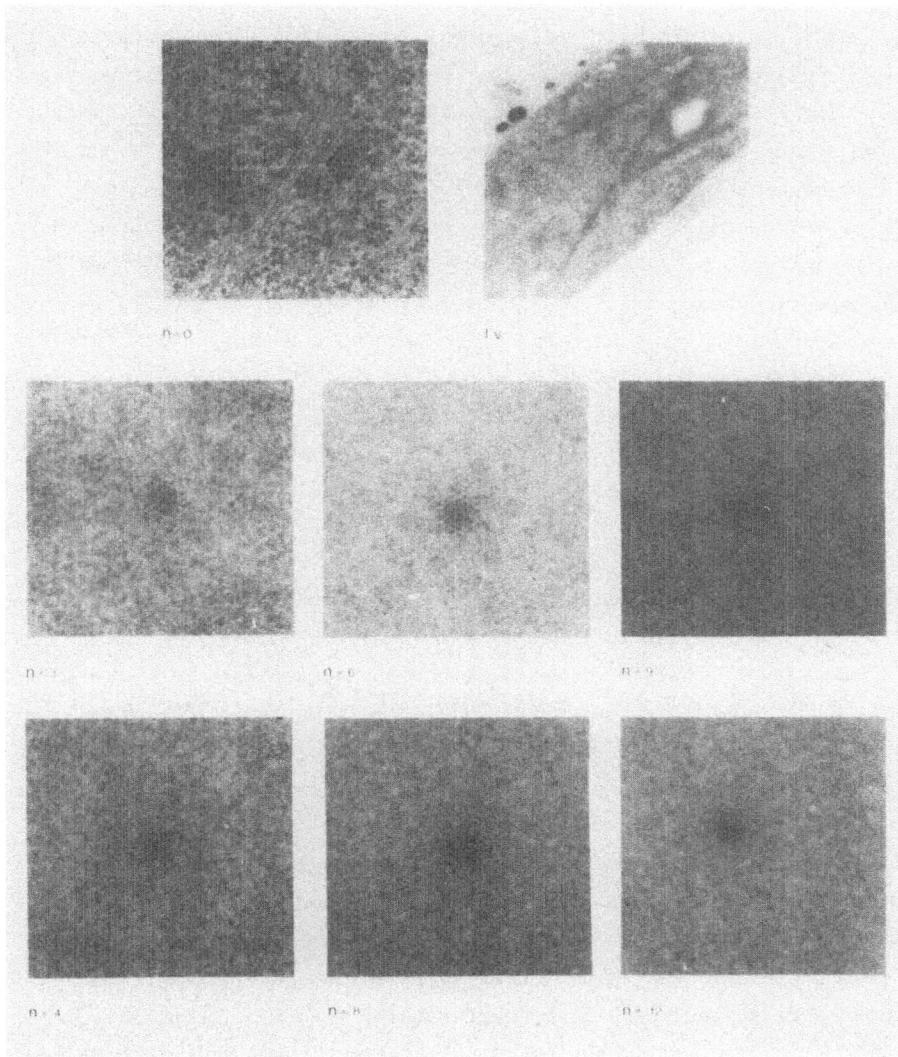

Fig. 7 Rotary photographic enhancement of the tubular arragement in plastic
of water stressed seedlings.
n)values as indicated; l.v : longitudinal view.

In support of this conclusion, we may recall the evidence given by Oquist et al (22) for a chlorophyll a-β carotene association in the chlorophyll a antenna of PS I, which is believed to function as a trap for excess excitation energy; Sigfridson (28), working with dessicated <u>Cladonia impexa</u> thalli observed a decrease of the energy transfer between chlorophyll b and other accessory pigments to chlorophyll a.

Figure 5 shows that low RH not only retards the appearance but also decreases the measurable activity of PS II. A mechanism of action on PS II for atmospheric water stress, can therefore be proposed as for PS I. PS II, however, seems to be even more sensitive than PS I as the effect is more severe. PS II is believed to be more complex than PS I (2). Its high sensitivity to RH, as compared to PS I, could therefore be due to at least two additional factors. First, water stress might impair the molecular assembly of the complex *in vivo*. Second, PS II might be subject to a higher risk of structural destabilization during the extraction and measuring procedure.

A number of ultrastructural changes in plastids seem to be an overall response to a decrease in water supply, whatever the mechanism of its induction : osmotic stress, salt stress or atmospheric water stress. On the one hand, a swelling of thylakoids was observed in chloroplasts of different plant species, subjected to osmotic stress (8, 14, 32). On the other hand, a large number of plastoglobuli was already described for water stressed plastids (24, 32). These phenomena are similar to our observations induced by atmospheric water stress.

Tubular aggregates occur in plastids of *Avena sativa* (10), *Triticum aestivum* (21) and *Sedum spec.* (5), grown under normal conditons. In *Mesembryanthenum*, however, tubular arrangements in the stroma of plastids were found when the plants were grown under salinity stress (27). All these structures were suggested to be an aggregate of fraction I protein. Lack of more precise information on the composition and the geometric configuration of the aggregates found by us in plastids of water stressed barley, prevents us from deciding whether our observations correspond to any of those reported in the literature.

6. ACKNOWLEDGMENTS

The authors wishes to thank Prof. Dr. H. Clijsters for critical discussion and Mrs. B. Vanaken for technical assitance. This work was supported by grant n° 2.9006.80 of the "Fonds voor Kollektief Fundamenteel Onderzoek" Brussels, Belgium.

7. REFERENCES

1. Alberte RS Fiscus W and Naylor AW (1975) The effects of water stress on the development of the photosythetic apparatus in Greening Leaves. Plant Physiol 55:317-321.
2. Boardman NK Anderson JM & Goodchild DJ (1978) Chlorophyll-protein complexes and structure of mature and developing chloroplasts. Current Topics in Bioenerg 8:35-159.
3. Bradford MM (1976) A rapid and sensitive method for the quantization of microgram quantities of proteins utilizing the principle of protein-dye binding. Anal Biochem 72:248-254.
4. Bradbeer JW Ireland HMM Smith JW Rest J and Edge HJW (1974) Plastid develop-ment in primary leaves of Phaseolus vulgaris. VIII Effects of transfer of dark-grown to centinuous illumination. New Phytol 73:263-290.
5. Brandaõ I and Salemo R (1974) Microtubules in chloroplasts of a higher plant J Submicrosc Cytol 6:381-390.
6. Bruinsma J (1963) The quantitative analysis of chlorophyll a and b in plant extracts. Photochem Photobiol 2:241-249.
7. Chua NH and Bennoun P (1975) Thylakoid membrane polypeptides of Chlamydomo-nas. Reinhardii : wild type and mutant strains deficient in photosystem II reaction center. Proc Natl Acad Sc 72(6):2175-2179.
8. a. Duysen ME and Freeman FP (1974) Effects of moderate water deficit (stress on wheat seedlings growth and plastid pigment development. Physiol Plant 31:262-266.
   b. Duysen ME and Freeman TP (1975) Partial restoration of the high rate of plastid pigment development and the ultrastructure of plastids in de-tached water-stressed wheat leaves. Plant Physiol 55:768-773.
9. Fairbanks G Steck TL and Wallach DFH (1971) Electrophoretic analysis of the major polypeptides of the human erythrocyte membrane. Biochemistry 10(13):2606-2616.
10. Gunning BES (1965) The five structure of chloroplast stroma following ade-hyde osmiumtetroxide fixation. J Cell Biol 24:79-93.
11. Henningsen KW and Boynten JE (1969) Macromolecular physiology of plastids. VIII The effect of a brief illumination on plastids of dark-grown barley leaves. J Cell Sci 5:757-793.
12. Høyer-Hansen G and Simpson (1977) Changes in the polypeptide composition of internal membranes of barley plastids during greening. Carlsberg Res Commun 42:379-389.
13. Huffaker RC Radin T Kleinkopf GE and Cox EL (1970) Effects of mild water stress on enzymes of nitrate assimilation and of the carboxylative phase of photosynthesis in barley. Crop Science 10(5):471-474.
14. Izawa S and Good NE (1966) Effects of salts and electron transport on the conformation of isolated chloroplasts. II Electron microscopy. Plant Physiol 41:544-552.

15. Kannangara CG Gough SP Hansen B Rasmutten JN and Simpson DJ (1977) A homogenizer with replaceable razor blades for bulk isolation of active barley plastids. Carlsberg Res Commun 42:431-439.

16. Machold O Simpson DJ and Møller BL (1979) Chlorophyll proteins of thylakoids from wild type and mutants of barley. Carlsberg Res Commun 44: 235-254.

17. Markham R Frey S and Hills GJ (1963) Methods for the enhancement of image detail and accentuation of structure in electron microscopy. Virology 20:88-102.

18. Neville DM (1971) Molecular weight determination of protein SDS complexes by gel electrophoresis in a discontinuous buffer system. J Biol Chem 246(20):6328-6334.

19. Nir I. and Poljakoff-Mayber A. (1967) Effect of water stress on the photochemical activity of chloroplasts. Nature 213:418-419.

20. Oben G and Marcelle R (1975) The effects of CCC and GA on some biochemical and photochemical activities of primary leaves of bean plants. Marcelle R (Ed) Environmental and Biological Control of Photosynthesis : 211-216.

21. Oliviera L (1975) On the morphology and nature of the plastid inclusions of leaf cells of a Triticale. J Submicrosc Cytol 7:271-280.

22. Öquist G (1980) On the role of β-cartene in the reaction center chlorophyll a antennae of photosystem I. Physiol Plant 50:63-70.

23. O'Toole JC Ozban JL and Wallace DH (1977) Photosynthetic response to water stress in Phaseolus vulgaris. Physiol Plant 40:111-114.

24. Pham Thi AT and Vieira Da Silva J (1980) Influence de la sécheresse sur l'ultrastructure mitochondriale chez le cotonnier. Quelques implications métaboliques. Z Pflanzenphysiol 100:351-358.

25. Plesnicar M and Bendall DS (1973) The photochemical activities and electron carriers of developing barley leaves. Biochem J 136:803-812.

26. Reynolds J (1963) The use of lead citrate at high pH as an electron opaque stain in electron microscopy. J Cell Biol : 17:208.

27. Salema R and Brandaõ I. (1978) Development of microtubules in chloroplasts of two halphytes forced to follow CAM. J Ultrastruct Res 62:132-136.

28. Sigfridsson B and Öquist G (1980) Preferential distribution of excitation energie into photosystem I of dessicated samples of the lichen Cladonia impexa and the isolated lichen alga Trebomixia pyriformis. Physiol Plant 49:329-335.

29. Spurr AR (1969) A low viscosity epoxy resin embedding medium for electron microscopy. J Ultrastruct Res 26,31.

30. Todd GW and Basler E (1965). Fate of various protoplasmic constituents in droughted wheat plants. Phyton 22:79-85.

31. Van Assche F Clijsters H and Marcelle R (1979) Photosynthesis in *Phaseolus vulgaris* L. as influenced by supra-optimal zinc nutrition. Marcelle R Clijsters H and Van Poucke M (Ed) Photosynthesis and plant development p 175-184.

32. Vieira Da Silva J (1976) Water stress, ultrastructure and enzymatic activity Lange O Kapper L and Schulze ED (Ed) Ecological studies, vol 19, Water and Plant Life : 207-224.

33. Virgin HI (1965) Chlorophyll formation and water deficit. Physiol Plant 18:994-1000.

34. Weier TE Sjoland RD and Brown DL (1970) Changes induced by low light intensities on the prolamellar body of 8-day, dark-grown seedlings. Am J Bot 57(3):276-284.

EFFECT OF WATER STRESS ON RATE OF PHOTOSYNTHESIS, TRANSPIRATION, AND CHLOROPLAST ULTRASTRUCTURE IN WILLOW LEAVES

E. VAPAAVUORI[1], A. NURMI[2] & E. KORPILAHTI[3]

[1]The Finnish Forest Research Institute, Suonenjoki Research Station, Suonenjoki, Finland; [2]Department of Botany, University of Helsinki, Finland; [3]Department of Silviculture, University of Helsinki, Finland

ABSTRACT

Formation of crystals, supposed to be Fraction I-protein, in the chloroplast stroma has been reported frequently in the literature (1). In our previous study with willow (*Salix sp.* 'aquatica gigantea')(2), under the strong light regime large crystal formations were observed with increasing water stress. In weak light these crystal structures could be seen in the well-watered controls, but seemed to disappear as water stress became severe. Since CO_2 assimilation at low $p(CO_2)$ has been thought to follow the kinetics of RuP_2-carboxylase-oxygenase and to be determined by the RuP_2 saturated rate of the enzyme (3), an experiment was designed to determine whether crystal inclusions observed in the chloroplasts are correlated with photosynthesis during water stress. The characteristics of CO_2 exchange and transpiration at 20 $^{\circ}C$ and at a quantum flux density of 550-700 $\mu E\ m^{-2}\ s^{-1}$ were examined in willow leaves (*Salix sp.* 'aquatica gigantea') at several concentrations of intercellular CO_2.

With decreasing leaf water potentials, CO_2 exchange became inhibited at intercellular partial pressures of CO_2 that were both rate-limiting and rate-saturating. This inhibition was increasingly caused by nonstomatal factors, especially at CO_2 concentrations equal to or higher than that of the ambient air. In moderately stressed plants (Ψ_L -0,74 MPa) the inhibition of CO_2 exchange was relieved but not totally overcome two days after the plants were rewatered, although the values for leaf water potential, resistance in the stomatal and boundary layers, and rate of transpiration had values similar to those for well-watered controls. Under severe stress (Ψ_L -0,94 MPa) the plants had recovered only slightly two days after rewatering. This slow recovery was caused mainly by nonstomatal factors and was indicated by high values for mesophyll resistance.

Under electron microscopy large grana stacks showed up in the chloroplast lamellae, a feature typical of low-light plants. The chloroplast lamellae remained fairly intact throughout the drying period, although the lamellae showed somewhat poorer contrast to electrons. In well-watered control plants and under mild water stress, large crystal formations were observed in the stroma area of the chloroplasts. The number of crystal formations seemed to decrease with increasing water deficit. In severely stressed plants few or none of these crystal formations were observed. This is similar to our previous results with willow grown in weak light (2).

Since willow (*Salix sp.* 'aquatica gigantea') is a typical heliophyte, low irradiance during the growing period of the plants might have been the primary stress factor for crystallization in our study. With inducement of another stress factor, in this case water stress, proteolytic degradation of Fraction I-protein might have increased in the leaves, causing the crystals to disappear and the rate of CO_2 uptake to decrease.

REFERENCES

1. Sprey, B. 1977. Lamellae-bound inclusions in isolated spinach chloroplasts. I. Ultrastructure and isolation. Z. Pflanzenphysiol. 83:159-179.
2. Vapaavuori, E. & A. Nurmi. 1982. Chlorophyll-protein complexes in *Salix sp.* 'aquatica gigantea' under strong and weak light II. Effect of water stress on the chlorophyll-protein complexes and chloroplast ultrastructure. Plant & Cell Physiol. 23, in press.
3. Farquhar, G. D., S. von Caemmerer & J. A. Berry. 1980. A biochemical model of photosynthetic CO_2 assimilation in leaves of C_3 species. Planta 149:78-90.

EFFECT OF THE SIMULATION PROCEDURES OF ROOT WATER UPTAKE ON THE ASSESMENT OF THE PLANT WATER STRESS

M. ALAERTS, M. BADJI and J. FEYEN[*]

ABSTRACT

Four root water extraction terms different in concept and varying complexity, used in soil water balance models, were compared, and their sensitivity was tested. The models were run under simplified initial and boundary conditions. The simulated total water extraction, reduction of transpiration rate and development of plant water stress, and water extraction patterns differed significantly. The complexity of the most sophisticated model appeared for the major part superfluous. The simplest one cannot be applied under all conditions.

SIMULATION OF WATER STRESS IN *BRACHIARIA RUZIZIENSIS* UNDER A HIGH EVAPORATIVE DEMAND

M. BADJI, M. ALAERTS and J. FEYEN[*]

ABSTRACT

An attempt to simulate the water potential in *Brachiaria Ruziziensis*, under high evaporative demand, was made using Ohm's law analog concept and the catenary concept including plant capacitance. Both models follow fairly well the measured plant water potential as long as the soil remains relatively moist. At high soil moisture deficits, both models tend to overestimate strongly the water stress. The concept taking into account the plant capacitance results, under adversed conditions, to slightly better computed water stress values, although not significantly better than the calculated results obtained with the single catenary concept.

[*] Laboratory of Soil and Water Engineering, K.U.Leuven, Kardinaal Mercierlaan 92, 3030   Leuven (Belgium)

RESPONSES OF CRASSULACEAN ACID METABOLISM (CAM) TO INCREASING AND
DECREASING WATER STRESS IN PLANTS IN THE SOUTHERN NAMIB DESERT

D. J. von Willert, E. Brinckmann, B. Scheitler, B. M. Eller[+]

Lehrstuhl Pflanzenökologie, Universität Bayreuth, D-8580 Bayreuth,
W.-Germany

[+]Institute of Plant Biology, University Zürich, CH-8008 Zürich,
Switzerland

ABSTRACT

CAM features in a variety of different succulents in response to
climatic changes between March 1977 and April 1981 were studied in the
southern Namib desert (Richtersveld). In 1977 and 1978 all investigated
succulents performed a CAM. After an extended period of insufficient rainfall which damaged the succulents between 80 and 100% most species did not
show a measurable CO_2 gas exchange or diurnal acid fluctuation, indicative
of CAM. But both features were restored quickly after an abundant rainfall
in March 1981.

Artificially increased or decreased water stress confirmed that
nocturnal CO_2 uptake and malate accumulation is correlated with water
supply and water content in the plant. The effects of the water status are
overruled by rapid changes in the water vapour pressure deficit during the
night. A sudden increase due to föhn-wind stopped or at least markedly
diminished CO_2 uptake and acid accumulation in CAM plants.

INTRODUCTION

From the available information in the literature the ecological
significance of CAM is difficult to define. Investigations of CAM plants
in their natural habitats favour a marked reduction of nocturnal
acidification with increasing drought (1, 2). Due to seasonal changes in
the climate, effects of increasing drought and temperature could not be
separated. On the other hand, laboratory experiments imply an adaptive
mechanism of CAM to drought and peak in the so-called induction of CAM by
water stress (for ref. see 3). Nevertheless, the contradictory and
confusing excess of different data seem to have at least one unifying
result. Nocturnal CO_2 uptake is less sensitive to changes in the
environment than photosynthetic CO_2 uptake.

In order to elucidate the contribution of CAM to the survival of
succulents in arid environments we investigated the responses of CAM plants
to increasing and decreasing water stress over several years in the
southern Namib desert.

MATERIAL AND METHODS

All measurements were done in the southern part of the Namib desert,
the Richtersveld, during March 1977, Sept. and Oct. 1978, March and April
1980 and Feb. to April 1981 close by the old copper mine Numees. CO_2 gas
exchange was measured by an infrared gas analyzer in an open system. Leaves
were harvested at sun set and sun rise. Samples were immediately dried at
95°C and afterwards extracted with hot water. In aliquots of these extracts
malate and citrate were determined enzymatically and proline by a
colorimetric method. The osmotic potential of the cell sap was determined
by a cryoscopic method. Air temperatures were measured by copper-constantan
thermocouples, dew point temperature by a dew point mirror. Water vapour
partial pressure deficit (VPD) was calculated from air and dew point
temperatures. For further details see ref. 4 and 5.

List of plants cited below:

Mesembryanthemaceae: *Brownanthus schlichtianum* (Sonder) V. Bittrich et
Ihlenfeldt, *Cheiridopsis sp.*, *Psilocaulon cf. subnodosum* (A. Berger)
N. E. Br., *Ruschia schneideriana* (A. Berger) L. Bolus, *Ruschia sp. 1*
(with red stems), *Ruschia sp. 2* (with lateral inflorescences),
Sphalmanthus trichotomus (L. Bolus) L. Bolus, *Stoeberia beetzii* (Dinter)
Dinter et Schwantes v. *arborescens* (Friedrich). Crassulaceae: *Crassula
perfoliata* L., *Crassula deceptrix* Schoenl., *Tylecodon reticulatus*
(Thunbg.) Portulacaceae: *Ceraria namaquensis* (Sond.) Pears. et Stephens.
Asteraceae: *Senecio longiflorus* (DC.) Sch. Bip. Liliaceae: *Aloe pearsonii*
Schoenl.

RESULTS AND DISCUSSION

The average annual rainfall in the Richtersveld is 64 mm and winter
is the main rain season. Nevertheless, rainfall is episodic and extended
rainless periods occur. Fig. 1 gives the rainfall data of the last 6
years that were recorded at Lekkersing in the Richtersveld. Although
these data will not reflect the rainfall situation at Numees which lies
about 100 km north of Lekkersing they demonstrate the severe drought period

that started in 1978. The insufficient and mostly offseason rainfall
since 1978 damaged the vegetation considerably. Especially succulents dried
off. Most of them were damaged to about 80% but some even completely
(*Prenia sladeniana*). The still living succulents had a very low water
content and their leaves were crumpled (fig. 3a). The slightly better rainfall in 1980 led to a weak increase in the water content but was not enough
to restore the vegetation. The recovery process was initiated by the
abundant rainfalls in 1981 and provided further rain will continue in
1982 (there have been 67 mm rain in March and April 1982).

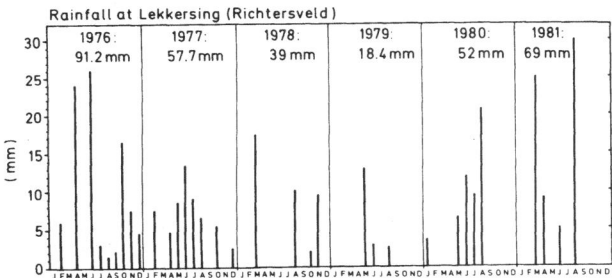

FIGURE 1. Monthly distribution and amount of rainfall in mm. Data from
1976 - 1980 were recorded at Lekkersing that for 1981 at Numees (Richtersveld).

From a first screening of the vegetation at Numees in 1977 we know
that all investigated succulents performed a CAM (6). This screening was
extended in 1978 and continued in 1980 and 1981. In 1981 measurements were
done prior to an abundant rainfall and thereafter. As is shown in fig. 2
for plants from different families representative for all investigated
plants malate and citrate fluctuations declined drastically with increasing drought from 1978 to 1980 and leaf water content was markedly
reduced. But both recovered a bit from 1980 to 1981 and considerably after
the abundant rainfall of March 1981 (fig. 3).

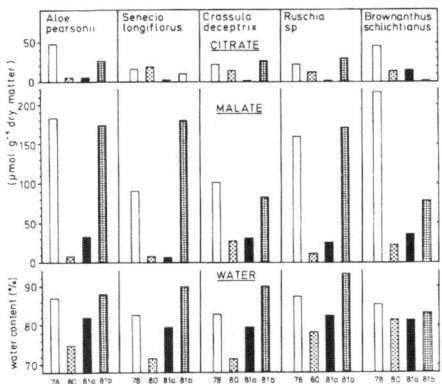

FIGURE 2. Overnight accumulation of citrate and malate and the water content in % of fresh weight in 5 different CAM-species. Data were obtained in Sept. 1978, March 1980, February 1981 and April 1981.

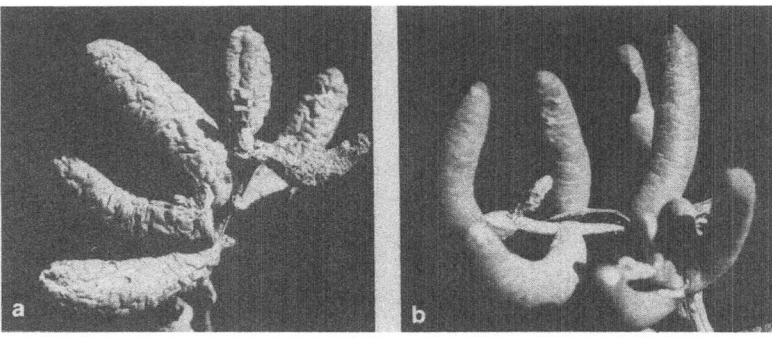

FIGURE 3. *Ruschia sp. 2* in April 1980 (a) and after the abundant rainfall in March 1981 (b).

From September to October 1978 we observed a steady decrease in nocturnal malate synthesis which runs parallel to a decreasing water availability in the soil (soil water potential changed from -8 to -44 bars). Corresponding results were obtained by setting a severe water stress. Cutting off the roots of a CAM exhibiting *Psilocaulon* plant in a good water condition led to a rapid decline in malate synthesis. Even the evening content of malate (stock pool) was continuously reduced (March 1981). In contrast, watering a rooted plant of the same species caused a rapid increase in malate synthesis over night and the stock pool recovered (April 1980). Fig. 4 compares the results of both the experiments.

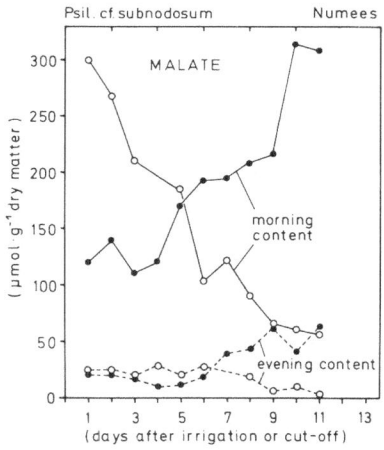

FIGURE 4. The evening and morning content of malate in *Psilocaulon* cf. *subnodosum* during the course of an irrigation experiment in April 1980 (●) and in a cut-off experiment in March 1981 (o).

During the course of these experiments the leaf water content changed. Fig. 5a shows that irrespective of leaf age, year and treatment (cut-off or irrigation) nocturnal malate synthesis can be directly correlated with leaf water content. A replot of the data from March 1981 over the solute water potential Ψ_π shows that the endogenous water availability governs malate synthesis at night (fig. 5b). If we accept proline as a water stress indicator then fig. 5d demonstrates that after cutting off the root *Psilocaulon* faces an increasing water stress which is answered by increasing proline and decreasing malate content. Finally, fig. 5c reveals that irrespective of the season, the year and the plant species nocturnal malate synthesis is correlated with the apparent leaf water content which from the above given evidence is a suitable indicator of water stress in the plant and easy to measure in the field. As can be seen on the graph most plants cease their nocturnal CO_2 fixation at about 82% leaf water while others (*Sphalmanthus*) are much more sensitive.

160

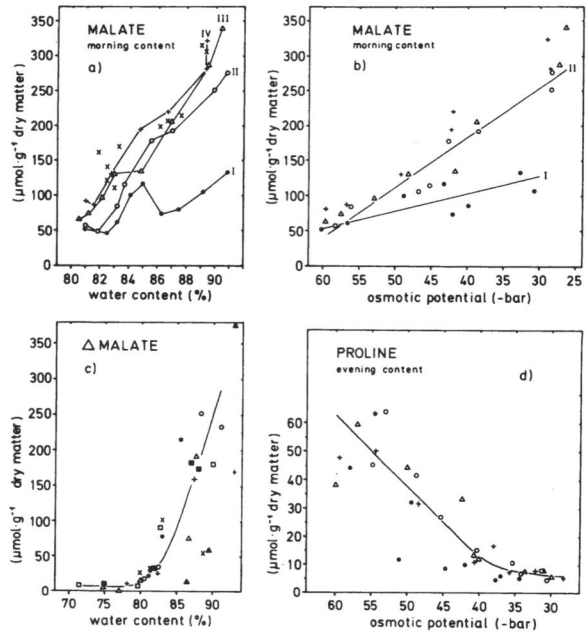

FIGURE 5. a) Correlation between the morning content of malate and the leaf water content in % of fresh weight in leaves of different age of *Psilocaulon* cf. *subnodosum*. Youngest leaf I (●), second leaf II (o), third leaf III (Δ), forth leaf IV (+). Data of the connected symbols were obtained in a cut-off experiment in March 1981, the crosses (✗) are data from an irrigation experiment in April 1980.

b) Correlation between the morning content of malate and the osmotic potential of the cell sap in leaves of different age of *Psilocaulon* cf. *subnodosum*. The malate content of the youngest leaf differs significantly from that in the following leaves.

c) Correlation between the overnight accumulation of malate and the apparent leaf water content in naturally growing (●) *Brownanthus schlichtianum*, (▲) *Sphalmanthus trichotomus*, (o) *Stoeberia beetzii*, (+) *Ruschia sp. 2*, (Δ) *Cheiridopsis sp.*, (✗) *Crassula deceptrix*, (□) *Senecio longiflorus*, (■) *Aloe paersonii*. Data were obtained in Sept. 1978, April 1980, Feb. 1981 and April 1981.

d) Correlation between the evening content of proline and the osmotic potential of the cell sap in leaves of different age of *Psilocaulon* cf. *subnodosum*. Leaf age as in 4a.

There can be no doubt that water supply to the soil via water content in the plant is one important factor which determines the expression of CAM in a plant. But even a good water supply from the soil does not necessarily guarantee a maximum nocturnal CO_2 fixation. For a well watered plant the overnight accumulation of malate still depends on the microclimatic conditions during the night, mainly temperature and relative humidity which are best expressed by the term VPD. There is, irrespective of leaf age, a good negative correlation between nocturnal malate synthesis and mean VPD during the night (fig. 6). Hence VPD overrules soil water supply and internal water availability and is a second important factor controlling nocturnal malate accumulation in CAM plants of the southern Namib.

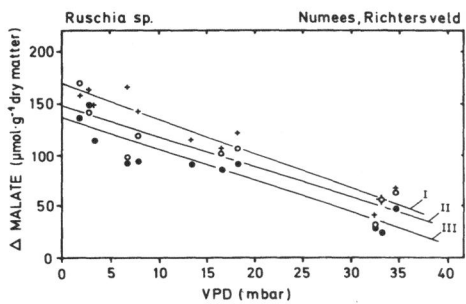

FIGURE 6. Correlation between the overnight accumulation of malate in leaves of different age of *Ruschia sp. 1* and the mean value of the water vapour partial pressure deficit of the night (20h00 - 7h00). I: youngest leaf.

Nocturnal CO_2 uptake of non-irrigated plants prior to the rainfall in March 1981 was also extremely sensitive to VPD as is shown by *Psilocaulon* in fig. 8. CO_2 uptake did not start unless a VPD of about 4 mbar has been reached. The night March 14/15 was a "normal" cool and humid night and net CO_2 uptake started at midnight. The following night was in the first part slightly warmer and drier and net CO_2 uptake started 2 hours later. At 4h30 föhn-wind (extremely hot and dry east wind) arose which after a lag period due to the delay in VPD increase in the measuring device stopped the CO_2 uptake suddenly.

With increasing VPD nocturnal CO_2 uptake of a non-irrigated *Psilocaulon* plant declined, irrigation disturbed that correlation temporary unless the internal water content of the plant has adjusted to the new environmental condition (fig. 7).

162

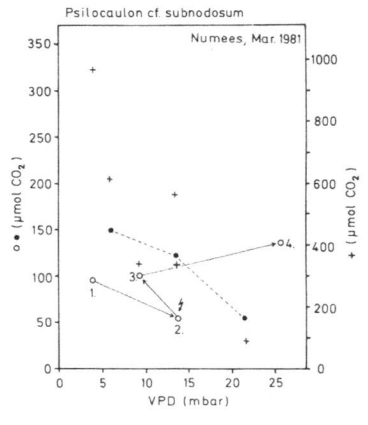
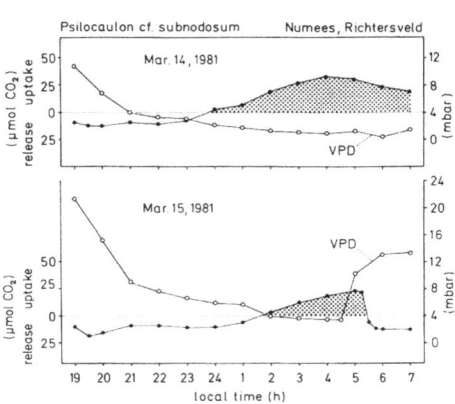

fig. 7   fig. 8

FIGURE 7. Correlation between the nocturnal CO_2 uptake of *Psilocaulon* cf. *subnodosum* and the mean VPD of the night (20h00 - 7h00) of a non-irrigated plant (●), a non-irrigated plant (o) that was irrigated after the second night (⚡) and of an irrigated plant (+).

FIGURE 8. Night trace of the CO_2 gas exchange of *Psilocaulon* cf. *subnodosum* in a "normal" cool and humid night and in a night with a föhn-wind arising at 4h30. VPD was calculated from air and dew point temperature.

The CAM exhibiting succulents of the Richtersveld belong to different life forms and strategies. Fig. 9 compares the effect of an abundant rain-fall on the overnight acid accumulation in some of the different life forms. *Ruschia* is a shrub with woody stems and succulent perennial leaves. *Brownanthus* has a succulent stem and drought deciduous succulent leaves. Both organs exhibit a CAM. Crassula is a perennial leaf succulent. *Tylecodon* and *Ceraria* have succulent but presumably photosynthetic not active stems and succulent deciduous leaves. As CAM is suggested to be a biochemical adaptation to water stress it seems paradoxal that CAM exhibiting deciduous leaves exist. This opens the discussion about the ecological significance of this special metabolism for the vegetation of the Namib desert. It is too early to give a final answer but it seems that nocturnal acid synthesis functions in water harvesting in its widest sense. If conditions do not allow water uptake (or translocation) acid accumulation does not occur or only very limited. In case of possible water uptake, acid accumulation is significantly stimulated. A more detailed paper on this subject is in preparation.

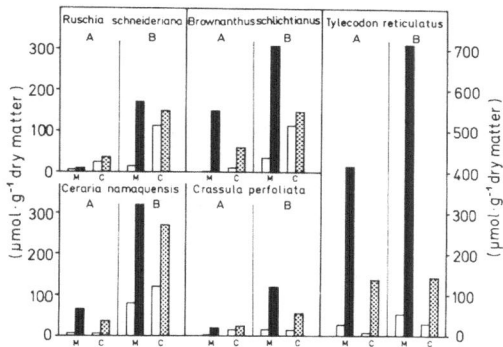

FIGURE 9. The effect of rainfall on the overnight accumulation of malate (M) and citrate (C) in 5 representative species. A: data prior to the rainfall, B: data 7 days after the rainfall in March 1981. Open columns: morning content.

ACKNOWLEDGEMENT
    This work was supported by the Deutsche Forschungsgemeinschaft, the Swiss National Science Foundation and the Department of Education of the South African Government. The generous help of A. P. Lötter (director, Octha Mine, Sendelingsdrift) during our stay in the field is thankfully acknowledged. We got further support by Mettler, Boehringer, Eppendorf and South African Airways. The work in the Richtersveld was kindly permitted by the Department of Internal Affairs and the Department of Nature and Environmental Conservation in Cape Town.

REFERENCES

1. Bartholomew B. 1973. Drought response in the gas exchange of *Dudleya farinosa* (Crassulaceae) grown under natural conditions. Photosynthetica 7, 114 - 120

2. Szarek SR. 1974. Physiological mechanisms of drought adaptation in *Opuntia basilaris* Engelm. Ph. D. Thesis, Univ. Calif. Riverside

3. Kluge M., Ting IP. 1978. Crassulacean acid metabolism. Ecol. Studies Vol. 30, Springer Verlag, Berlin, Heidelberg, New York

4. von Willert DJ., Brinckmann E., Scheitler B., Thomas DA., Treichel S. 1979. The activity and malate inhibition/stimulation of PEP carboxylase in CAM plants in their natural environment. Planta 147, 31 - 36

5. von Willert DJ., Eller BM., Brinckmann E., Baasch R. 1982. CO_2 gas exchange and transpiration of *Welwitschia mirabilis* Hook. fil. in the central Namib desert. Oecologia, in press

6. von Willert DJ., Brinckmann E., Schulze E-D. 1979. Ecophysiological investigations of plants in the coastal desert of southern Africa. Ion content and CAM. In: Ecological processes in coastal environments, Blackwell Sci. Pub. Oxford

LOW AND HIGH TEMPERATURE INFLUENCES ON CACTI

PARK S. NOBEL

Department of Biology and Laboratory of Biomedical and Environmental Sciences, University of California, Los Angeles, CA 90024

ABSTRACT

The low and high temperature extremes leading to tissue damage were examined for various species of cacti. For *Opuntia bigelovii*, uptake of a neutral red stain by its chlorenchyma cells was 50% inhibited by a 1-hour treatment at $-7^\circ C$ or $53^\circ C$, indicating the temperature limits leading to impairment of membrane function. The low and high temperature treatments leading to 50% inhibition of nocturnal acid accumulation by this CAM succulent were $-4^\circ C$ and $44^\circ C$, respectively. Cold hardening leading to a decrease in the half-inactivation temperature for stain uptake averaged $0.5^\circ C$ per $10^\circ C$ decrease in environmental temperature for nine species of cacti. Heat hardening was much greater, $4.3^\circ C$ increase per $10^\circ C$ increase in environmental temperature. More negative osmotic potentials were correlated with greater heat sensitivity for *O. bigelovii*, e.g., a $1.0^\circ C$ decrease in the high temperature for half-inactivation of stain uptake occurred per 0.6 MPa decrease in osmotic potential. For *Coryphantha vivipara*, the low temperature for half-inactivation of stain uptake was essentially uninfluenced as the photoperiod was varied from 3 to 21 hours.

INTRODUCTION

Low and high temperature tolerances of plants are often related to the seasonal temperature extremes of their habitats (11, 12). Although intermediate temperatures help set the photosynthetic rate and net productivity, the extreme temperatures are generally more crucial for the limits of plant distribution. For instance, only three species of the 65 arborescent ceroid cacti in the Sonoran desert (*Carnegiea gigantea, Lophocereus schottii,* and *Stenocereus thurberi*) occur further north than where frost occurs (14, 20, 22). For plants native to hot habitats, physiological processes may become disrupted when temperatures exceed about $50^\circ C$. The highest nonlethal temperatures recorded for tissues of higher plants are $62^\circ C$ to $65^\circ C$ for various species of *Opuntia* (9, 10, 13).

Emphasis here will be on the effect of temperature extremes on the metabolism and viability of cacti. Cacti exhibit Crassulacean acid metabolism (CAM), where nocturnal CO_2 uptake leads to an increase in tissue acidity as the CO_2 is incorporated into malate and other organic acids. Studies with cacti have indicated that accumulation of a stain (neutral red) by the chlorenchyma cells is a useful index of membrane integrity and hence cell viability (7, 17), as the stain is accumulated in the vacuoles of living cells only. Such studies were expanded here to include several species of cacti and examined high and low temperature hardening in response to changing environmental temperatures, the influence of tissue osmotic potential on high temperature tolerance, and the effect of photoperiod on low temperature tolerance.

MATERIALS AND METHODS

Plant material was collected in the field at the following locations: *Carnegiea gigantea* (Engelm.) Britton & Rose from central Arizona at $33°46'N$, $112°41'W$, 500 m elevation; *Coryphantha vivipara* (Nutt.) Britton & Rose var. *deserti* (Engelm.) W. T. Marshall from southern Nevada at $36°40'N$, $116°1'W$, 1110 m; *Ferocactus acanthodes* var. *lecontei* (Engelm.) Lindsay from southern California at $33°38'N$, $116°24'W$, 840 m; *F. covillei* Britton & Rose from southern Arizona at $31°59'N$, $111°39'W$, 1160 m; *F. viridescens* (Nutt.) Britton & Rose from coastal southern California at $33°13'N$, $117°22'W$, 40 m; *F. wislizenii* from southern Arizona at $32°21'N$, $111°2'W$, 850 m; *Lophocereus schottii* (Engelm.) Britton & Rose var. *schottii* from southern Arizona at $31°55'N$, $112°57'W$, 410 m; *Opuntia bigelovii* Engelm. from southern California at $33°38'N$, $116°24'W$, 850 m; and *Stenocereus thurberi* (Engelm.) Buxb (= *Lemaireocereus thurberi* (Engelm.) Britton & Rose var. *thurberi*) from northern Sonora, Mexico at $30°40'N$, $112°6'W$, 350 m. The plants were transferred in native soil to environmental chambers maintained at day/night air temperatures of $30°C/20°C$. Photosynthetically active radiation (PAR) averaging 360 µmol m^{-2} s^{-1} at the stem surface was provided for 12 hours each day, and the plants were watered weekly with 1/10 Hoagland no. 1 solution plus micronutrients so that the soil water potential near the roots was generally -0.3 ± 0.1 MPa.

Stems were cooled to subzero temperatures in a deep freezer (Revco ULT-385A) or heated by immersing them in water baths at carefully controlled temperatures. Air and chlorenchyma temperatures represent the average for three copper-constantan thermocouples 150 µm in diameter. Stems were maintained at the test temperature for one hour and then samples were removed to study accumulation of 50 µM neutral red (3-amino-7-dimethyl-amino-2-methylphenazine (HCl)) in 1 M mannitol (7, 17, 19). Stain accumulation was determined for at least 400 cells from the center of the chlorenchyma approximately 1 mm below the stem surface; sections 100 µm thick were examined with a phase-contrast microscope at approximately 500X. To study cold or heat hardening, the day/night temperatures in the environmental chamber were decreased or increased in $10°C$ steps at approximately 10-day intervals.

To examine temperature influences on diurnal acidity changes, the acid level was measured at the beginning and end of the dark period using three 0.97-cm^2 stem samples following a temperature treatment at the beginning of the dark period. Samples were ground with sand, boiled in 80 ml distilled water for 5 minutes, and then titrated to an endpoint of pH 6.8 with 5 mM NaOH (8). Nocturnal acid accumulation refers to the value at the end of the dark period minus the value at the end of the temperature treatment. Osmotic potentials of chlorenchyma were determined using a Wescor C-52 leaf chamber and a HR-33T dew-point microvoltmeter. To obtain various osmotic potentials, the rate of tissue water loss was accelerated by cutting the stem into segments.

RESULTS AND DISCUSSION

Temperature effects on staining and nocturnal acid accumulation

Figure 1 indicates that uptake of neutral red and nocturnal accumulation of acid decrease at high and low temperature extremes for the jointed cylindropuntia, *Opuntia bigelovii*. The change in acidity is more sensitive to temperature than is the uptake of stain, especially at high temperatures. This is consistent with previous findings (1, 3, 11, 12), where disruption of membrane integrity occurs at more extreme temperatures than does inhibition of a multicomponent metabolic process.

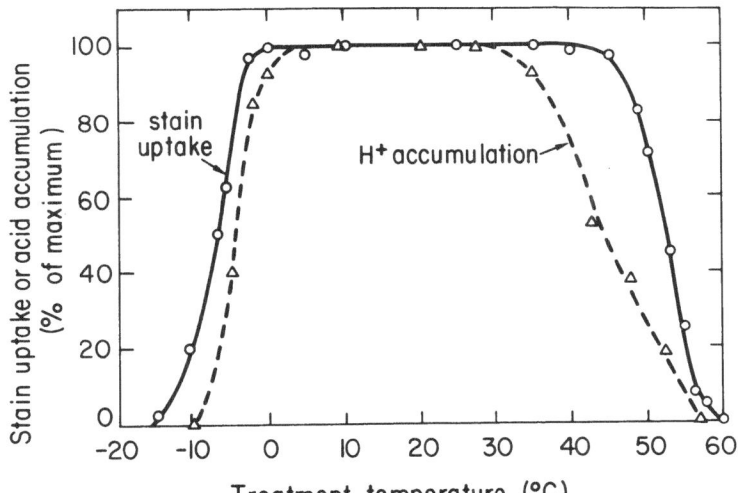

FIGURE 1. Uptake of stain or nocturnal increase in acidity for *Opuntia bigelovii* following a 1-hour treatment at the indicated temperature. A maximum of 79% of the chlorenchyma cells took up neutral red, and the maximal increase in tissue acidity during the dark period following the temperature treatment was 0.16 mol m^{-2}.

Stems of cacti exhibit supercooling (cooling below the equilibrium freezing point) as their temperatures are lowered to about -5°C (17, 19, 25). Following such supercooling, the tissue temperature usually rises, which represents the latent heat released as the extracellular water freezes (6, 12). For *Coryphantha vivipara*, such an exothermic reaction represents the freezing of about 10% of the stem water (17). Based on the temperatures involved, such extracellular ice formation apparently does not impair the subsequent ability of the chlorenchyma cells to take up stain or to accumulate acid. As the cooling continues following the exothermic reaction, intracellular water presumably leaves the cell and crystallizes onto the extracellular ice that has already been formed.

Such intracellular dehydration, which accounts for the shrunken appearance of the protoplasts observed for *C. vivipara* (17) and *O. bigelovii* as the freezing progressed, apparently is responsible for the low temperature damage.

Freezing tolerance may reflect the ability to tolerate desiccation. No persisting adverse effect on gas exchange by *C. vivipara* was caused by a single night at -10°C, although death occurred near -15°C (17). The dehydration leading to a freezing point depression of -15°C corresponds to an osmotic potential of about -17 MPa (-170 bar), which is about 16 times more negative than the osmotic potential of a hydrated stem of *C. vivipara* in the field. If we assume that the water content varies inversely with the absolute value of the osmotic potential, a loss of 94% of the cellular water would be necessary to give a freezing point depression of 15°C. In agreement with this, a drought dehydration in which 91% of the stem water was lost led to the death of *C. vivipara* (17). For *O. bigelovii* (Fig. 1) death occurred at about -11°C, which would correspond to a loss of 95% of the water from a hydrated stem in the field. The subzero temperature leading to cellular dehydration that in turn leads to disruption of cellular integrity occurs 2°C to 5°C lower than for a comparable disruption in nocturnal acid accumulation for *O. bigelovii* (Fig. 1).

The basis for the damage to cacti at high temperatures is somewhat harder to assess. Membrane impairment at the plasmalemma or tonoplast level apparently led to 50% inhibition of stain uptake at 53°C for *O. bigelovii* (Fig. 1). Nocturnal acid accumulation was half-inactivated about 9°C lower and inhibition was already quite apparent at 40°C. Some of the decrease reflects a lowered nocturnal stomatal conductance (7), and enzyme inactivation at elevated temperatures presumably also plays a role (1). In any case, high temperature extremes induce much greater differences between the half-inactivation temperatures for stain uptake and acid accumulation than do low temperature extremes (Fig. 1).

Survey of half-inactivation temperatures and hardening

Next, the high and low temperature tolerances were examined for stain uptake by nine species of cacti (Table 1). Excluding *C. vivipara*, the low temperature extreme averaged $-6.9 \pm 0.9^\circ$C for 50% inhibition of stain uptake. For *C. vivipara*, it was -19.3°C. This species occurs in Manitoba and Alberta, Canada (50°N), as well as at high elevations in Wyoming (up to 3000 m), and hence it can be subjected to quite low temperatures (the other species generally do not occur at latitudes above 35°N or at such high elevations). The high temperature extreme where stain uptake was inhibited 50% averaged 56.0 ± 1.8 (Table 1). Relations between high temperature tolerances and distributions are not as obvious for these cacti as the low temperature tolerances. Also, magnitudes of tolerances of high and low temperatures are not necessarily related for cacti (r = -0.13), in agreement with studies on species of potato (22).

Table 1. Summary of high and low temperatures where staining was reduced 50% for plants maintained at day/night air temperatures of 30°C/20°C. Cold hardening (change in temperature for 50% reduction in staining) and heat hardening were averages for sequential 10°C changes in day/night temperatures down to 0°C/-10°C for cold hardening and up to 50°C/40°C for heat hardening. Temperatures were changed at 10-day intervals until the plants became necrotic.

Species	Low temperature (°C)	Cold hardening (°C/10°C decrease)	High temperature (°C)	Heat hardening (°C/10°C increase)
Carnegiea gigantea	-7.6	-0.5	56.1	3.6
Coryphantha vivipara	-19.3	-1.7	56.4	3.9
Ferocactus acanthodes	-7.4	-0.3	58.8	3.8
F. covillei	-6.2	0.0	55.5	5.8
F. viridescens	-5.1	-0.3	56.0	5.4
F. wislizenii	-7.4	-0.3	58.0	4.5
Lophocereus schottii	-7.2	-0.3	56.0	4.1
Opuntia bigelovii	-6.5	-0.8	52.6	3.0
Stenocereus thurberi	-8.0	-0.3	55.0	4.7

Many plants exhibit hardening, where the tolerance of specific processes to temperature extremes increases as the environmental temperatures gradually change seasonally or experimentally (12). Such hardening may represent the development of tolerances to increased intracellular dehydration or, perhaps, to the protection of certain membrane-bound proteins. Also, changes in membrane permeability possibly reflecting changes in lipid composition could occur during the hardening process (23). This was examined for the nine species of cacti by increasing or decreasing the day/night air temperature stepwise (Table 1). The low temperature at which stain uptake was inhibited by 50% decreased 0.5 ± 0.5°C for each 10°C lowering of the air temperature and the high temperature increased 4.3 ± 0.9°C for each 10°C rise (Table 1). Thus, the high temperature tolerance averaged nearly 10-fold greater capability for acclimation than did the low temperature tolerance. The marked acclimation in high temperature tolerance also occurs for nocturnal acid accumulation, e.g., its half-inactivation temperature increased 6.1°C per 10°C increase in environmental temperature for *O. bigelovii* (7). Also, increased heat hardening was somewhat correlated with reduced cold hardening (r = -0.52).

The non-lethal range of subzero temperatures survived by cacti varies widely among species. For two species from warm desert regions below 1200 m in southern California (*O. bigelovii* and *O. ramosissima*), half-inactivation of stain accumulation by chlorenchyma cells occurs at $-7^{\circ}C$ to $-4^{\circ}C$ and tissue necrosis sets in $3^{\circ}C$ to $5^{\circ}C$ below such cold treatment temperatures (19). For three species that can range up to 3000 m in southeastern Wyoming (*O. polyacantha, Pediocactus simpsonii, and C. vivipara*), half-inactivation of stain uptake occurs at $-17^{\circ}C$ to $-20^{\circ}C$ (19). Thus, cold tolerance and distribution of cacti appear to be related, as would be expected.

The optimal temperature for nocturnal CO_2 uptake increased an average of $7.7^{\circ}C$ as six species of cacti and three species of agaves were changed from day/ night air temperatures of $10^{\circ}C/10^{\circ}C$ to $30^{\circ}C/30^{\circ}C$, or a shift of $3.8^{\circ}C$ per $10^{\circ}C$ change in environmental temperature (21). For most non-CAM plants, the optimal temperature for CO_2 uptake increases $3^{\circ}C$ to $4^{\circ}C$ as growth temperature is increased $10^{\circ}C$. The acclimation of cold tolerance was much less than this index and that of heat tolerance was somewhat more. However, even the acclimation of the low temperature extreme could have important ecological consequences, since the minimum air temperatures at constant elevation can change about $-0.8^{\circ}C$ per degree increase in latitude and $-0.5^{\circ}C$ to $-0.6^{\circ}C$ per 100 m increase in elevation (15, 24). The high temperature acclimation of *O. bigelovii* is apparently necessary for the survival of mature plants at certain field sites (7). Since it propagates primarily by rooting of readily detached stem segments, the high temperature acclimation may also be crucial for the survival of such propagules exposed to the high temperatures of an unshaded soil surface.

Osmotic potential effects on heat tolerance

For certain species, more negative osmotic potentials of the chlorenchyma are correlated with greater heat tolerance, i.e., heat stability increases when the plants are subjected to water deficits (4, 12, 15). Here, the effect of osmotic potential on the high temperature for half-inactivation of stain uptake by the chlorenchyma cells of *O. bigelovii* was examined.

Cacti are well adapted to hot, arid areas, and yet do not tend to have very negative osmotic potentials. For instance, osmotic potentials of cacti range from -0.4 MPa to -1.8 MPa in the field (28). Also, such plants differ from most C_3 and C_4 plants by having considerably more water storage capacity per unit surface area. Hence, the changes in osmotic potential occur over longer time periods for cacti than for C_3 and C_4 plants. In fact, the osmotic potentials for certain uprooted cacti change very little over a period of months, as some dry weight is lost by respiration and some water is lost by the greatly restricted transpiration, so that the water content relative to solute content

decreases only slightly. Indeed, to achieve appreciable rates of water loss
from cacti the stem generally must be sectioned, as was done in the present
experiments.

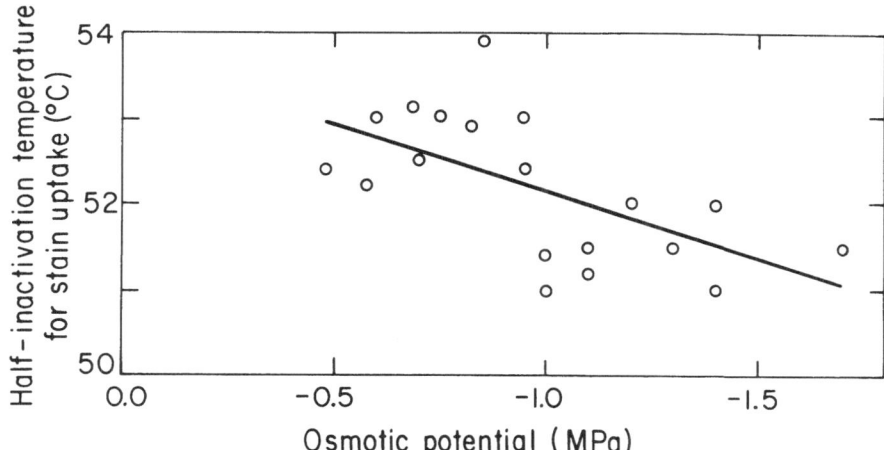

FIGURE 2. Effect of osmotic potential of the chlorenchyma on the heat tolerance
of *O. bigelovii*. Temperatures where the uptake of neutral red was decreased 50%
were determined graphically for desiccation periods of 1 to 3 months. The indicated regression line is: half-inactivation temperature in $^{\circ}C = 53.7 \pm 1.56 \times$
osmotic potential in MPa ($r^2 = 0.37$).

Figure 2 indicates that as the osmotic potential decreases, *O. bigelovii*
becomes slightly _more_ heat sensitive. The half-inactivation temperature for the
uptake of neutral red decreased $1.0^{\circ}C$ for each 0.6 MPa (6 bars) decrease in
osmotic potential. As the tissue became drier, stomata opened less fully and
for a shorter portion of the night. For instance, for a night at $20^{\circ}C$ the maximal stomatal conductance was 0.143 mm s^{-1} for an osmotic potential of -0.64 MPa
decreasing to 0.054 mm s^{-1} at -0.97 MPa and 0.012 mm s^{-1} at -1.32 MPa. This
decrease in stomatal opening would greatly restrict nocturnal CO_2 uptake and
acid accumulation. In fact, the gradual decrease in osmotic potential and net
loss of dry weight over a period of months could lead to some senescence of the
tissue, which could make it more vulnerable to high temperatures. This time
factor together with the response shown in Figure 2 is quite different from the
ameliorating effect that lower osmotic potentials can have on the high temperature tolerance of non-succulent C_3 and C_4 plants. In summary, desiccation
apparently does not enhance heat tolerance for *O. bigelovii*. Also, the heat
acclimation in response to increasing environmental temperatures (Table 1) is
not due to changes in osmotic potential, since the latter does not change in
intact stems over the course of the acclimation.

172

Photoperiod effect on cold tolerance

As the final topic, the influence of day (or night) length on tolerance of low temperatures will be considered. The progressive lowering of environmental temperatures from late summer through the autumn to the winter can lead to cold hardening, as discussed above. This period is also accompanied by progressively shorter days (longer nights). For many species, this decrease in photoperiod leads to a cold hardening where lower temperatures are tolerated (12). This is apparently a phytochrome-mediated process in many species and can be even more important than temperature-inducing hardening (5). Photoperiod, however, has no effect on the cold-hardening of herbaceous biennials and some other herbaceous plants whose growth and development is unaffected by day length (12). Thus, the effect of photoperiod on the half-inactivation temperature for stain uptake by the chlorenchyma cells of *C. vivipara* was examined. This species was chosen since it has the greatest degree of cold hardening so far observed for cacti (Table 1; also ref. 19).

Figure 3 indicates that the cold tolerance of *C. vivipara* was essentially unchanged as the photoperiod was reduced from 21 to 3 hours. Under field conditions the maximum change in daylength (sunrise to sunset) for *C. vivipara* at 37°N in southern Nevada is from 14 hours 40 minutes at the summer solstice to 9 hours 22 minutes at the winter solstice. Hence, photoperiod would not be expected to have any significant influence on its cold tolerance.

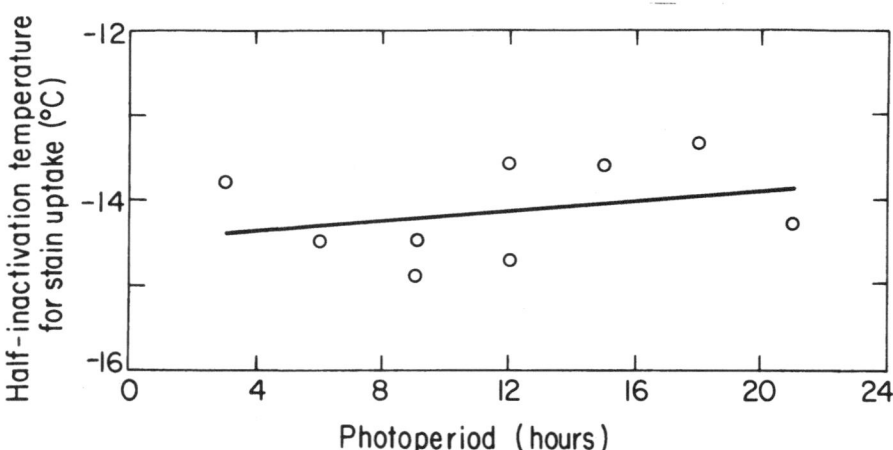

FIGURE 3. Influence of photoperiod on the low temperature tolerance of *Coryphantha vivipara*. Plants were maintained at a PAR of 360 μmol m^{-2} s^{-1} for 10 days at each photoperiod (day/night air temperatures were 10°C/10°C). The indicated regression line is: half-inactivation temperature for stain uptake in $^\circ$C = $-14.5 + 0.030 \times$ photoperiod in hours ($r^2 = 0.09$).

The concomitant effect of photoperiod on nocturnal acid accumulation can be determined by the changes in total daily PAR. Specifically, CAM plants reach 90% of light saturation for a total daily PAR of 20 to 25 mol m^{-2} day $^{-1}$, whic is the average value in the field for unshaded vertical stems on clear days (1 18, 20). The 21-hour photoperiod would here lead to 27 mol m^{-2} day $^{-1}$, and th 3-hour photoperiod would lead to a total daily PAR of 4 mol m^{-2} day $^{-1}$ (essentially the light compensation point for cacti, refs. 18, 20). Due to the range in PAR involved -- from light compensation to over 90% of saturation of nocturnal CO_2 uptake -- the major effect of changes in photoperiod on productivity would generally be by the variation in total daily PAR.

Acknowledgements. The author gratefully acknowledges the contributions of Brigitte Didden-Zopfy to Figure 2 and the technical assistance of Terry Hartsock. Financial support was provided by Department of Energy Contract DE-AM03-76-SF00012 and National Science Foundation grant DEB-81-00829.

REFERENCES

1. Alexandrov VYa. 1964. Cytophysiological and cytoecological investigations of heat resistance of plant cells toward the action of high and low temperature. Quart Rev Biol 39: 35-77.
2. Alexandrov VYa, Lomagin AG, Feldman NL. 1970. The responsive increase in thermostability of plant cells. Protoplasma 69: 417-458.
3. Berry JA, Fork DC, Garrison S. 1975. Mechanistic studies of thermal damage to leaves. Carnegie Inst Wash Yearbook 74: 751-759.
4. Biebl R. 1962. Protoplasmatische Ökologie der Pflanzen. I. Wasser und Temperatur. Protoplasmatologia XII. Vienna, Springer-Verlag.
5. Biebl R. 1967. Kurtztag-Einflüsse auf arktische Pflanzen während der arktischen Langtage. Planta 75: 77-84.
6. Burke MJ, Gusta LV, Quamme HA, Weiser CJ, Li PH. 1976. Freezing and injury in plants. Ann Rev Plant Physiol 27: 507-528.
7. Didden-Zopfy B, Nobel PS. 1982. High temperature tolerance and heat acclimation of Opuntia bigelovii. Oecologia 52: 176-180.
8. Hartsock TL, Nobel PS. 1976. Watering converts a CAM plant to daytime CO_2 uptake. Nature 262: 574-576.
9. Huber B. 1932. Einige Grundfragen des Wärmehaushalts der Pflanzen. I. Die Ursache der hohen Sukkulenten-Temperaturen. Ber Deutsche Bot Ges 50: 68-76
10. Konis E. 1950. On the temperature of Opuntia joints. Palestine J Bot, Jerusalem Ser, 5: 46-55.
11. Larcher W. 1980. Physiological plant ecology, 2nd ed. Berlin, Springer-Verlag.
12. Levitt J. 1980. Responses of plants to environmental stresses, 2nd ed., Vol. 1. Chilling, freezing, and high temperature stresses. New York, Academic Press.
13. MacDougal DT, Working EB. 1921. A new high temperature record for growth. Carnegie Inst Wash Yearbook 20: 47-48.
14. Nobel PS. 1980. Morphology, surface temperatures, and northern limits of columnar cacti in the Sonoran Desert. Ecology 61: 1-7.
15. Nobel PS. 1980. Influences of minimum stem temperatures on ranges of cacti in southwestern United States and central Chile. Oecologia 47: 10-15.
16. Nobel PS. 1980. Interception of photosynthetically active radiation by cacti of different morphology. Oecologia 45: 160-166.

17. Nobel PS. 1981. Influence of freezing temperatures on a cactus, *Coryphantha vivipara*. Oecologia 48: 194-198.
18. Nobel PS. 1981. Influences of photosynthetically active radiation on cladode orientation, stem tilting, and height of cacti. Ecology 62: 982-990.
19. Nobel PS. 1982. Low temperature tolerance and cold hardening of cacti. Ecology, in press.
20. Nobel PS. 1982. Interaction between morphology, PAR interception, and nocturnal acid accumulation for cacti, In: Ting IP, Gibbs M (eds.). Crassulacean Acid Metabolism, Rockville, Maryland, American Society of Plant Physiologists, pp. 260-277.
21. Nobel PS, Hartsock TL. 1981. Shifts in the optimal temperature for nocturnal CO_2 uptake caused by changes in growth temperature for cacti and agaves. Physiol Plant 53: 523-527.
22. Palta JP, Chen HH, Li PH. 1981. Relationship between heat and frost resistance of tuber-bearing Solanum species: Effect of cold acclimation on heat resistance. Bot Gaz 142: 311-315.
23. Pike CS, Berry JA. 1980. Membrane phospholipid phase separation in plants adapted to or acclimated to different thermal regimes. Plant Physiol 66: 238-241.
24. Smith WK, Geller GN. 1979. Plant transpiration at high elevations: theory, field measurements, and comparisons with desert plants. Oecologia 41: 109-122.
25. Steenbergh WF, Lowe CH. 1976. Ecology of the saguaro: I. The role of freezing weather in a warm-desert plant population. In: Research in the parks. National Park Service Symposium Series, No. 1, Washington, D.C., U.S. Government Printing Office, pp. 1-242.
26. Steenbergh WF, Lowe CH. 1977. Ecology of the saguaro: II. Reproduction, germination, establishment, growth and survival of the young plant. In: National Park Service Monograph Series, No. 8, Washington, D.C., U.S. Government Printing Office.
27. Turnage WV, Hinckley AL. 1938. Freezing weather in relation to plant distribution in the Sonoran desert. Ecol Mon 8: 529-550.
28. Walter H, Stadelman E. 1974. A new approach to the water relations of desert plants. In: Brown GW (ed.) Desert Biology, Vol. 2. New York, Academic Press, pp. 214-302.

REGULATION OF CARBON METABOLISM IN MESEMBRYANTHEMUM CRYSTALLINUM

J. G. FOSTER, K. WINTER, and G. E. EDWARDS

United States Department of Agriculture, Appalachian Soil and Water Conserva-
tion Research Lab, P.O. Box 867, Beckley, West Virginia 25801 USA; Botanik II
der Universität, Mittlerer Dallenbergweg 64, D-8700, Würzburg, West Germany;
Department of Botany, Washington State Univ., Pullman, Washington 99164 USA.

ABSTRACT

Assays of enzyme activity in fractions obtained after Percoll density gradient certrifugation of protoplast extracts from *Mesembryanthemum crystallinum*
performing exclusively C_3 photosynthesis or Crassulacean acid metabolism
(CAM) established the cytoplasmic location of phosphoenolpyruvate (PEP) carboxylase and NADP-malic enzyme, the mitochondrial location of NAD-malic enzyme,
and the chloroplastic location of pyruvate,Pi dikinase. The activity of these
enzymes was elevated in the CAM tissue, and net synthesis of PEP carboxylase
protein increased during CAM induction. This compartmentation and regulation
of enzymes of the carbon pathway implicate the transport of pyruvate and PEP
across the chloroplast envelope during the metabolism of malate in the light
in the CAM tissue.

When analyzed by SDS polyacrylamide gradient gel electrophoresis, chloroplast envelopes, prepared from *M. crystallinum*, operating in either the C_3
or CAM mode, spinach (C_3) and the obligate CAM plant *Kalanchoë daigremontiana*, exhibited a prominent 29 kilodalton polypeptide which has been identified as the phosphate translocator in spinach. Like chloroplast envelopes from
K. daigremontiana, those from CAM-*M. crystallinum* had an intensely stained
32 kilodalton polypeptide which occurred only in limited amounts in the C_3
tissue examined. These data suggest that a 32 kilodalton polypeptide of the
chloroplast envelope may be a translocator protein which is required in CAM,
but not in C_3 photosynthesis.

INTRODUCTION

Early studies on photosynthetic carbon metabolism in isolated chloroplasts
of C_3 plants demonstrated a requirement for inorganic phosphate (Pi) [1,2].
Subsequent observations that organic phosphates could relieve the inhibition
of photosynthesis caused by excessive amounts of Pi in the medium [3] led to
the discovery of a carrier protein which mediates stoichiometric uptake of Pi

in exchange for triose phosphate [14]. Further, compartmentation of the enzymes which catalyze reactions associated with fixation of CO_2 into carbohydrate in C_3 plants necessitates transport of photosynthetic intermediates across the chloroplast envelope [24]. In spinach, a 29 kD phosphate translocator located in the chloroplast envelope facilitates export of carbon, in the form of triose phosphate, from the chloroplast for its metabolism to sucrose in the cytoplasm [8,9]. Control of this transport *in vivo* can therefore influence both the rate of photosynthesis and the partitioning of carbon between starch synthesis in the chloroplast and sucrose synthesis in the cytoplasm [12,21,24].

The capacity for transport of pyruvate and PEP across the C_3 chloroplast envelope is minimal [7,13], but in C_4 plants cytoplasmic PEP, which accepts CO_2 during the initial fixation reaction in the mesophyll cells, is synthesized in the chloroplast from pyruvate that originated outside the organelle [6]. A specific translocator for pyruvate and a phosphate translocator(s) which catalyzes the exchange of PEP with Pi and phosphoglyceric acid (PGA) with triose phosphate have been demonstrated in chloroplasts from these cells [16,17]. While it is uncertain whether the phosphate translocator of C_4 mesophyll chloroplasts is modified to recognize PEP as well as PGA, Pi, and triose phosphate or whether there are two translocators, one catalyzing exchange of PEP for Pi and the other exchange of PGA for triose phosphate, similar transport requirements are expected for malic enzyme-type CAM plants in which photosynthetic events characteristic of C_4 plants are separated temporally in a single cell type. Thus, the manner in which enzymes of carbon metabolism are compartmentalized in these cells will contribute to the regulation of the different phases of CAM as well as determine the requirements for metabolite transport between organelles and the cytoplasm.

The capacity of *Mesembryanthemum crystallinum* to perform Crassulacean acid metabolism when subjected to water stress and the ease with which intact organelles can be isolated from the leaf tissue [25] make this plant a desirable species with which to study the regulation of carbon metabolism. Since comparative analysis of *M. crystallinum* operating in either the C_3 or CAM mode are not complicated by species variations, differences in carbon metabolism in the two tissue types can be evaluated in relation to differences in enzyme levels while differences in the polypeptide composition of the chloroplast envelopes can be interpreted in terms of differences in carrier-mediated transport.

MATERIALS AND METHODS

Plant culture, protoplast isolation, sample preparation, and analytical procedures have been described in detail elsewhere [10,25]. Envelope membranes
were isolated from intact chloroplasts of spinach (obtained from protoplast extracts) and C_3-M. crystallinum (prepared mechanically) according to methods
described by Douce [4] except that the chloroplast lysis medium contained 2 mM
EDTA instead of 4 mM $MgCl_2$. CAM chloroplasts were prepared mechanically in
medium containing 450 mM sorbitol, 50 mM HEPES-NaOH, pH 7.5 and 0.2% (w/v) BSA.
For K. daigremontiana, the homogenization medium also contained 5 mM dithiothreitol and 2% (w/v) PVP-40. CAM chloroplasts were lysed in medium containing
2 mM EDTA (M. crystallinum) or 4 mM $MgCl_2$ (K. daigremontiana) and 150 mM sucrose in 10 mM Tricine-NaOH, pH 7.5. Alternatively, lysis of CAM chloroplasts
was accomplished by freezing chloroplasts in lysis medium containing 600 mM
sucrose, then thawing on ice. Polypeptides of the purified chloroplast envelopes were resolved on 0.75 mm-thick discontinuous SDS polyacrylamide 7.5-15%
gradient slab gels [10,25]. After electrophoresis, gels were stained with
silver [20] and scanned spectrophotometrically at 450 nm. Envelope protein
applied to the gel was estimated according to Markwell et al. [19].

RESULTS AND DISCUSSION

Enzyme regulation during CAM induction

The efficiency of the photosynthetic process depends upon a sequence of integrated metabolic events which include the photochemical reactions, the enzy-
mology of carbon assimilation, and transport of photosynthetic intermediates
between subcellular compartments. The inherent potential of the plant and restrictions imposed by the environmental conditions determine the performance of
a plant at any given time. Exposure of the halophyte M. crystallinum to environmental conditions which cause water stress induces plants exhibiting C_3
photosynthesis to perform Crassulacean acid metabolism [26]. Plants stressed by
growth in saline culture medium display characteristics typical of obligate CAM
species. Substantial CO_2 uptake and malic acid synthesis in the dark followed
by deacidification in the subsequent light period occur, and these responses
become increasingly pronounced with increasing salinity of the culture medium
[10]. In contrast to obligate CAM plants which always possess the metabolic
machinery necessary to perform CAM in expanded leaves, inducible CAM plants such
as M. crystallinum presumably develop the requisite biochemical apparatus in
response to an environmental signal.

Table 1. Activities of enzymes of carbon metabolism in protoplast extracts from *M. crystallinum* performing C3 photosynthesis or Crassulacean acid metabolism.

Enzyme	Activity, mkat (kg, Chl)$^{-1}$	
	C$_3$	CAM
RuBP carboxylase	110	110
NADP-glyceraldehyde-3-phosphate dehydrogenase	100	120
Pyruvate,Pi dikinase	ND[1]	8
PEP carboxylase	6	280
NADP-malic enzyme	1	18
NAD-malic enzyme	3	31

[1]ND: nondetectable

Assays of ribulose bisphosphate (RuBP) carboxylase and NADP-glyceraldehyde-3-phosphate dehydrogenase in extracts from C$_3$- and CAM-*M. crystallinum* reveal that activities of these two enzymes of the C$_3$ pathway are relatively unaffected by the osmotic stress imposed upon the plant, Table 1. Pyruvate,Pi dikinase activity was present in protoplast extracts from CAM *M. crystallinum* (Table 1) but was 40% lower than the sum of the activities of this enzyme in the supernatant [0 mkat(kg Chl)$^{-1}$] and chloroplast pellet [12 mkat (kg Chl)$^{-1}$] obtained by differential centrifugation of the protoplast extract. Failure to detect activity in either the supernatant or the pellet fractions after differential centrifugation of C$_3$ protoplast preparations indicates that absence of detectable pyruvate,Pi dikinase activity in the C$_3$ protoplast extract (Table 1) is not due to destabilizing effects of cytoplasmic components. Marked increases in the activities of PEP carboxylase and the decarboxylating enzymes NADP-malic enzyme and NAD-malic enzyme in CAM-*M. crystallinum* over those observed in C$_3$-*M. crystallinum* (Table 1) further attest to specific biochemical responses to the stress-induced shift in the mechanism of carbon metabolism and are consistent with other observations with this species [11,15].

Whether PEP carboxylase protein is newly synthesized or simply modified during the induction of CAM in *M. crystallinum* is important in understanding the cellular control of the enzyme and the development of CAM. Figure 1 provides evidence that the induction of CAM is accompanied by a specific increase in at least eight of the polypeptides detected on SDS polyacrylamide gels. The most striking difference between tissue operating in the C$_3$ or CAM mode occurred in polypeptide 2. This 98 kD polypeptide was identified as the subunit of PEP carboxylase by comparison with the corresponding band from partially purified PEP carboxylase from the same tissue; thus, it was concluded that increased net synthesis of PEP carboxylase protein occurs during CAM induction in this species.

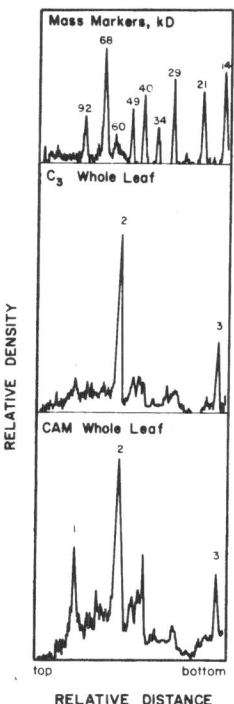

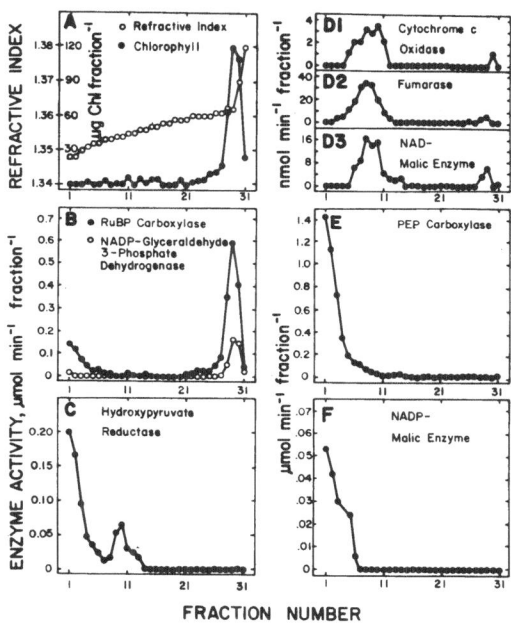

FIGURE 1. Densitometric scans of a Coomassie-stained poly-acrylamide gradient slab gel of leaf extracts from C₃- and CAM-*M. crystallinum*. Dissociated protein, applied to the gel in 50 μl, represented equal leaf areas: C₃-*M. crystallinum*, 16 μg protein; CAM-*M. crystallinum*, 24 μg protein; mass markers (M), 1 μg protein each. Peak 1 designates the subunit (98 kD) of PEP carboxylase; peaks 2 and 3 are the 55 kD and 14 kD subunits, respectively, of RuBP carboxylase.

FIGURE 2. Distribution of chlorophyll (A) and activity of several enzymes (B-F) following centrifugation of a of protoplast extract from CAM-*M.crystallinum* on a Percoll density gradient. Fractions are numbered from top to bottom of the gradient.

Enzyme localization in M. crystallinum

There has been considerable controversy over the compartmentation of enzymes of carbon metabolism in CAM species [18,22]. Successful preparation of intact organelles from mesophyll protoplasts of M. crystallinum was a key factor leading to the definitive localization of enzymes in this study. Data from typical Percoll gradient certrifugations using protoplast extracts of CAM-M. crystallinum are presented in Figure 2. As indicated by the distribution of chlorophyll and the chloroplast marker enzymes, NADP-glyceraldehyde-3-phosphate dehydrogenase and RuBP carboxylase, minimal breakage of chloroplasts occurred and intact chloroplasts formed a distinct band in the lower part of the centrifuge tube. Hydroxypyruvate reductase activity was found to peak in fraction 10, but was also observed at the top of the gradient, probably indicating some breakage of peroxisomes. No activity of the mitochondrial markers, fumarase and cytochrome c oxidase, was found at the top of the gradient. There was a peak of activity of these enzymes around fraction 9, and some activity was associated with the chloroplasts. Evidently there was no breakage of mitochondria, but some of these organelles comigrated with the chloroplasts during centrifugation. The distribution of NAD-malic enzyme on the gradient was identical with that of the mitochondrial marker enzymes, while restriction of PEP carboxylase and NADP-malic enzyme to the top of the gradient suggests cytoplasmic localization of these enzymes, tacitly presupposing that they are not of vacuolar origin. The extrachloroplastic location of the malic enzymes, localization of pyruvate,Pi dikinase in the chloroplast and the restriction of PEP carboxylase to the cytoplasm suggest a complicated flow of carbon during deacidification in malic enzyme-type CAM species. Pyruvate formed upon decarboxylation of malate must enter the chloroplast for conversion to PEP. PEP would then leave the chloroplast for further metabolism to sucrose [5,23,25]. Gluconeogenic conversion of PEP to triose phosphate in the cytoplasm would provide a transport metabolite which could reenter the chloroplast, in exchange for Pi, via a phosphate translocator.

Polypeptide composition of the chloroplast envelope

Differences in chloroplast functions in M. crystallinum tissue performing either C_3 photosynthesis or Crassulacean acid metabolism suggest differences may exist in the carrier proteins of the chloroplast envelope. It was therefore reasonable to consider whether chloroplast envelopes of malic enzyme-type CAM plants differ from C_3 plants in their polypeptide composition. As shown by the scans of silver-stained SDS polyacrylamide gradient gels in Figure 3,

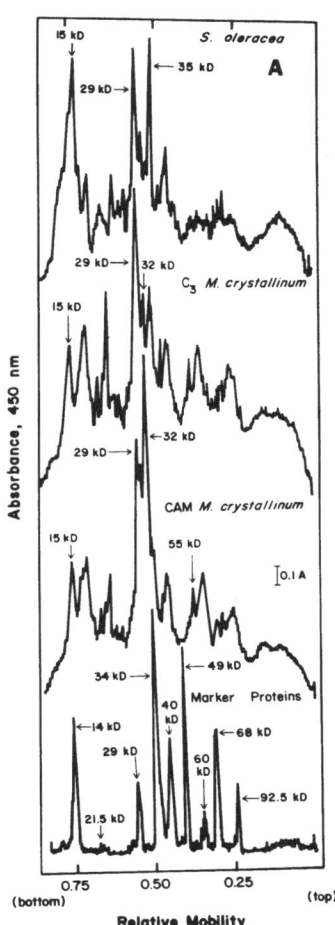
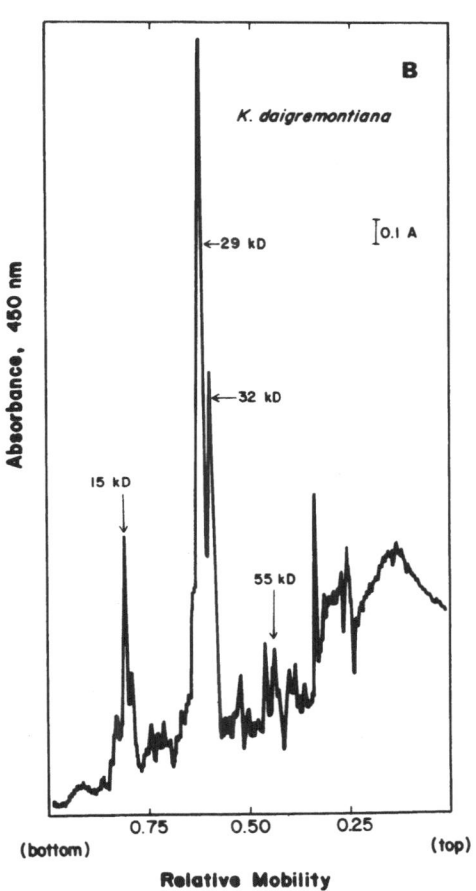

FIGURE 3. Polypeptide profile of chloroplast envelopes from spinach and *M. crystallinum* operating in the C₃ or CAM mode (panel A) and *K. daigremontiana* (panel B). Polyacrylamide gradient slab gels were stained with silver, dried on porous cellophane and scanned spectrophometrically at 450 nm. Each profile represents 10 µg of envelope protein or 275 ng of each of nine mass markers.

chloroplast envelopes, isolated from the C_3 plant spinach, from *M. crystallinum* operating in either the C_3 or CAM mode and from the obligate CAM plant *K. daigremontiana* have a major 29 kD polypeptide. This polypeptide, identified as the phosphate translocator in spinach [8,9], may possess similar transport activity in all species. Occurrence of an intensely stained 32 kD polypeptide in chloroplast envelopes from *M. crystallinum* after induction of Crassulacean

182

acid metabolism and from *K. daigremontiana* suggests that this polypeptide may be a translocator protein which is required for CAM but not for C_3 photosynthesis.

It is not presently known whether pyruvate and/or PEP as well as Pi, triose phosphate and PGA are transported by a 29 kD polypeptide in the chloroplast envelopes from these CAM species. Direct analysis of the transport properties of these chloroplasts and isolation and functional characterization of the prominent polypeptides of the envelope membranes are needed in order to confirm proposed mechanisms of intracellular metabolic transport during the metabolism of carbon in CAM species.

ACKNOWLEDGEMENTS

This research was supported by the College of Agriculture and Life Sciences University of Wisconsin, Madison; by USDA Competitive Research Grants 5901-0410-8-0088-9 and 59-2531-0-516-0 and by NSF Grant PCM 81-07953.

REFERENCES
1. Cockburn W Baldry CW and Walker DA (1967) Oxygen evolution by isolated chloroplasts with carbon dioxide as the hydrogen acceptor. A requirement for orthophosphate or pyrophosphate. Biochim Biophys Acta 131: 594-596.
2. Cockburn W Baldry CW and Walker DA (1967) Some effects of inorganic phosphate on O_2 evolution by isolated chloroplasts. Biochim Biophys Acta 143: 614-624.
3. Cockburn W Walker DA and Baldry CW (1968) Photosynthesis by isolated chloroplasts. Reversal of orthophosphate inhibition by Calvin cycle intermediates. Biochem J 107: 89-95.
4. Douce R and Joyard J (1979) Structure and function of the plastid envelope Adv Bot Res 7: 1-16.
5. Edwards GE Foster JG and Winter K (1982) Activity and Intracellular compartmentation of enzymes of carbon metabolism in CAM plants. In Ting IP and Gibbs M, eds. Crassulacean acid metabolism, pp. 92-111. Rockville: American Society of Plant Physiologists.
6. Edwards GE and Huber SC (1981) The C_4 pathway. In Hatch MD and Boardman NK, eds. The biochemistry of plants a comprehensive treatise vol 8 photosynthesis, pp. 238-281. New York: Academic Press.
7. Fliege R Flügge UI Werden K and Heldt HW (1978) Specific transport of inorganic phosphate, 3-phosphoglycerate and triosephosphates across the inner membrane of the envelope in spinach chloroplasts. Biochim Biophys Acta 502: 232-247.
8. Flügge UI and Heldt HW (1976) Identification of a protein involved in phosphate transport in chloroplasts. FEBS Lett 68: 259-262.
9. Flügge UI and Heldt HW (1981) The phosphate translocator of the chloroplast envelope. Isolation of the carrier protein and reconstitution of transport. Biochim Biophys Acta 638: 296-304.
10. Foster JG Edwards GE and Winter K (1982) Changes in levels of phosphoenolpyruvate carboxylase with induction of Crassulacean acid metabolism in *Mesembryanthemum crystallinum*. Plant Cell Physiol 23: in press.

11. Greenway H Winter K and Lüttge G (1978) Phosphoenolpyruvate carboxylase during development of Crassulacean acid metabolism and during a diurnal cycle in *Mesembryanthemum crystallinum*. J Exp Bot 29: 547-559.
12. Heber U and Heldt HW (1981) The chloroplast envelope: structure, function and role in leaf metabolism. Ann Rev Plant Rhysiol 32: 139-168.
13. Heber U and Krause GH (1971) Transfer of carbon, phosphate energy and reducing equivalents across the chloroplast envelope. Proc II Internat Cong Photosynthesis, pp. 1023-1033. The Hague: Junk.
14. Heldt HW and Rapley L (1970) Specific transport of inorganic phosphate, 3-phosphoglycerate and dihydroxyacetone phosphate, and of dicarboxylates across the inner membrane of spinach chloroplasts. FEBS Lett 10: 143-148.
15. Holtum JAM and Winter K (1982) Activity of enzymes of carbon metabolism during the induction of Crassulacean acid metabolism in *Mesembryanthemum crystallinum*. Planta: in press.
16. Huber SC and Edwards GE (1977) Transport in C_4 mesophyll chloroplasts. Characterization of the pyruvate carrier. Biochim Biophys Acta 462: 583-602.
17. Huber SC and Edwards GE (1977) Transport in C_4 mesophyll chloroplasts. Evidence for an exchange of inorganic phosphate and phosphoenolpyruvate. Biochim Biophys Acta 462: 603-612.
18. Kluge M and Ting IP (1978) Crassulacean acid metabolism: analysis of an ecological adaptation. Berlin: Springer-Verlag.
19. Markwell MAK Haas SM Bieber LL and Tolbert NE (1978) A modification of the Lowry procedure to simply protein determination in membrane and lipoprotein samples. Anal Biochem 87: 206-210.
20. Oakley BR Kirsch DR and Morris NR (1980) A simplified ultrasensitive silver stain for detecting proteins in polyacrylamide gels. Anal Biochem 105: 361-363.
21. Robinson SP and Walker DA (1979) The site of sucrose synthesis in isolated leaf protoplasts. FEBS Lett 107: 295-299.
22. Schnarrenberger C Gross D Bulkhard C and Herbert M (1980) Cell organelles from Crassulacean acid metabolism (CAM) plants. II. Compartmentation of enzymes of the Crassulacean acid metabolism. Planta 147: 477-484.
23. Spalding MH Schmitt MR Ku SB and Edwards GE (1979) Intracellular localization of some key enzymes of Crassulacean acid metabolism in *Sedum praealtum*. Plant Physiol 63: 738-743.
24. Usuda H and Edwards GE (1980) Localization of glycerate kinase and some enzymes for sucrose synthesis in C_3 and C_4 plants. Plant Physiol 65: 1017-1022.
25. Winter K Foster JG Edwards GE and Holtum JAM (1981) Intracellualr localization of enzymes of carbon metabolism in *Mesembryanthemum crystallinum* exhibiting photosynthetic characteristics of a C_3 or a Crassulacean acid metabolism plant. Plant Physiol 69: 300-307.
26. Winter K and von Willert DJ (1972) NaCl-induzierter Carssulaceensaure-stoffwechsel bei *Mesembryanthemum crystallinum*. Z Pflanzenphysiol 67: 166-170.

WELWITSCHIA MIRABILIS HOOK. FIL. - A CAM PLANT?
ECOPHYSIOLOGICAL INVESTIGATIONS IN THE CENTRAL NAMIB DESERT

D. J. VON WILLERT, E. BRINCKMANN, R. BAASCH, B. M. ELLER+

Lehrstuhl Pflanzenökologie, Universität Bayreuth, D-8580 Bayreuth,
W-Germany;
+ Institute of Plant Biology, University of Zürich, CH - 8008 Zürich,
  Switzerland

ABSTRACT

The diurnal course of CO_2 gas exchange, $^{14}CO_2$ incorporation, malate, citrate and proline content, transpiration, and energy budget of *Welwitschia mirabilis* were measured in the central Namib desert in order to decide which CO_2 fixation pathway is performed by this gymnosperm.

The CO_2 gas exchange of *Welwitschia* is that of a C_3-plant under arid conditions. Young leaf parts show a two-peaked pattern of photosynthetic CO_2 uptake, older ones show only the morning peak. Irrigation improves photosynthetic CO_2 uptake considerably but does not change the constant CO_2 release during the night. $^{14}CO_2$ incorporation experiments supported these findings.

The contents of malate and citrate are high but do not undergo diurnal fluctuations in terms of a CAM. The proline content virtually indicates a severe water stress but transpiration is exceptionally high and helps to dissipate about 12% of the absorbed energy.

INTRODUCTION

The criteria for the existence of a CAM are: net CO_2 uptake in the night, diurnal fluctuations in malate, citrate or other acids with increasing content during night, a low daytime transpiration due to stomatal closure, a pronounced water store capacity (succulence) and $\delta^{13}C$ values in the range between C_3 and C_4 plants. For *Welwitschia mirabilis* only some of these criteria have been shown to occur. With greenhouse plants a weak diurnal fluctuation in total acidity and a better $^{14}CO_2$ incorporation at night than during the day have been reported (1). Stomatal resistance and CO_2 uptake of plants growing in the Namib desert implied a C_3 pathway of photosynthesis (2) while the $\delta^{13}C$ values fulfilled the CAM requirements (3).

The lack of sufficient information concerning continuously recorded diurnal courses of CO_2 gas exchange together with determination of $^{14}CO_2$ incorporation, malate and citrate content and simultaneous measurement of transpiration of *Welwitschia* in its natural habitat encouraged us to perform detailed investigations in order to decide to which type of CO_2 fixation pathway *Welwitschia* belongs.

MATERIAL AND METHODS

All measurements have been made in the central Namib desert (Welwitschia Vlakte) with naturally growing plants in August and September 1981. As the methods and equipment used for our investigations are described in detail elsewhere (4) we will here only briefly list them up. Transpiration was estimated by a weighing method. CO_2 gas exchange was measured on intact growing leaves with an infrared gas analyzer in an open system. Malate and citrate content was estimated enzymatically from extracts of leaf samples; proline was determined in aliquots of the same extracts by a colorimetric method. Microclimatic data (radiation, temperatures, relative humidity) were continuously recorded.

RESULTS AND DISCUSSION

Welwitschia mirabilis has only two perennial leaves that emerge from a groove around the margin of the stem (fig. 1). Total leaf length is a function of growing at the base and dying at the tip. Depending on the climatic conditions the age of a leaf will increase from 0 at the base to 7 or more years at the tip. The average growth rate is about 10 - 15 cm per year. Along the leaf gradients in the distribution of inorganic ions, organic acids, aminoacids, water content, physiological and biochemical reactions occur.

FIGURE 1. *Welwitschia mirabilis* in the central Namib desert

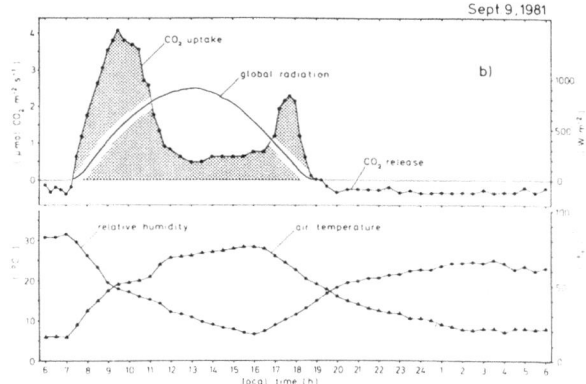

FIGURE 2. CO_2 gas exchange of a non-irrigated (a) and an irrigated (b)
Welwitschia mirabilis plant. Relative humidity and air temperature were
measured at ground level.

The CO_2 gas exchange of a young part of the leaf (fig. 2a) is
characterized by a pronounced CO_2 uptake peak in the morning with a rapid
decline due to increasing air temperature. During the rest of the day this
leaf part operates at the compensation point. Throughout the night there is
only a continuous respiratory CO_2 release.

Irrigation known to evolve or enhance nocturnal CO_2 fixation in CAM
plants (5, 6) had no effect on the CO_2 gas exchange pattern during the
night but improved the photosynthetic CO_2 uptake considerably (fig. 2b).
Maximum rate of net photosynthesis was higher and evening peak of CO_2
uptake evolved.

It is well established that in many CAM plants the tendency to CAM
increases with leaf age (7). Measurements of the CO_2 gas exchange over a
range of different leaf ages along the axis of a 85 cm long leaf did not
show any sign of net CO_2 uptake at any time during the night. These
findings were substantially supported by the results of $^{14}CO_2$ fixation
experiments. $^{14}CO_2$ was only incorporated when supplied in the morning.

Incorporation at night was in the same order of magnitude as for C_3 plants (4). Maximum rate of net photosynthesis as well as the CO_2 balance depends on leaf age as it is illustrated in fig. 3a+b. It is obvious that the old part of the leaf lives at the expense of the younger ones. A good reason for *Welwitschia* to do this might be that an improved water supply rapidly enhances photosynthetic CO_2 uptake (fig. 2b) and old leaf parts will obtain a positive carbon balance and hence support growth under temporary improved environmental conditions (rainfall).

Although we could not detect any CAM feature for *Welwitschia* the possibility of refixation of parts of the respiratory CO_2 at night still exists. For that reason the malate, citrate and isocitrate contents in the leaf at the beginning and end of the night were determined. With the exception of isocitrate which is present only in small amounts the other acids show rather high concentrations depending significantly on leaf age (fig. 4). No clear cut diurnal oscillations in terms of CAM were detected. All differences were proved to be statistically insignificant.

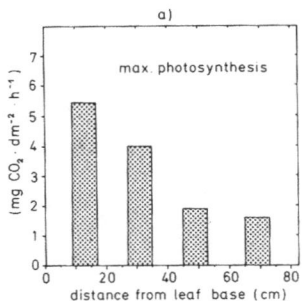 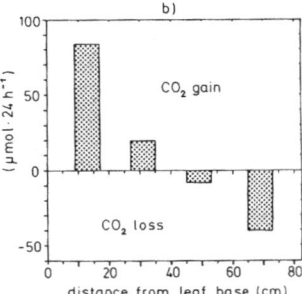

FIGURE 3. Maximum rate of photosynthesis (a) and CO_2 balance (b) along the leaf of *Welwitschia mirabilis*

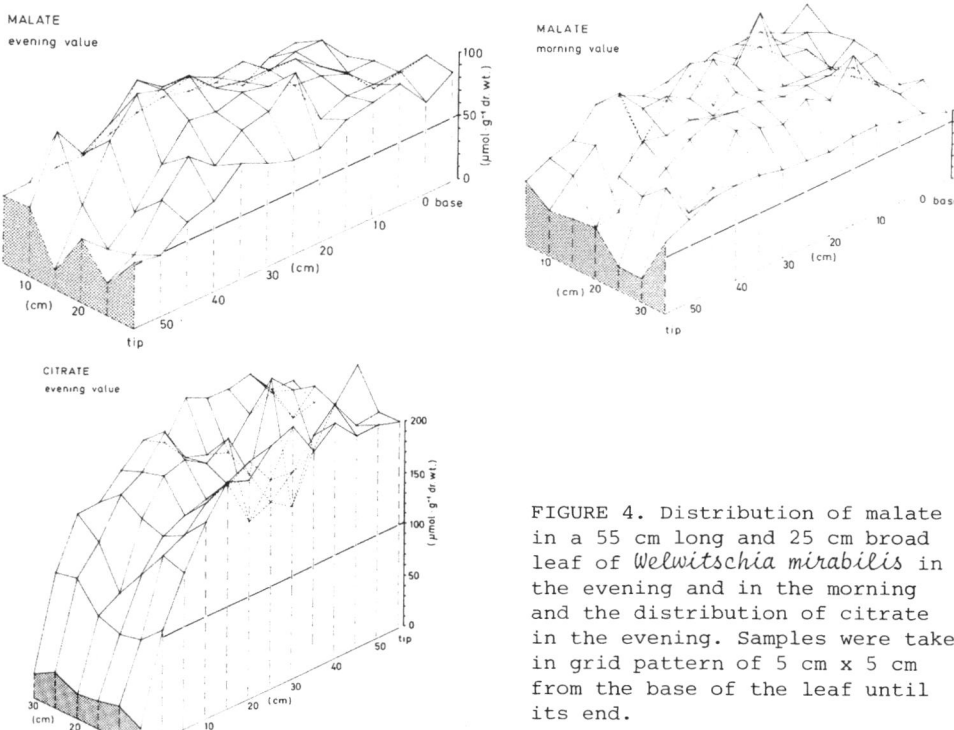

FIGURE 4. Distribution of malate in a 55 cm long and 25 cm broad leaf of *Welwitschia mirabilis* in the evening and in the morning and the distribution of citrate in the evening. Samples were taken in grid pattern of 5 cm x 5 cm from the base of the leaf until its end.

Thus, all efforts to verify CAM in *Welwitschia mirabilis* failed. Nevertheless, the $\delta^{13}C$ values are in the range of -17.77 to -19.64% and hence in the same range as previously reported and interpreted by means of a CAM (3). Possible reasons for such low ^{13}C contents in a C_3 plant are discussed elsewhere (4) but lack of experimental proof.

It is amazing that *Welwitschia*, living in an environment favouring CAM, covers its CO_2 demand only by daytime CO_2 fixation which consequently results in a marked water loss due to transpiration. In fact, even after a drought period of several years an average Welwitschia plant replaces 1l of water lost by transpiration during a day. We observed maximum transpiration rates at noon of 2.6 mmol $m^{-2}s^{-1}$. A calculation of the leaf conductance from the transpiration dates revealed a two-peaked pattern of stomatal opening which coincides with the peaks of the CO_2 uptake. For several plants especially halophytes proline has been shown to be correlated with salt and water stress (cf. 8, 9). If we take proline as an indicator for an existing water stress then *Welwitschia* has to endure

a tremendous stress condition. Its proline content is by far the highest
ever reported for higher plants and fulfills all requirements of a water
stress indicator. Irrigation lowers its content, a föhn wind condition
(severe hot east wind throughout day and night) raises it (fig. 5), and
under normal climatic conditions (hot and dry days, cool and humid nights)
proline exhibits diurnal fluctuations with the evening value being highest.

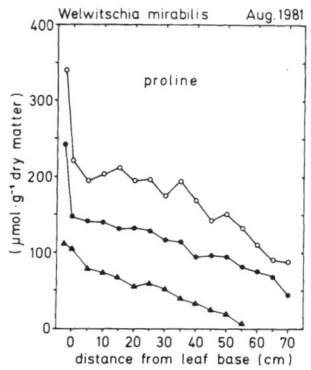

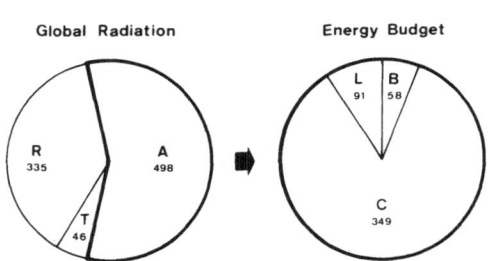

FIGURE 5. Proline content along the
leaf of *Welwitschia mirabilis* (o)
during föhn-wind conditions, (•)
prior to föhn-wind and (▲) with an
improved water supply

FIGURE 6. Global radiation
fluxes and energy budget of
Welwitschia mirabilis leaves
(6 cm from the margin). Mean
values in W m^{-2} around noon
(Sept. 8, 1981)

Generally desert plants have small leaves. From an energy budget
point of view this provides an efficient convective energy transfer.
Welwitschia is an exception of this generalization. Hence it seems worth
to answer the question how *Welwitschia* prevents lethal leaf temperatures.
Neglecting energy exchange from metabolism and storage of sensible heat
the energy budget is A = L + B + C, where A is the energy input from
absorbed global radiation and L, B and C are the energy outputs by net
thermal (longwave) radiation, transpiration and convection, respectively.
Considering earlier reported data (10) fig. 6 gives the values for the
energy fluxes resulting from reflection, transmission and absorption of
879 W m^{-2} global radiation (mean value between 11h30 and 14h00, Sept.
8, 1981) and how the absorbed energy is dissipated. The most important
factor in the energy balance is the high reflectivity of 28.2% of the
inciding global radiation which compensates the bad convective energy
dissipation due to the unfavourable size of the leaf. Transpiration

plays an important role in the energy output. From 11.6% (mean value in fig. 6) to 23.7% (peak value) of the absorbed solar radiation is dissipated by transpiration which should help cooling the leaf below lethal temperatures.

It is paradoxical that one way (perhaps the only one) to survive with big leaves in a desert environment is by means of a high transpiration. The stomata once forced to be open at daytime to prevent lethal leaf temperatures will then provide CO_2 for photosynthesis. Only a deep reaching rooting system and a sufficient water availability in the deeper soil layers can meet the water demand and guarantee the survival even during extended drought periods as is the case with this in all respects curious gymnosperm *Welwitschia mirabilis*.

This work was supported by the Deutsche Forschungsgemeinschaft and the Swiss National Science Foundation. The generous help of Mettler (loan of an electronic balance) is thankfully acknowledged. The work on *Welwitschia mirabilis* in the Welwitschia-Vlakte was kindly permitted by the Department van Landbou en Natuurbewaring in Windhoek.

REFERENCES

1. Dittrich P, Huber W 1974. Carbon dioxide metabolism in members of the Chlamydospermae. In M Avron (ed), Proceedings of the Third International Congress on Photosynthesis. Elsevier, Amsterdam 1974
2. Gaff DF 1972. Drought resistance in *Welwitschia mirabilis* Hook. fil. Dinteria 7, 3 - 7
3. Schulze E-D, Ziegler H, Stichler W 1976. Environmental control of Crassulacean acid metabolism in *Welwitschia mirabilis* Hook. fil. in its range of natural distribution in the Namib desert. Oecologia 24, 323 - 334
4. von Willert DJ, Eller BM, Brinckmann E, Baasch R 1982. CO_2 gas exchange and transpiration of *Welwitschia mirabilis* Hook. fil. in the central Namib desert. Oecologia (in press)
5. Hanscom Z, Ting IP 1978. Irrigation magnifies CAM-photosynthesis in *Opuntia basilaris* (Cactaceae). Oecologia 33, 1 - 15
6. von Willert DJ, Brinckmann E, Scheitler B, Schulze E-D, Thomas DA, Treichel S 1980. Ökophysiologische Untersuchungen an Pflanzen der Namib-Wüste. Naturwissenschaften 67, 21 - 28
7. von Willert DJ 1979. Vorkommen und Regulation des CAM bei Mittags- blumengewächsen (Mesembryanthemaceae). Ber. Deutsch. Bot. Ges. 92, 133 - 144
8. Treichel S 1979. Der Einfluß von NaCl auf den Prolinstoffwechsel bei Halophyten. Ber. Deutsch. Bot. Ges. 92, 73 - 85
9. Kinzel H 1982. Pflanzenökologie und Mineralstoffwechsel. Verlag Eugen Ulmer
10. Schulze E-D, Eller BM, Thomas DA, von Willert DJ, Brinckmann E 1980. Leaf temperature and energy balance of *Welwitschia mirabilis* in its natural habitat. Oecologia 44, 258 - 262

LOW TEMPERATURE STRESS AND MEMBRANE LIPID PHASE IN THE BLUE-GREEN ALGAE

N. MURATA, H, WADA, T. OMATA and T. ONO
Department of Biology, University of Tokyo, Komaba, Meguro-ku, Tokyo 153, Japan

INTRODUCTION

The blue-green algal cells, which are similar to the chloroplasts of the eukaryotic organisms from the viewpoint of membrane structure, can be regarded as model systems in the study on the temperature effect on plants. We have investigated by various methods the lipid phase of the thylakoid membranes and the cytoplasmic membranes of a blue-green alga, *Anacystis nidulans*, grown at different temperatures (1-4), and have suggested that the chilling injury (irreversible damage of physiological activities at the chilling temperature) is associated with the transition of lipid phase of the cytoplasmic membranes (4-6).

The present investigation was designed to examine validity of this mechanism for the chilling injury in the blue-green algae. The changes in lipid phase with temperature in the membranes from *A. nidulans* (sensitive to chilling) and *Anabaena variabilis* (resistant to chilling) were measured by the spin probe method.

MATERIALS AND METHODS

A. nidulans and *A. variabilis* (strain M-3) obtained from the Algal Collection of the Institute of Applied Microbiology, University of Tokyo, were photoautotrophically grown as described previously (5). Growth temperatures were 28 and 38°C in *A. nidulans*, and 22 and 38°C in *A. variabilis*. The activity of the photosynthetic oxygen evolution was measured by following the changes in oxygen concentration in the cell suspension with a calibrated Clark-type oxygen electrode (5). For the chilling treatment of the algal cells, the cell suspension in a test tube was immersed for 60 min in a water bath, the temperature of which was set at a designated level (5).

In the preparation of the thylakoid membranes from *A. nidulans* and *A. variabilis*, the cell suspension was passed through the French pressure cell, and the thylakoid membranes were isolated by centrifuging the homogenate on a sucrose density gradient. In the preparation of the cytoplasmic membranes from *A. nidulans*, the cells were treated with lysozyme and disrupted with the French press (7). The cytoplasmic membranes were isolated by floatation centrifugation of the homogenate on a sucrose density gradient.

The ESR spectrum of nitroxide-free radical was measured with an ESR spectrometer (Nihon Denshi, Model JES-PE-3X). A spin probe, N-oxyl-4',4'-dimethyloxazolidine derivative of 16-ketostearic acid (16-SAL), was added to a dense suspension of the membranes.

RESULTS

The ESR spectrum of 16-SAL added to the membrane suspension showed a signal of the nitroxide radical in a highly immobilized state, suggesting that the spin probe was all bound to the membranes. The rotational correlation time (τ_c), which is inversely related to the rate of rotational diffusion of the nitroxide group, was calculated from the ESR spectrum, and its logarithms were plotted against the reciprocal of the absolute temperature (Figs. 1 and 2).

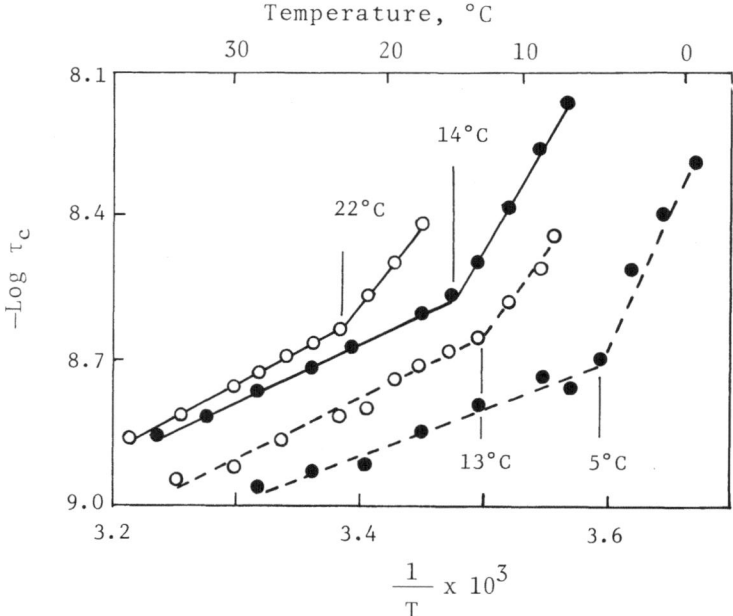

FIGURE 1. Rotational correlation time (τ_c) of the ESR spectrum of 16-SAL in the membranes from *A. nidulans* plotted against the reciprocal of absolute temperature. O——O ; Thylakoid membranes from 38°C-grown cells. ●——● ; Thylakoid membranes from 28°C-grown cells. O---O ; Cytoplasmic membranes from 38°C-grown cells. ●---● ; Cytoplasmic membranes from 28°C-grown cells.

In the membranes from *A. nidulans* (Fig. 1), the plot followed two straight lines. Break points appeared at 14 and 23°C in the thylakoid membranes from cells grown at 28 and 38°C, respectively, and 5 and 13°C in the cytoplasmic membranes from cells grown at 28 and 38°C, respectively. These temperatures for break points in the thylakoid membranes were close to those for the onset of phase separation in the same membranes, that were detected by the X-ray diffraction (16 and 26°C, respectively, in Ref. 1) and the fluorescent probe method (13 and 25°C, respectively, in Ref. 2). The temperatures for the break points in the cytoplasmic membranes were close to those for the onset of phase separation in the same membranes, that were detected by the freeze-fracture electron microscopy (5 and 16°C, respectively, in Ref. 4). The results in Fig. 1 and in the previous study suggest that the two types of membranes respond to the temperature in different manners; the thermotrophic phase transition from the liquid crystalline to the phase separation state in the thylakoid membranes takes place at a higher temperature than that in the cytoplasmic membranes does. They also demonstrate that the temperature for

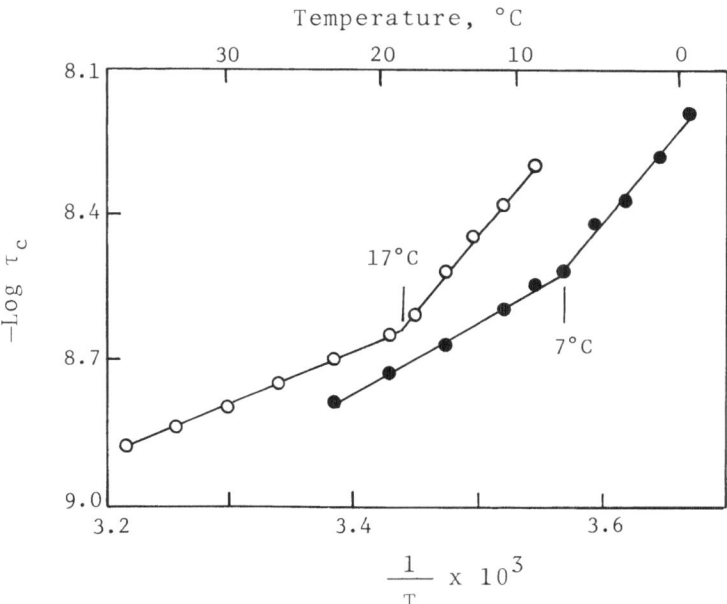

FIGURE 2. Rotational correlation time (τ_c) of the ESR spectrum of 16-SAL in the thylakoid membranes from *A. variabilis* plotted against the reciprocal of absolute temperature. O——O ; Thylakoid membranes from 38°C-grown cells. ●——● ; Thylakoid membranes from 22°C-grown cells.

196

the onset of phase separation in both types of the membranes depends on the growth temperature.

A similar experiment was done in the thylakoid membranes from the chilling-resistant alga, *A. variabilis*. In Fig. 2, the logarithms of τ_c are plotted against the reciprocal of the absolute temperature. Break points were at 7 and 17°C in the membranes from cells grown at 22 and 38°C, respectively. This indicates that the thylakoid membranes were in the liquid crystalline state above these temperatures and in the phase separation state below them.

The Arrhenius plots of the photosynthetic oxygen evolution in *A. nidulans* and *A. variabilis* are shown in Fig. 3. The break points appeared at 14 and 22°C in *A. nidulans* grown at 28 and 38°C, respectively, and at 7 and 15°C in *A. variabilis* grown at 22 and 38°C, respectively. These temperatures were equal, or very close, to those for the onset of phase separation of the

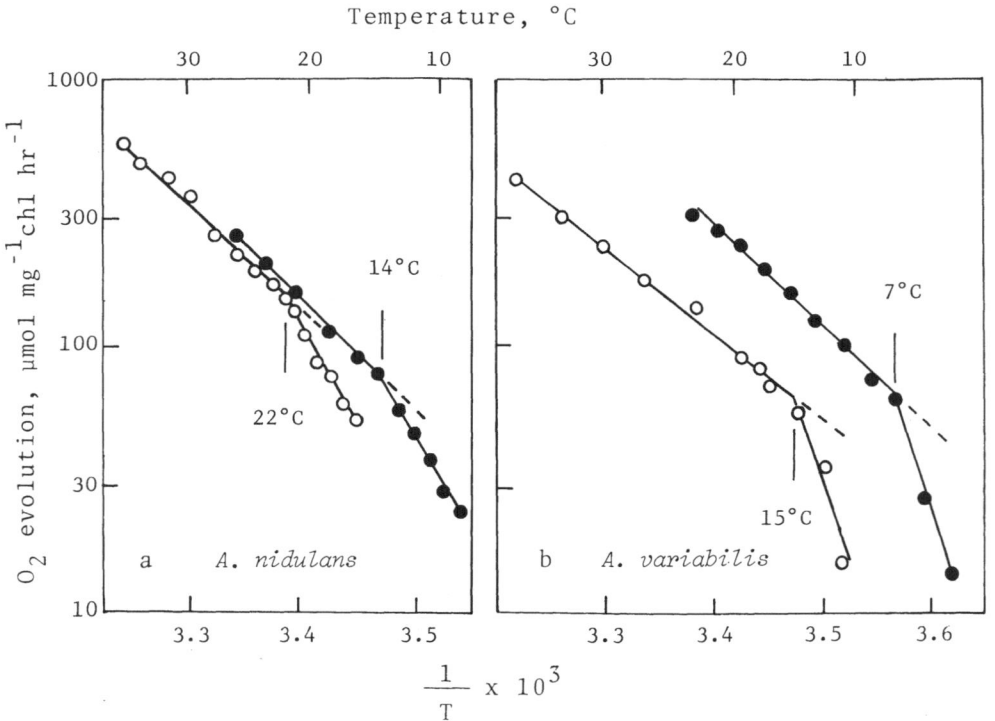

FIGURE 3. The Arrhenius plot of the photosynthetic oxygen evolution in *A. nidulans* and *A. variabilis*. a) *A. nidulans*. ○——○; Cells grown at 38°C. ●——●; Cells grown at 28°C. b) *A. variabilis*. ○——○; Cells grown at 38°C. ●——●; Cells grown at 22°C.

thylakoid membranes from the corresponding algal cells. The apparent activation energy of photosynthesis calculated from the inclination of the straight lines in Fig. 3 changed at the break point from 16 to 32 Kcal mol^{-1} in *A. nidulans* and from 16 to 57 Kcal mol^{-1} in *A. variabilis*. Large values for the apparent activation energy of photosynthesis such as 32 and 57 Kcal mol^{-1} suggest that photosynthesis virtually diminishes below the break points, and that the algal cells become photosynthetically inactive when the thylakoid membranes are in the phase separation state.

The irreversible damage (chilling injury) of photosynthetic oxygen evolution in *A. nidulans* was investigated by exposing the algal cells to low temperatures (Fig. 4). The irreversible decline of photosynthetic activity began at 5 and 15°C in cells grown at 28 and 38°C, respectively. These temperatures were equal, or very close, to those for the onset of phase separation in the cytoplasmic membranes. In *A. variabilis*, on the other hand, no irreversible damage of photosynthetic activity was observed in the temperature region between 0 and 20°C.

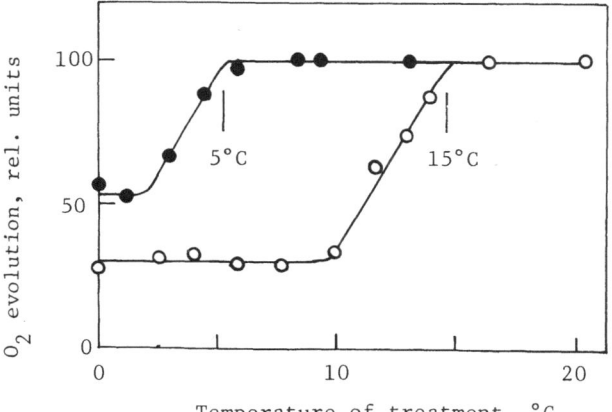

FIGURE 4. Effect of low temperature treatment on the activity of photosynthetic oxygen evolution in *A. nidulans*. The cells were treated at designated temperatures for 60 min, and then the photosynthetic oxygen evolution was measured after the cell suspension was warmed to the growth temperature. o——o ; Cells grown at 38°C. ●——● ; Cells grown at 28°C.

DISCUSSION

The temperatures for the onset of phase separation of the membrane lipids, for the break points of the Arrhenius plot of photosynthesis, and for the beginning of the irreversible damage of photosynthesis in the chilling-sensitive alga, *A. nidulans*, are summarized in Table 1. The break point of the Arrhenius plot appeared at the temperature for the onset of phase separation of the thylakoid membranes. The irreversible damage of photosynthesis began to appear at the temperatures for the onset of phase separation of the cytoplasmic membranes. These findings indicate that in a temperature region near the growth temperature the thylakoid membranes and the cytoplasmic membranes are both in the liquid crystalline state. Under these conditions, the algal cells are fully active. In a temperature region near 0°C, both types of the membranes are in the phase separation state. Under these conditions, the cells are irreversibly damaged. Between these regions, there exists a certain temperature region in which the cytoplasmic membranes are in the liquid crystalline state, and the thylakoid membranes are in the phase separation state. Under these conditions, the photosynthetic activity is only reversibly suppressed.

TABLE 1. Temperatures for the onset of phase separation, for the break points of the Arrhenius plot of photosynthesis, and for the beginning of irreversible damage of photosynthesis in *A. nidulans* and *A. variabilis*.

	Temperature (°C)			
	A. nidulans grown at		*A. variabilis* grown at	
	28°C	38°C	22°C	38°C
Onset of phase separation				
Thylakoid membranes	14	23	7	17
Cytoplasmic membranes	5	13	<0[*]	<0[*]
Photosynthesis				
Break points in the Arrhenius plot	14	22	7	15
Beginning of irreversible damage	5	15	<0	<0

[*]Taken from the results by the freeze-fracture electron microscopy in Ref. 4.

In the chilling-resistant alga, *A. variabilis* (Table 1), the break point of the Arrhenius plot of photosynthesis was very close to the temperature for the onset of phase separation of the thylakoid membranes. The cytoplasmic membranes are in the liquid crystalline state above 0°C. This fact can be related to the resistivity of this alga to chilling; no irreversible damage of photosynthesis was observed above 0°C. The results in *A. nidulans* and *A. variabilis* suggest that the chilling injury is closely associated with the phase separation of the cytoplasmic membranes but not the thylakoid membranes.

CONCLUSION

In the blue-green algae, *A. nidulans* and *A. variabilis*, grown at different temperatures, the temperatures for the onset of phase separation was higher in the thylakoid membranes than in the cytoplasmic membranes. When the cells are exposed to the temperature at which both types of the membranes are in the phase separation state, the photosynthetic activity is irreversibly damaged. When the cells are exposed to the temperature at which the thylakoid membranes are in the phase separation state and the cytoplasmic membranes are in the liquid crystalline state, the photosynthetic activity of the algal cells was only reversibly suppressed.

REFERENCES

1. Tsukamoto Y, Ueki T, Mitsui T, Ono T. Murata N. 1980. Biochim. Biophys. Acta 602: 673–675.

2. Murata N, Ono T. 1981. Photosynthesis (Ed. G. Akoyunoglou), pp. 473–481. Balaban International Science Services, Philadelphia.

3. Murata N, Troughton JH, Fork DC. 1975. Plant Physiol. 56: 508–517.

4. Ono T, Murata N. 1982. Plant Physiol. 69: 125–129.

5. Ono T, Murata N. 1981. Plant Physiol. 67: 176–181.

6. Ono T, Murata N. 1981. Plant Physiol. 67: 182–187.

7. Murata N, Sato N, Omata T, Kuwabara T. 1981. Plant & Cell Physiol. 22: 855–866.

INACTIVATION AND PROTECTION OF ISOLATED THYLAKOID MEMBRANES DURING FREEZING

K.A. SANTARIUS and Ch. GIERSCH

Botanisches Institut, Universität Düsseldorf, Universitätsstrasse 1,
D-4000 Düsseldorf, Federal Republic of Germany

ABSTRACT

In freezing experiments with thylakoid membranes isolated from spinach leaves (*Spinacia oleracea L. cv. Monatol*) the effect of the solute concentration on membrane damage and cryoprotection was studied.

Calculation of the final concentration of solutes in the unfrozen part of the system revealed that at a given freezing temperature membrane inactivation appeared to be predominantly a function of the osmolality of cryotoxic compounds reached in the surroundings of the membranes. However, during freezing in the presence of extremely low solute concentrations, e.g. 1-15 mmolal NaCl and appropriate low amounts of carbohydrates, mechanical damage caused by the large amount of ice formed was likely to contribute to freezing injury. Moreover, freezing of thylakoids at various temperatures exhibits that in addition to the injurious effect of membrane-toxic compounds other factors were involved in membrane inactivation.

Cryopreservation is mainly due to nonspecific colligative action of the compounds, i.e. cryoprotectants diminish the concentration of membrane-toxic solutes in the surroundings of the membranes; simultaneously, they reduce the amount of ice formed and, thus, enlarge the volume of the unfrozen solution which also contains the membranes. In addition, to some extent a specific noncolligative-type mechanism plays a role in cryoprotection of biomembranes.

1. INTRODUCTION

In plant cells, biomembranes are known to be extremely frost-sensitive (2-6). It is a common view that membrane damage is mainly due to increase in the concentration of membrane-toxic solutes which accompanies dehydration in the course of ice formation. In addition, membrane inactivation is dependent on the freezing temperature, on the duration of freezing,

and on the freezing and thawing conditions (6, 10, 20). For cryopreservation, synthesis and accumulation of cryoprotectants plays a major role.
These compounds decrease, on the one hand, the concentration of membrane-
toxic solutes during freezing (7, 9, 13), but, on the other hand, they
also exert a specific noncolligative effect (17-19).

In this paper it is investigated to which extent elevated concentrations of cryotoxic solutes reached during freezing of isolated thylakoids
in the surroundings of the membranes affect membrane inactivation and
protection compared to other factors.

2. MATERIAL AND METHODS

Thylakoids isolated from leaves of greenhouse- or field-grown spinach
(*Spinacia oleracea* L. *cv. Monatol*) were used as a model system for biomembranes. The isolation procedure was described recently (16, 18). Freezing
at different temperatures took place in the absence or presence of various
concentrations of the cryotoxic NaCl and of low molecular weight carbohydrates as cryoprotectants. For slow and fast freezing, small samples
(0.5 ml each) kept in glass tubes were either transferred into a deep
freezer or in a temperature-bath, respectively, both previously adjusted
to the respective freezing temperatures. After thawing in a water bath at
room temperature, the functional integrity of the membranes was estimated
by measuring the activity of cyclic photophosphorylation with phenazine
methosulfate as cofactor (16). Chlorophyll concentrations, chloride contents and freezing point depressions of solutes were determined as outlined
earlier (19). Calculation of the final electrolyte concentration in the
unfrozen part of the system which was in equilibrium with ice at a given
freezing temperature was adapted from Lineberger and Steponkus (7) as
recently described in detail (19). As deviations from the ideal osmotic
response of the solutes are considerable at high solute concentrations,
calculated final concentrations were corrected for the non-ideal thermodynamic behaviour of the solutes.

3. RESULTS AND DISCUSSION

3.1. Membrane inactivation

When a suspension of membranes is frozen in the presence of solutes to
temperatures which are above eutectic crystallization, water is converted
to ice, and the concentration of the compounds in the unfrozen part of

the system, which also contains the membranes, is increased. As already shown earlier, loss of function of thylakoid membranes during a freeze-thaw cycle depends on the molal ratio of cryoprotective to cryotoxic solutes present in the membrane suspension (1, 16, 17, 19). In Fig. 1A it is demonstrated that the more sugar was added to thylakoids prior to freezing the more salt was necessary to obtain a comparable degree of membrane inactivation during freezing. If damage is primarily due to increase in the concentration of membrane-toxic compounds such as electrolytes, one would expect that a comparable degree of membrane inactivation can be observed at comparable final concentrations of electrolytes. Indeed, if membrane survival was plotted against the final concentration of NaCl predicted in the surroundings of the membranes, the differences shown in Fig. 1A are no longer evident (Fig. 1B). This clearly suggests that at a given freezing temperature membrane damage seems to be due predominantly to the high electrolyte concentration reached during freezing.

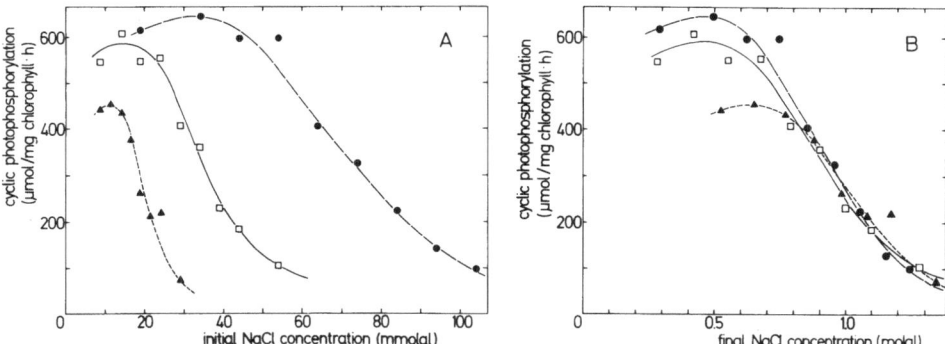

FIGURE 1. The effect of freezing on isolated thylakoid membranes in the presence of 100 (-----▲-----), 200 (————□————) and 400 mmolal D(+)-glucose (— —●— —), respectively, and various concentrations of NaCl. After fast freezing, membranes were kept for 3-4 h at -15.5°C. Membrane survival was plotted as function of the NaCl concentration prior to freezing (A) and of the final molality of NaCl predicted in the surroundings of the thylakoids during freezing (B).

However, if freezing took place in the presence of extremely low solute concentrations, e.g. 1-5 mmolal NaCl and appropriate low sucrose levels, membrane survival was nearly independent of the NaCl concentration present in the membrane suspension before freezing (Fig. 2A). Below an initial NaCl concentration of about 15 mmolal, membrane inactivation was apparently no longer determined by the final molality of NaCl reached during freezing

(Fig. 2B): the lower the initial salt level the lower the final concentration of NaCl at which a comparable degree of membrane inactivation
was observed. Thus, at extremely low solute concentrations, in addition
to the increase in the concentration of membrane-toxic solutes other
factors apparently contribute to membrane damage during freezing. It is
suggested that under these conditions most of the water is converted to
ice, and the membranes become highly compressed in the extremely low
volume of the unfrozen part of the system so that mechanical damage caused
by ice crystals may take place (see ref. 19). Comparable results were
recently obtained by Mazur and coworkers (11, 12) during slow freezing
of erythrocytes.

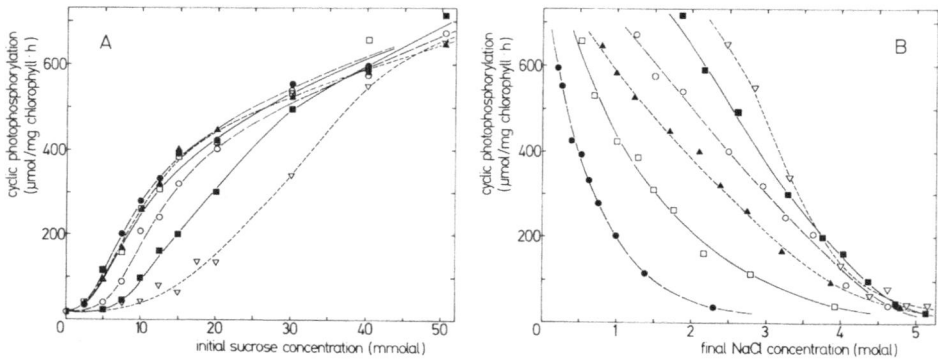

FIGURE 2. The effect of freezing on isolated thylakoid membranes in the
presence of various concentrations of sucrose and NaCl. After slow freezing, membranes were kept for 3-4 h at $-23^{O}C$. Membrane survival was plotted
as function of the sucrose concentration prior to freezing (A) and of the
final molality of NaCl predicted in the surroundings of the thylakoids
during freezing (B). NaCl concentration prior to freezing (in mmolal):
— —●— — 1.14; ———□——— 3.12; -----▲----- 6.0; — —O— — 10.3;
———■——— 16.8; -----▽----- 25.8.

In a previous paper it was shown that the rate of inactivation of cyclic
photophosphorylation of isolated thylakoids during freezing above eutectic
crystallization was highly dependent on the freezing temperature. Inactivation showed a maximum at about $-12^{O}C$ (20). Therefore, the dependence of
membrane damage on the final electrolyte concentration reached in the
unfrozen part of the system was investigated at various freezing temperatures. Mild freezing up to about -12° resulted in increase of membrane
injury with decreasing freezing temperature (Fig. 3A). In contrast, at
freezing temperatures below this value, membrane inactivation decreased

with lowering the temperature (see also Fig. 7A). If membrane survival
is plotted as function of the final NaCl concentration reached at the
respective freezing temperature in the unfrozen part of the system, it
is obvious that thylakoid damage occurs at comparable electrolyte levels
under mild freezing conditions (Fig. 3B; see also Fig. 1B). However, at
more severe freezing, the final salt concentration in the surroundings
of the membranes at which a comparable degree of injury was observed
increased with decreasing temperature (see also Fig. 7B).

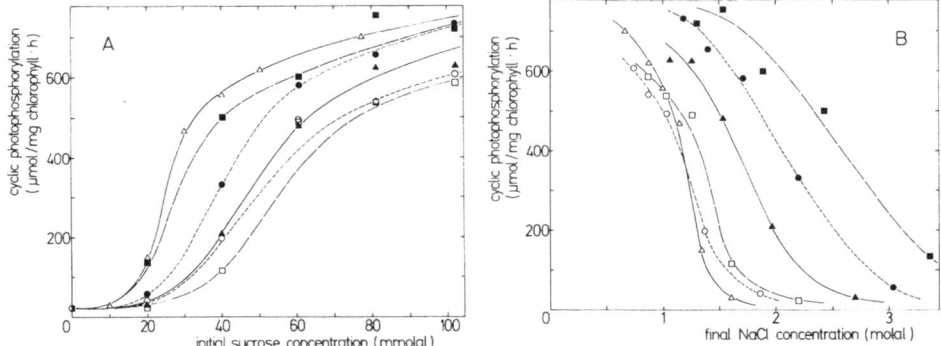

FIGURE 3. The effect of freezing temperature on isolated thylakoid membranes which were suspended in solutions containing 26.5 mmolal NaCl and
various concentrations of sucrose. After fast freezing and storage at the
respective freezing temperature for 3-4 h, membrane survival was plotted
as function of the sucrose concentration prior to freezing (A) and versus
the final molality of NaCl predicted in the surroundings of the thylakoids
during freezing (B). Freezing temperatures (in °C): ———△——— -7;
-----O----- -10; — —□— — -12; ———▲——— -15; -----●----- -17;
— —■— — -19.

From Fig. 3 it is evident that the effect of freeze-induced increase
in electrolyte concentrations on membrane damage shows pronounced dependence on the temperature: under mild freezing conditions, the degree of
membrane inactivation seems to be primarily a function of the final molality of NaCl in the unfrozen part of the system, i.e. the increase in
electrolyte concentration with decreasing temperatures seems to be the
dominating factor. However, this interpretation should be considered with
some caution: membrane inactivation by NaCl at 0°C occurs at electrolyte
concentrations which are clearly lower than final concentrations causing
damage at moderate freezing conditions (Fig. 4), even if the simultaneously occurring increase of the concentration of the carbohydrate during

206

freezing is taken into account; in this case differences are somewhat
smaller than given in Fig. 4 but still considerably (compare also Figs. 3B
and 8). Thus, besides salt concentration, other factors contribute to
freezing injury. This becomes also evident at temperatures below about
$-12^{O}C$ where a given degree of membrane inactivation occurs at increasing
final salt concentrations with decreasing temperatures. These findings
cannot be explained yet. In earlier investigations it has been suggested
that in addition to membrane damage caused by the temperature-dependent
increase in electrolyte concentration, a temperature-dependent loss in
membrane activity takes place which decreases with lowering the temperature (20). However, preliminary results seem to indicate that at temperatures around $0^{O}C$, i.e. in the absence of freezing, membrane damage increases with lowering the temperature. This suggests that additionally
other, yet unknown effects might be responsible for differences observed
in temperature-dependent membrane inactivation.

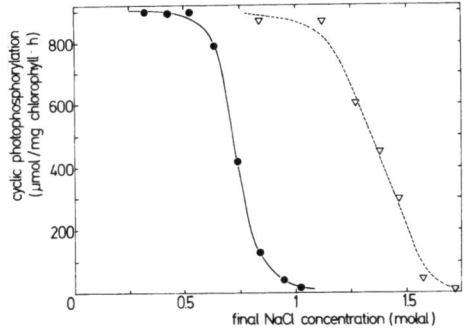

FIGURE 4. The effect of freezing and storage at $0^{O}C$ on isolated thylakoid
membranes which were suspended in solutions containing 300 mmolal glycerol
and various concentrations of NaCl. Thylakoids were kept for 6 h at
$0^{O}C$ (————●————) and $-11.6^{O}C$ (-----▽-----), respectively. Membrane
survival was plotted as function of the molality of NaCl in the surroundings of the thylakoids at the respective temperatures.

Other factors which contribute to thylakoid damage in the course of a
freeze-thaw cycle are the rates of cooling (Fig. 5) and thawing (see
ref. 1), the time for which the membranes were stored at the respective
freezing temperatures (Fig. 6) and differences in the toxicity of the
solutes (see, e.g., refs. 4, 5, 14). Loss of thylakoid function during
freezing seems to be primarily due to changes in protein structure and
in the overall permeability of the membranes, as discussed previously in
detail (4, 5; see also ref. 8).

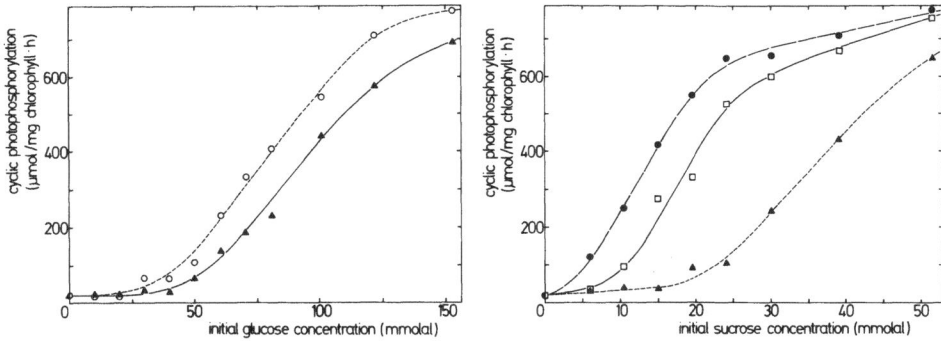

FIGURE 5. The effect of the cooling rate on isolated thylakoid membranes which were suspended prior to freezing in solutions containing 25 mmolal NaCl and D(+)-glucose concentrations as indicated on the abscissa. After slow (-----O-----) and fast freezing (———▲———), membranes were kept for 5-6 h at -22°C.

FIGURE 6. The effect of freezing time on isolated thylakoid membranes which were suspended prior to freezing in solutions containing 25 mmolal NaCl and sucrose concentrations as indicated on the abscissa. After slow freezing, membranes were kept for 3 (— —●— —), 5.5 (———□———) and 22 h (-----▲-----) at -22°C.

3.2. Membrane protection

It is amply documented in the literature that addition of cryoprotectants such as carbohydrates, certain carbonic acids and amino acids and even high molecular weight compounds such as proteins, dextrans and polyvinyl pyrrolidone to cellular membrane systems prior to freezing results in partial or complete cryopreservation (see refs. 2, 3, 16, 17).

The predominant mechanism of cryoprotection is the colligative action of the solutes, i.e. the nonspecific prevention of an increase in the concentration of membrane-toxic compounds in the course of the freeze-thaw cycle. This was first outlined by Lovelock (9) for erythrocytes and since that time confirmed for various cells and membrane systems (e.g., refs. 7, 11-13, 19). In Fig. 1 it was shown that at a given freezing temperature, membrane survival seems to be predominantly a function of the final electrolyte concentration reached in the unfrozen part of the system; a rise in the initial molal ratio of cryoprotective to cryotoxic solutes results in decrease of the final concentration of the membrane-toxic compounds in the unfrozen solution which is in equilibrium with ice.

However, comparison of the protective efficiency of various cryopro-

208

tectants revealed conspicuous differences which cannot be explained by
the colligative concept. Considerable differences in protective quali-
ties of glucose and sucrose which are evident when membrane preservation
is plotted as function of the sugar concentration (Fig. 7A), became not
abolished when cryoprotection was related to the final concentration of
membrane-toxic solutes reached in the unfrozen part of the system during
freezing (Fig. 7B). This is valid for various freezing temperatures and
was recently confirmed for several low molecular weight carbohydrates (19).
The findings suggest participation of a noncolligative-type mechanism in
cryopreservation.

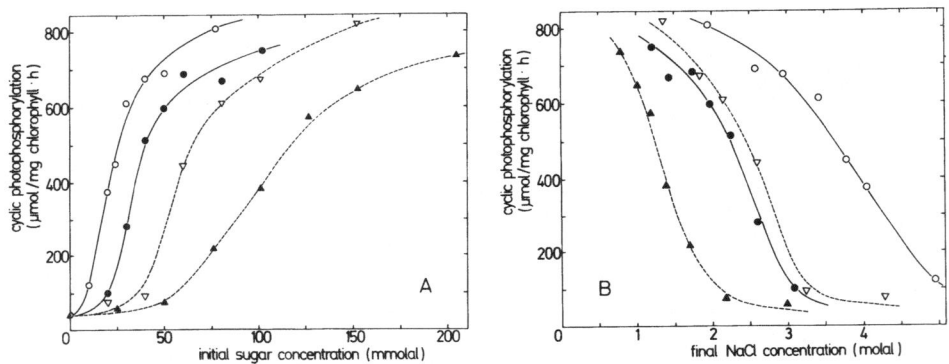

FIGURE 7. The effect of freezing on isolated thylakoid membranes which
were suspended prior to freezing in solutions containing 27.5 mmolal NaCl
and various concentrations of sucrose (——— O,● ———) and D(+)-glucose
(----- ▽,▲ -----). After fast freezing, membranes were kept for 3-4 h at
-17°C (●,▲) and -23°C (O,▽), respectively. Membrane survival was
plotted as function of the sugar concentration prior to freezing (A) and
as function of the final molality of NaCl predicted in the surroundings
of the thylakoids during freezing (B).

Moreover, a number of carbohydrates such as glucose and galactose,
various pentoses or deoxy-hexoses which - on a molal basis - exhibit
comparable activity-concentration profiles even at high concentrations,
differed considerably in their protective capacity during a freeze-thaw
cycle (17, 18). Even in the absence of freezing, when colligative dilution
of membrane-toxic solutes is not possible, carbohydrates stabilize biomembranes against the deleterious effect of high electrolyte concentrations
(15-18). In Fig. 8 it is shown that glucose and galactose which exhibit
considerable differences in cryoprotective efficiency (17, 18), also pro-

tect thylakoids to a different extent against the deleterious effect of high NaCl concentrations at $0^{\circ}C$, although the concentration of the membrane-toxic electrolytes in the surroundings of the membranes is not reduced by the sugars under these conditions. Also these effects cannot be explained by the colligative concept.

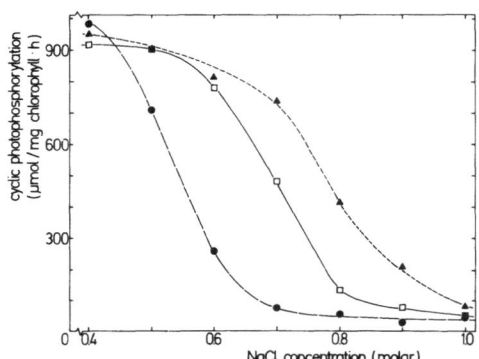

FIGURE 8. Protective effect of different hexoses on isolated thylakoid membranes which were suspended for 3-4 h at $0^{\circ}C$ in NaCl solutions of various concentrations indicated on the abscissa. -----▲----- 1 M D(+)-galactose; ————□———— 1 M D(+)-glucose; —— ——●—— —— without sugar.

Thus, in addition to the salt-buffering action and reduction in the amount of ice formed at the respective freezing temperature, cryoprotectants exert specific noncolligative effects which also play a role in membrane preservation during freezing. The possible nature of noncolligative cryoprotection was recently discussed (18).

ACKNOWLEDGMENTS

We are grateful to Miss Britta Dietzel and Miss Margrit Meyer for competent technical assistance, and to Herbert Vetter for capable assistance with the numerical calculations which were carried out on the TR 445 of the Rechenzentrum der Universität Düsseldorf.

REFERENCES

1. Heber U, Santarius KA (1964) Loss of adenosine triphosphate synthesis caused by freezing and its relationship to frost hardiness problems. Plant Physiol 39: 712-719.
2. Heber U, Santarius KA (1973) Cell death by cold and heat and resistance to extreme temperatures. Mechanisms of hardening and dehardening. In Precht H, Christophersen J, Hensel H, Larcher W, eds. Temperature and life, pp. 232-263. Berlin/Heidelberg/New York: Springer.

3. Heber U, Santarius KA (1976) Water stress during freezing. In Lange OL, Kappen L, Schulze E-D, eds. Water and plant life. Ecological Studies, Vol 19, pp. 253-267. Berlin/Heidelberg: Springer.

4. Heber U, Schmitt JM, Krause GH, Klosson RJ, Santarius KA (1981) Freezing damage to thylakoid membranes in vitro and in vivo. In Morris GJ, Clarke A, eds. Effects of low temperatures on biological membranes, pp. 263-283. London/New York/Toronto/Sydney/San Francisco: Academic Press.

5. Heber U, Volger H, Overbeck V, Santarius KA (1976) Membrane damage and protection during freezing. In Fennema O, ed. Proteins at low temperatures. Advan Chem Ser 180: 159-189.

6. Levitt J (1980) Responses of plants to environmental stresses, 2nd Ed, Vol 1. New York/London/Toronto/Sydney/San Francisco: Academic Press.

7. Lineberger RD, Steponkus PL (1980) Cryoprotection by glucose, sucrose, and raffinose to chloroplast thylakoids. Plant Physiol 65: 298-304.

8. Lineberger RD, Steponkus PL (1980) Effects of freezing on the release and inactivation of chloroplast coupling factor 1. Cryobiology 17: 486-494.

9. Lovelock JE (1953) The mechanism of the protective action of glycerol against haemolysis by freezing and thawing. Biochim Biophys Acta 11: 28-36.

10. Mazur P (1969) Freezing injury in plants. Annu Rev Plant Physiol 20: 419-448.

11. Mazur P (1981) Fundamental cryobiology and the preservation of organs by freezing. In Karow AM, Pegg DE, eds. Organ preservation for transplantation, 2nd Ed, pp. 143-175. New York/Basel: Marcel Dekker, Inc.

12. Mazur P, Rall WF, Rigopoulos N (1981) Relative contributions of the fraction of unfrozen water and of salt concentration to the survival of slowly frozen human erythrocytes. Biophys J 36: 653-675.

13. Meryman HT, Williams RJ, Douglas MSJ (1977) Freezing injury from "solution effects" and its prevention by natural or artificial cryoprotection. Cryobiology 14: 287-302.

14. Santarius KA (1969) Der Einfluß von Elektrolyten auf Chloroplasten beim Gefrieren und Trocknen. Planta 89: 23-46.

15. Santarius KA (1971) The effect of freezing on thylakoid membranes in the presence of organic acids. Plant Physiol 48: 156-162.

16. Santarius KA (1982) Cryoprotection of spinach chloroplast membranes by dextrans. Cryobiology 19: 200-210.

17. Santarius KA (1982) The mechanism of cryoprotection of biomembrane systems by carbohydrates. In Li PH, Sakai A, eds. Plant cold hardiness and freezing, Vol 2. New York: Academic Press (in press).

18. Santarius KA, Bauer J (1982) Cryopreservation of spinach chloroplast membranes by low molecular weight carbohydrates I. Evidence for cryoprotection by a noncolligative-type mechanism. Cryobiology (in press).

19. Santarius KA, Giersch Ch (1982) Cryopreservation of spinach chloroplast membranes by low molecular weight carbohydrates II. Discrimination between colligative und noncolligative protection. Cryobiology (in press).

20. Santarius KA, Heber U (1970) The kinetics of the inactivation of thylakoid membranes by freezing and high concentrations of electrolytes. Cryobiology 7: 71-78.

LOW TEMPERATURE EFFECTS ON PHOTOSYNTHESIS IN CONIFERS

GUNNAR ÖQUIST

Department of Plant Physiology, University of Umeå, S-901 87 Umeå, Sweden

ABSTRACT

In the Northern temperate region, and at high altitudes, conifers are exposed to quite severe climatic conditions during the winter; i.e. low temperatures below zero, often in combination with bright sunlight. This winter stress inhibits photosynthesis and is considered to be mediated by photo-oxidation that occurs under conditions when photosynthesis is largely inhibited by low temperatures. As briefly discussed in this communication the effects of winter stress on photosynthesis of *Pinus silvestris* are reflected in a number of ways in the function and organization of chloroplasts. Particular attention is paid to identifying the sites of inhibition in the photosynthetic electron transport chain. Analysis of electron transport and fluorescence kinetic studies at room temperature have shown that winter stress inhibits electron transport on both the oxidizing and on the reducing side of photosystem II. The inhibition is most pronounced on the reducing side after the primary electron acceptor. The electron transport over photosystem I seems to be relatively resistant to winter stress although a partial inhibition is also evident here. This may partly be related to the about 40% decrease of the fraction of absorbed energy (α) that is transferred directly to photosystem I, as studied at 77 $^\circ$K by fluorescence kinetics.

INTRODUCTION

As conifers can acquire deep frost hardiness and retain their needles for several years, they are of particular interest in studies of the effects of low, including subzero, temperatures on photosynthesis. Conifers kept under prolonged frost hardening conditions (short days and a few degrees above zero) show a reduced capacity for net photosynthesis (2). This depression was studied in more detail in frost hardened *Pinus silvestris* (17). It was shown that frost hardening caused a reduction only in the rate of light saturated net photosynthesis, whereas the quantum yield under light

limiting conditions was unaffected. The partial inhibition of light saturated photosynthesis in frost hardened pine is also expressed in a reduced
electron transport capacity over plastoquinone (13) and in a decreased
activity of ribulose-1,5-bisphosphate carboxylase *in vitro* (6).

In the Northern temperate region, and at high altitudes, conifers are
exposed to quite severe climatic conditions during the winter; i.e. subzero
temperatures often in combination with bright sunlight. This winter stress
inhibits photosynthesis to a much greater extent than observed during prolonged frost hardening at low, above zero, temperatures. Complete inhibition
of the potential for net photosynthesis has for example frequently been
observed in conifers exposed to severe winter conditions (8, 20). Tranquillini
(21) showed that the rate of net photosynthesis of *Pinus cembra* depended on
the degree of frost during the proceeding night, and several days or weeks
at above zero temperatures were needed for the recovery of photosynthesis
in frost treated *Pseudotsuga menziesii* and *Pinus ponderosa* (19).

Photosynthetic electron transport studies in chloroplasts isolated from
needles collected in a natural stand of *Pinus silvestris* showed that the
potential for electron transport varied approximately in parallel to that
reported for seasonal variations in the capacity for net photosynthesis (10,
12). The dominating site for inhibition, partial or complete, was ascribed
to the function of the plastoquinone pool linking the two photosystems,
whereas the partial photosystems, particularly photosystem I, was less
affected (10, 14).

Winter inhibition of photosynthesis in conifers also occurs in parallel
with a partial chlorophyll bleaching in the needles (22). Results from
sodium dodecyl sulphate polyacryl amide gel electrophoresis (SDS-PAGE) of
chlorophyll-protein complexes, fluorescence analyses at 77 $^{\circ}$K and analyses
of chlorophyll $\underline{a}/\underline{b}$ ratios have made us conclude that winter stress of *Pinus
silvestris* causes a more pronounced partial destruction of the reaction
center antennae, containing only chlorophyll $\underline{a}$, than of the light-harvesting
chlorophyll $\underline{a}/\underline{b}$ complex (10, 15). Comparative SDS-PAGE analyses of the polypeptide distribution pattern of chloroplast thylakoids revealed no major
differences between active and winter stressed pine (16). Winter stress,
however, caused a marked decrease in the level of unsaturation of monogalactosyl diglyceride in the thylakoids of pine (12). As the saturation
level of monogalactosyl diglyceride correlated very well with the seasonal
variation in the electron transport capacity over plastoquinone it was hypo-

thesised that the structure of the acyl lipids affects the function of the lipophilic plastoquinone pool.

Martin et al. (9) showed that winter inhibition of the photosynthetic electron transport occurred much faster and to a greater extent when pine was exposed to freezing temperatures at high than at low irradiances. From this and other experiments on chlorophyll bleaching (15) we have concluded that exposure of pine to freezing temperatures (but above the frost killing temperature) in light causes inhibition of photosynthesis through photo-inhibition and eventually photo-oxidation of membrane components (11). The effect of low temperature therefore primarily consists in creating conditions for photo-oxidation; i.e. slowing down the rate of photosynthesis so that the photosynthetic apparatus becomes overexcited to such an extent that the protective mechanisms can no longer prevent photodynamic damages from occuring.

In this communication I give some further data on the sites of winter stress induced electron transport inhibition and on the excitation energy distribution to the two photosystems of pine.

MATERIALS AND METHODS

Current year needles were collected from an about 20 year-old natural stand of *Pinus silvestris* L. (Umeå N63°50' E20°20', altitude 15 m) on April 16, 1982. This is the period of the year when, on this latitude, the effects of winter stress are most pronounced; recovery begins in the beginning of May (10). The needles were used for analyses immediately and also after storage in plastic bags in darkness at -18 °C until repetition of the experiments. No effects of this storage were observed in the studied parameters. Active, secondary needles were collected from one-year-old seedlings of *Pinus silvestris* grown on nutrient solutions in climate chambers as described before (14).

Chloroplasts were isolated according to Martin et al. (9) and the photo-reactions $H_2O \rightarrow 2,6$ dichlorophenol indophenol (DPIP; PSII), $H_2O \rightarrow NADP$ (PSII + PSI) and Na-ascorbate/DPIP $\rightarrow$ NADP (PSI) were studied as before (14). Chlorophyll was determined in 80% acetone (1). Fluorescence kinetics of PSII was studied at 685 nm at room temperature or of PSII (693 nm) and PSI (729 nm) concomitantly at 77 °K. A three-branched fiber optic based spectro-fluorometer constructed basically according to (4) was used.

RESULTS AND DISCUSSION

Table I shows the light saturated electron transport properties of active and winter stressed pine chloroplasts. The results essentially confirm earlier findings, although the extent of winter inhibitions may vary from year to year (9,10,12,14). The over all electron transport from H_2O to NADP is inhibited more severely than are the partial reactions of PSI and PSII; PSI is generally less affected than is PSII. It is thus obvious that the primary site of winter inhibition must be somewhere in the electron transport chain linking PSII to PSI. As no inhibition was found in the electron transport between cytochrome $\underline{f}$ and P700 (14) it is concluded from the electron transport studies that the primary target for winter inhibition of photosynthetic electron transport is associated with the function of the plastoquinone pool. In fact, fluorescence kinetic studies with and without the photosynthesis inhibitor DCMU (Fig. 1), shows that winter stress like DCMU inhibits the reoxidation of the primary electron acceptor Q of PSII; i.e. the inhibition is on the reducing side of PSII.

Table 1. Light saturated photoreduction rates of DPIP and NADP ($\mu mol \cdot mg^{-1}$ $Chl \cdot h^{-1}$) in chloroplasts isolated from active and winter stressed *Pinus silvestris*. The rates were measured at room temperature.

Plants	$H_2O \rightarrow$ DPIP (PSII)	Asc/DPIP $\rightarrow$ NADP (PSI)	$H_2O \rightarrow$ NADP (PSII + PSI)
Active	70.1	29.0	23.4
Winter stressed	37.0	18.8	7.5
Active/Stressed	1.9	1.5	3.1

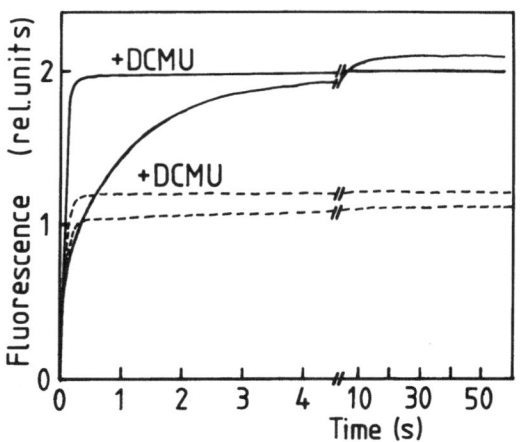

FIGURE 1. Fluorescence kinetics (±DCMU) at 685 nm (half band width 6.6 nm) of chloroplasts isolated from active and winter stressed *Pinus silvestris*. Excitation was with a broad band filter, 380–590 nm (peak at 525 nm), 230 μmol quanta m^{-2} s^{-1}.

———— active
– – winter stressed

As winter stress largely is thought to be mediated by photoinhibition, and eventually photo-oxidation, at subzero temperatures unfavourable for photosynthesis (11), it is of interest to compare the sites of winter stress inhibition with those observed under other photoinhibitory conditions. Photoinhibition of shade plants, CO_2-depleated plants and chilled plants (18) all differ from winter stressed pine by having the most severe inhibition on the oxidizing rather than on the reducing side of PSII. Whether the partial winter inhibition observed on the oxidizing side of PSII, and the stronger and sometimes complete inhibition on the reducing side of PSII of pine, depend on photoinhibition above or below freezing or on the duration of the photoinhibitory treatment, is currently under investigation in our laboratory.

As our earlier studies of winter stress effects on photosynthesis in pine also revealed partial bleaching of chlorophyll (10,15) it was of interest to investigate how winter stress affects energy distribution to the two photosystems. For this purpose the energy distribution model of Butler (3) and coworkers, based on the characteristics of the specific PSII (693 nm) and PSI (729 nm) emission bands at 77 $^{\circ}$K, has been applied. Fig. 2 shows the PSI fluorescence (F729) as a function of the PSII fluorescence (F693) as the fluorescence of whole needles or isolated chloroplasts increased from F_o (open traps) to F_m (closed traps) upon exciting mainly chlorophyll b at 77 $^{\circ}$K. The straight lines obtained demonstrate that the variable (F_v) part of F729 is due entirely to energy transfer from PSII to PSI; i.e. "spillover" (7). The results of winter stress on the fluorescence yield changes during PSII-trap closure were similar for whole needles and isolated chloroplasts. The problem with reabsorption of the short wavelength fraction of the fluorescence light was minimized by using a very diluted chloroplast suspension (3 μg chlorophyll/ml and an optical pathlength of 1 mm).

Winter stress quenched F_v for both PSII (2.8 times) and PSI (1.9 times) considerably, whereas F_o decreased less (1.3 and 1.4 times for PSII and PSI, respectively). The quenching of F_v is attributed to a winter stress induced partial disorganization on the oxidizing side of PSII so that the reaction center P680 accumulates in its oxidized state, thereby quenching the variable fluorescence normally obtained upon reduction of the primary electron acceptor Q. This type of quenching has been called reaction center quenching (3) and it is typical for photoinhibited leaves damaged on the

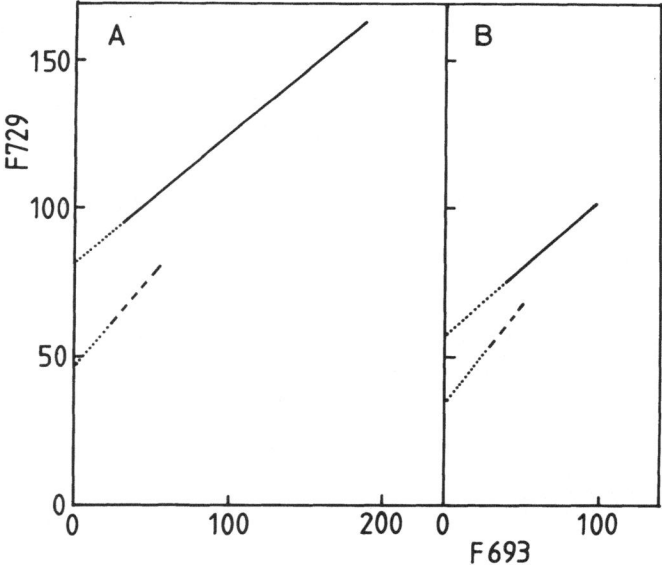

FIGURE 2. F693 (half band width 6.6 nm) vs. F729 (half band width 13 nm) of needles (A) and chloroplasts (B) of active (—) and winter stressed (- -) *Pinus silvestris*. The dotted lines are extrapolations. Excitation was on chlorophyll b mainly, using a 478 interference filter with a half band width of 10 nm; 8 mol quanta m^{-2} s^{-1}. The equations of the lines were for active and winter needles y = 0.44x + 80.5 (r^2=0.997) and y = 0.62x + 46.2 (r^2=0.993), respectively, and for active and winter stressed chloroplasts y = 0.45x + 57.5 (r^2=0.991) and y = 0.67x + 34.8 (r^2=0.954), respectively.

oxidizing side of PSII, probably very close to the reaction center (5). Note that the winter stress induced inhibition on the reducing side of PSII is not revealed in these experiments since the electron transport becomes inhibited after Q at 77 ^{O}K.

The level of F_o at 729 nm is composed of two parts; one part is due to the fraction of absorbed energy (α) transferred directly to PSI and another part is due to the energy transfer from PSII to PSI. A relative value of α is obtained by extrapolating the straight lines of Fig. 2 back to the Y-axis (3,7). It was found that winter stress decreased α with about 40% in both needles and isolated chloroplasts. This finding is consistent with an earlier observation that winter stress causes the formation of a chlorophyll fraction that no longer can photo-oxidize P700 of PSI (14), which findings may be related to the partial winter inhibition of photosystem I (Table 1).

It can furthermore be derived that the slopes of the functions shown in

Fig. 2 are approximately equal to F729/F693 = $k_{T(II \rightarrow I)}$ x $\emptyset F729/k_{F693}$ (3).
How the winter stress induced increase of the slope (1.5 times) depends
on these three factors is currently under investigation.

REFERENCES

1. Arnon DI. 1949. Copper enzymes in isolated chloroplasts. Polyphenol-
   -oxidase in *Beta vulgaris*. Plant Physiol 24:1-15.
2. Bauer H, Larcher W and Walker RB. 1975. Influence of temperature stress
   on CO_2-gas exchange. In Cooper JP, ed. Photosynthesis and productivity
   in different environments, pp. 557-586. Cambridge University Press.
3. Butler WL. 1978. Energy distribution in the photochemical apparatus
   of photosynthesis. Ann Rev Plant Physiol 29:345-378.
4. Fork DC, Ford GA and Catanzaro B. 1979. Measurements with a microprocessor-
   -based fluorescence spectrophotometer made on the blue-green alga *Anacystis nidulans* above and below the phase transition temperature. Carnegie
   Inst Year Book 78:196-199.
5. Fork DC, Öquist G and Powles SB. 1981. Photoinhibition in bean: a fluorescence analysis. Carnegie Inst Year Book 80: in press.
6. Gezelius K and Hallén M. 1980. Seasonal variation in the ribulose bisphosphate carboxylase activity in *Pinus silvestris*. Physiol Plant 48:
   88-98.
7. Kitajima M and Butler WL. 1975. Excitation spectra for photosystem I
   and photosystem II in chloroplasts and the spectral characteristics of
   the distribution of quanta between the two photosystems. Biochim Biophys
   Acta 408:297-305.
8. Linder S and Troeng E. 1980. Photosynthesis and transpiration of 20-year-
   -old Scots pine. In Persson T, ed. Structure and function of northern
   coniferous forests - an ecosystem study, Ecological Bulletins 32:165-181.
   Swedish Natural Science Research Council, Stockholm.
9. Martin B, Mårtensson O and Öquist G. 1978. Effects of frost hardening
   and dehardening on photosynthetic electron transport and fluorescence
   properties in isolated chloroplasts of *Pinus silvestris*. Physiol Plant
   43:297-305.
10. Martin B, Mårtensson O and Öquist G. 1978. Seasonal effects on photo-
    synthetic electron transport and fluorescence properties in isolated
    chloroplasts of *Pinus silvestris*. Physiol Plant 44:102-109.
11. Öquist G. 1981. Chloroplast structure and photosynthetic efficiency.
    In Johnson CB, ed. Physiological processes limiting plant productivity,
    pp. 53-80. Butterworths.
12. Öquist G. 1982. Seasonally induced changes in acyl lipids and fatty
    acids of chloroplast thylakoids of *Pinus silvestris*. A correlation be-
    tween the level of unsaturation of monogalactosyl diglyceride and the
    rate of electron transport. Plant Physiol 69:869-875.
13. Öquist G and Hellgren NO. 1976. The photosynthetic electron transport
    capacity of chloroplasts prepared from needles of unhardened and hardened
    seedlings of *Pinus silvestris*. Pl Sci Lett 7:359-369.
14. Öquist G and Martin B. 1980 Inhibition of photosynthetic electron trans-
    port and formation of inactive chlorophyll in winter stressed *Pinus
    silvestris*. Physiol Plant 48:33-38.
15. Öquist G, Mårtensson O, Martin B and Malmberg G. 1978. Seasonal effects
    on chlorophyll-protein complexes isolated from *Pinus silvestris*. Physiol
    Plant 44:187-192.

218

16. Öquist G, Martin B, Mårtensson O, Christersson L and Malmberg G. 1978. Effects of season and low temperature on polypeptides from thylakoids isolated from chloroplasts of *Pinus silvestris*. Physiol Plant 44:300-306.
17. Öquist G, Brunes L, Hällgren J-E, Gezelius K, Hallén M and Malmberg G. 1980. Effects of artificial frost hardening and winter stress on net photosynthesis, photosynthetic electron transport and RuBP carboxylase activity in seedlings of *Pinus silvestris*. Physiol Plant 48:526-531.
18. Osmond CB. 1981. Photorespiration and photoinhibition. Some implications for the energetics of photosynthesis. Biochim Biophys Acta 639: 77-98.
19. Pharis RP, Hellmers H and Schuurmans E. 1970. Effects of subfreezing temperatures on photosynthesis of evergreen conifers under controlled environments. Photosynthetica 4:273-279.
20. Pisek A and Winkler E. 1958. Assimilationsvermögen und Respirationen der Fichte (*Picea excelsia* Link.) in verschiedener Höhenlage und der Zirbe (*Pinus cembra* L.) an der alpinen Waldgrenze. Planta 51:518-543.
21. Tranquillini W. 1957. Standortsklima, Wasserbilanz und CO_2-Gaswechsel junger Zirben (*Pinus cembra* L.) an der alpinen Waldgrenze. Planta 49: 612-661.
22. Tranquillini W. 1964. The physiology of plants at high altitudes. Ann Rev Plant Physiol 15:345-360.

PLANT REACTION TO HEAT STRESS AT LOW OXYGEN AND HIGH CO_2-CONCENTRATION

F. LENZ, P. KORNKAMHAENG and H.-G. LEVIN
Institut für Obstbau und Gemüsebau der Universität Bonn, D-5300 Bonn, FRG

1. ABSTRACT

To eliminate virus from fruit trees, plants have to be maintained several weeks at temperatures between 37-44°C for heat treatment.

In particular, heat sensitive fruit species die off after a few days at such extreme condition. However, reducing photo-inhibition and photorespiration by altering O_2 and CO_2 levels allow plants to survive for longer periods.

Measurements of photosynthesis, respiration and water consumption of heat treated plants grown at different O_2 and CO_2 concentration are described.

2. INTRODUCTION

For elimination of virus from fruit trees plants are kept for about 3-6 weeks at temperatures of 37-39°C (2). By this, mobility and synthesis of virus is reduced. Therefore new growing shoot tips can be free of virus (7). Unfortunately, plants, particularly of prunus species, are sensitive to heat (10) and die off after a few days of treatment. Moreover, there are heat resistent virus in these species for elimination of which even temperatures of 40°C and higher for at least 3 weeks would be required. During such treatment plants should not only survive but also produce new shoots, which might be virus free.

It was suggested by Kriedemann et al. 1976 that for maintaining plants at extreme high temperatures CO_2 should be enriched for increasing photosynthetic rates. Therefore in a preliminary experiment with Capsicum annuum at high temperatures and later with small plants of Prunus avium, CO_2 and O_2 levels were altered to maintain a positive carbon balance even at 40°C.

3. MATERIAL AND METHODS

3.1. Plants

For the expermiments, young plants of Capsicum annuum, var. 'Bell Boy' and Prunus avium, var. 'F 12/1' were used.

The <u>Capsicum</u> plants were 25 cm in size. Fruit set was prevented by flower removal. The <u>Prunus avium</u> plants were 40 cm high with 5 side branches. All plants were grown in 5 l plastic containers filled with quartz sand (0.7-1.2 mm) and watered three times daily with 300 ml of a modified Hoagland solution.

3.2. Heat treatment

For heat treatment 4 plants were placed into each of three growth chambers (90x80x80 cm) in which temperature, light (HQ IL-400 W lamps), humidity, CO_2 and O_2 were controlled. The conditions for the experiments are indicated in the graphs. The plants were continuously supplied with nutrient solution through a drip system. Surplus solution drained back into a storage tank. From the water level in a calibrated glass tube outside the tank it was possible to calculate the water consumption of the plants. The plants were kept up to 4 weeks at 38°C and 40°C respectively.

3.3. Measurements

CO_2 uptake under illumination was monitored by automatic compensation and dark respiration was calculated from the rate of CO_2 increase in the growth chamber (5). At the end of the experiment the total leaf area was measured with a Li-cor area meter. The plants were separated into roots, stems, leaves for determination of dry matter. Starch contents in leaves and stems were analysed according to (4).

4. RESULTS AND DISCUSSION

4.1. CO_2-Effect

In a first experiment the effect of high CO_2-levels (900 ppm CO_2) as compared to 350 ppm was used to prolong plant life at 38°C.

Before heat treatment plants were adapted to higher temperatures 25°, 30° and 35°C respectively for about 4 hours.

In Fig. 1, CO_2 effects are shown on photosynthesis and dark respiration of the green pepper plants grown at 38°C. At 350 ppm CO_2-uptake steeply declined after 4 days while dark respiration remained at similar levels.

After 9 days, respiration was higher than CO_2-uptake. Initially the photosynthetic rates of plants were considerably higher at 900 ppm CO_2. However after five days of treatment CO_2-uptake became similar to the control plants.

Parallel to photosynthesis, water consumption of plants quickly declined after four to five days (Fig. 2). Despite the larger leaf area, plants at 900 ppm CO_2 used less water than control plants.

It can be assumed that stomatal sensitivity to increased CO_2-levels as observed in many plant species (6, 9, 11) also functioned at the high temperatures resulting in higher stomatal resistance. As indicated in Fig. 3, the plants had high starch contents in leaf blades at the end of the experiment, particularly at the high CO_2 level.

It also appeared that leaf symptoms with CO_2 enrichment were similar to those after ethylene treatment. Unfortunately ethylene was not measured but recent findings with sunflower plants (1) indicate that high CO_2 concentrations enhance ethylene production. Heat treatment may have even further promoted ethylene release and this may be a reason for the early decline of the green pepper plants.

Plants grown in CO_2 enriched atmosphere produced more total dry matter as compared to control plants due to the initially increased photosynthetic rates (Fig. 4).

4.2. Reduction of O_2-level

In a second experiment, the O_2-level was changed at heat treatment to reduce photorespiration and by this to prolong a positive carbon balance of the plants.

In Fig. 5, it is again demonstrated that at a normal atmosphere, high temperature resulted in a rapid decrease of photosynthesis and a negative CO_2-balance of the plants after 14 days.

In contrary, CO_2-uptake remained at a relatively high level throughout the experiment when O_2 was reduced to 5 %.

The dark respiration of plants was not clearly affected and altered by the O_2-treatment.

Plants of low oxygen level had produced considerably more dry matter than control plants (Fig. 6).

4.3. CO_2 and O_2 effects

Both enrichment of CO_2 and lowering of the O_2 level resulted in higher CO_2-uptake of plants, at least for a short period of time.

Therefore, in a third experiment with plants of Prunus avium, effects of a controlled atmosphere - 900 ppm CO_2, 5 % O_2 - as compared to normal atmosphere were measured on photosynthesis and dark respiration at 40°C.

The plants were adapted to 40°C by increasing the temperature from 25°C to

40°C by 5°C daily. From the beginning on plants grown at 900 ppm CO_2 and 5 % O_2 had considerably higher photosynthetic rates than the control plants.

It is shown in Fig. 7 that under normal atmospheric conditions photosynthetic rates of plants decreased rapidly and the plants died off after a few days. Treated plants stayed alive for 4 weeks and even produced side shoots. The photosynthetic rates dropped to a level of about 2.5 mg, but recovered after 16 days of treatment.

Herewith it was shown that by CO_2-enrichment and reduced level of oxygen heat sensitive plants like Prunus avium can be maintained during heat treatment for virus elimination. Moreover, growth can be stimulated even at the extreme high temperature of 40°C. From experiments with green pepper plants it can be concluded that the reduction of O_2-level to 5 % may be more important for plant survival.

Both enrichment of CO_2 and low oxygen concentration suppress photorespiration, which is supposed to be increased by the high temperature (8). But moreover temperature related inhibition of photosynthesis (3) may be strongly reduced by lowering the oxygen concentration.

REFERENCES

1. Bassi PK, Spencer MS (1982) Effect of carbon dioxide and light on ethylene production in intact sunflower plants. Plant Physiol 69 : 1222–1225.
2. Baumann G. (1973) Wichtige Viruskrankheiten des Kern- und Steinobstes - Erkennung und Verhütung. Berlin und Hamburg, Paul Parey.
3. Björkman A, Badger MR, Armond PA (1980) Response and adaptation of photosynthesis to high temperatures. In Turner NC and Kramer PK eds. Adaptation of plants to water and high temperature stress. pp 233–249. New-York : John Wiley and Sons.
4. Claussen W (1975) Untersuchungen über den Einfluss der Frucht auf die Netto-Photosyntheseraten und den Saccharose- und Stärkestoffwechsel der Blätter und Wurzeln von Auberginen (Solanum melongena L.).
5. Daunicht HJ (1970) Ein Verfahren zur exakten automatischen Photosynthese-kompensation. Ber Dt Bot Ges 83 : 499.
6. Heath OVS (1948) Control of stomatal movement by a reduction in the normal carbon dioxide content of the air. Nature (London) 1961 : 179–181
7. Kassanis B (1980) Therapy of virus-infected plants and the active defence mechanism. Outlook on Agriculture 10 : 288–292.
9. Mansfield TA, Davies WJ (1981) Stomata and stomatal mechanisms. In : Paleg LG and Aspinall D eds. The physiology and biochemistry of drought resistance in plants. pp 315–346. New-York, Academic Press.
10. Naumann G (1978) Virus diseases in fruit growing and possible ways of controlling them. Plant research and development 7 : 108–118.
11. Raschke K (1975) Stomatal action. Ann Rev Plant Physiol 26 : 309–340.

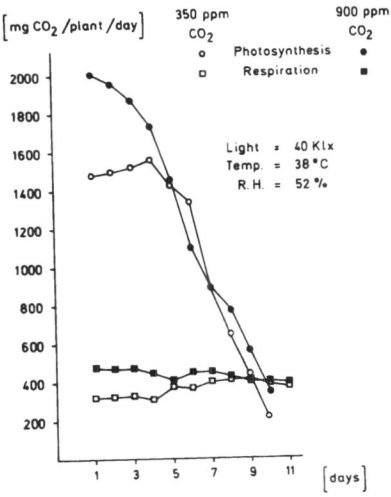

Figure 1. Effect of CO_2-enrichment on CO_2-uptake and respiration of green pepper plants during heat treatment.

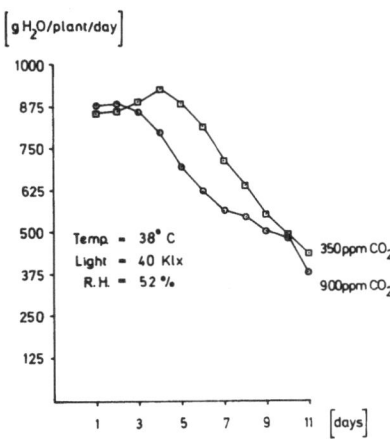

FIGURE 2. Water consumption of green pepper plants as affected by CO_2-concentration during heat treatment.

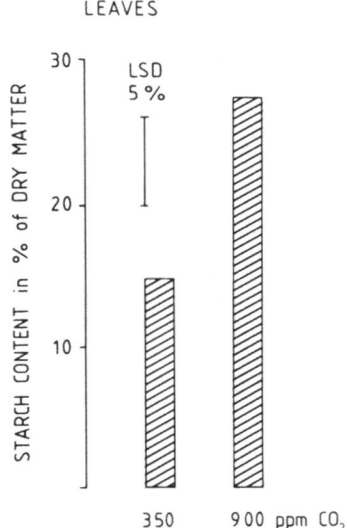

FIGURE 3. Starch content of green pepper leaves as affected by CO_2-treatment.

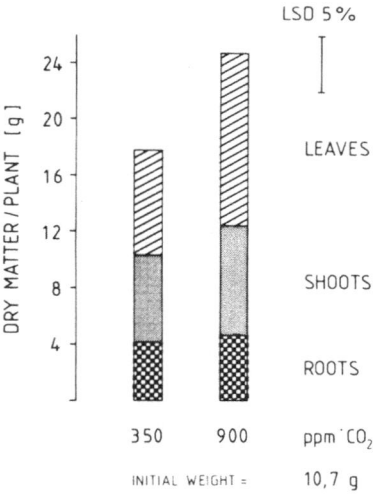

FIGURE 4. Dry matter of green pepper plants as affected by CO_2-concentration.

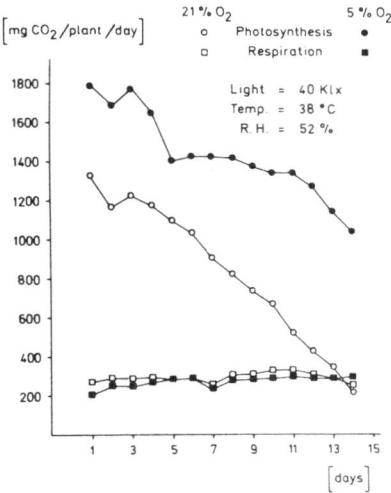

FIGURE 5. Effect of low oxygen concentration on CO_2-uptake and respiration of green pepper plants during heat treatment.

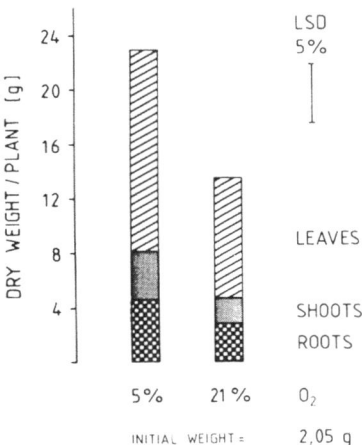

FIGURE 6. Dry matter of green pepper plants as affected by O_2-concentration.

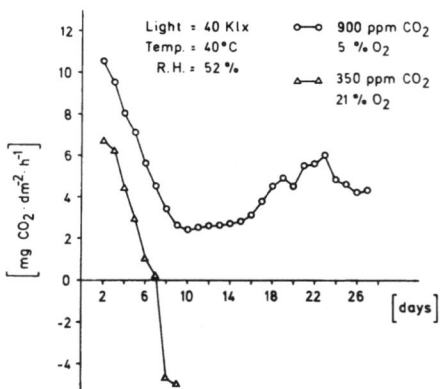

FIGURE 7. Photosynthetic rates of <u>Prunus avium</u> plants during heat treatment as affected by controlled atmosphere.

CHILLING-INDUCED INHIBITION OF PHOTOSYNTHESIS IN TOMATO

DONALD R. ORT AND BJORN MARTIN
Department of Botany, USDA/ARS
University of Illinois, Urbana, Illinois, 61801

INTRODUCTION

Photosynthesis, in a fairly diverse group of chilling-sensitive plants, is inhibited for an extended period by brief exposure to low temperature ($>0°<12°$). Dramatic reductions in the rate of whole plant photosynthesis have been documented in numerous plants (1-5) including several of the commercially most significant crop plants of temperate North America. Although it is clear that chilling of thermophilic plants in the dark results in inhibition of photosynthesis which is often severe and persistent, there has been no general agreement in the literature regarding the site of this inhibition (1,2,4,6,7,8) (i.e. at the level of stomatal control of gas exchange or at the level of the chloroplast). As a consequence a consensus regarding the underlying physiological and biochemical basis of the inhibition has not been possible.

It is clearly important to identify the contributions which partial stomatal closure and impaired chloroplast activities make to the overall inhibition of photosynthesis by chilling. In this paper we have summarized results obtained from recent experiments conducted on attached tomato leaves (*Lycopersicon esculentum* Mill. cv Rutgers and Floramerica) which enabled us to demonstrate that the effects of dark chilling on stomatal CO_2 conductance are relatively minor compared to the more direct effects of chilling on chloroplast activities.

There is a good deal of experimental evidence indicating that chloroplasts isolated from prechilled leaves of sensitive plants have impaired electron transport. Kaniuga and associates (9,10) have identified water oxidation as the chill-labile process in tomato chloroplasts supporting similar conclusions drawn earlier by others (e.g. 11-14) for a variety of plants. Consequently, it is a widely held belief that the reduced capacity for water oxidation observed in isolated chloroplasts is a major cause of the reduced photosynthesis measured in attached prechilled leaves (14). However, in this paper we have summarized three lines of evidence that have lead us to the conclusion that a significant reduction in the capacity of water oxidation does not occur in tomato due to prechilling and is not a primary element of the inhibition of photosynthetic CO_2 fixation observed in attached leaves. Furthermore, the data demonstrate that no photosystem II reaction prior to the reduction of plastoquinone can account for the inhibition observed in attached leaves.

MATERIALS AND METHODS

Conditions for the growth of the tomato plants and conditions for the chilling treatment are detailed elsewhere (15). The rate of light-saturated CO_2 fixation was measured in a closed compensating system at the specified ambient CO_2 level as described earlier (15). The quantum yield for photosynthetic CO_2 fixation in attached leaves was measured at 1500 μl/l ambient CO_2 with the same equipment as was used for measurement of light-saturated rates (16). All light intensities used for quantum yield determinations were at or below the light compensation point. Absorbed light was calculated by subtracting transmitted and reflected light from the incident light.

Procedure for the isolation of chloroplasts (intact naked lamellae) from fully expanded tomato leaves is given in reference (16).

Measurement of the concentration of active Photosystem II centers was accomplished by exciting the chloroplasts with a series of 100 saturating single turnover flashes from a xenon lamp (6 μsec half width, 3 Hz). The acidification of the medium was measured with a glass electrode (Orion 91-03)[1] and a Keithley 610C electrometer. Calibration was performed for each reaction mixture by adding a known amount of HCl as an internal standard.

RESULTS AND DISCUSSION

Figure 1B shows that the rates of both photosynthesis (trace 2) and transpiration (trace 3) measured at atmospheric CO_2 levels are dramatically reduced by a period of dark chilling (1ºC for 16 h) between days 1 and 2. It is this correlation between declining photosynthetic rates and declining transpiration rates that has led numerous investigators (1,2,4) to suggest that increased stomatal resistance to CO_2 flux is the cause of the chilling impairment of photosynthesis. However, by elevating the ambient CO_2 concentration to 1500

[1]Mention of a trademark, proprietary product, or vendor does not constitute a guarantee or warranty of the product by the U.S. Department of Agriculture or the University of Illinois and does not imply its approval to the exclusion of other products or vendors that may also be suitable.

$\mu l \cdot l^{-1}$, a saturating CO_2 concentration at the site of CO_2 reduction was attained and even under these conditions of saturating CO_2, photosynthesis was inhibited by more than 35% during the day following the chilling treatment (trace 1).

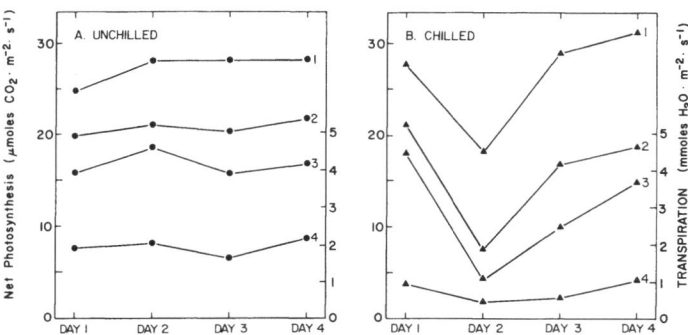

FIGURE 1. The effect of chilling on the rate of net photosynthesis and transpiration in attached tomato leaves. Measurements from a leaf prior to chilling are presented in part A, whereas in part B an attached leaf was chilled at 1°C for 16 hrs in the dark between day one and two. <u>Trace 1</u>, Net photosynthesis at 1500 $\mu l \cdot l^{-1}$ ambient CO_2; <u>Trace 2</u>, Net photosynthesis at 300 $\mu l \cdot l^{-1}$ ambient CO_2; <u>Trace 3</u>, transpiration at 300 $\mu l \cdot l^{-1}$ ambient CO_2; <u>Trace 4</u>, transpiration at 1500 $\mu l \cdot l^{-1}$ ambient CO_2. The relative humidity was 70% and the irradiance was 1800 $\mu E \cdot m^{-2} \cdot s^{-1}$.

Since photosynthesis remained depressed even at saturating levels of CO_2 it is clear that a direct inhibition of chloroplast activity accounted for a large fraction of the chilled-induced decline in photosynthesis, in fact larger than did the partial stomatal closure (Fig. 1B, cf traces 1 and 2). In order to quantitatively distinguish between the stomatal and non-stomatal contributions to the chilling impairment of photosynthesis, we investigated the dependence of photosynthesis on the intercellular CO_2 concentration (15). The reasoning behind this experiment was that any stomatal contribution to the inhibition at normal atmospheric CO_2 concentration can be overcome if any decreases in the intercellular CO_2 concentration caused by chilling are reversed by increasing the ambient CO_2 level. In other words, any remaining inhibition of photosynthesis, after readjustment of the intercellular CO_2 concentration to the level existing prior to chilling, must be due to direct impairment of one or more processes in the chloroplast. The arrows in Figure 2 show that, when the ambient CO_2 level was 300 $\mu l \cdot l^{-1}$, the intercellular CO_2 concentration decreased from 250 $\mu l \cdot l^{-1}$ to 180 $\mu l \cdot l^{-1}$ due to chilling. Even after the intercellular CO_2 level was increased to 250 $\mu l \cdot l^{-1}$ in the chilled leaf, the rate of photosynthesis remained depressed by 37%. Figure 2 (inset) also shows that the internal CO_2 concentration necessary for half maximal photosynthesis is the same before and after chilling with an apparent K_m of 170 to 180 $\mu l \cdot l^{-1}$ CO_2.

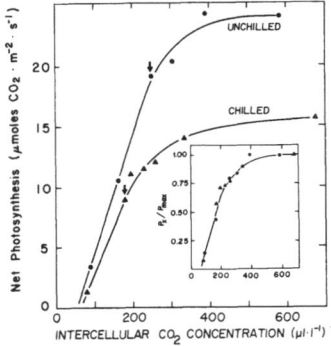

FIGURE 2. The dependence of photosynthesis of an attached chilled ($\triangle$) and an attached unchilled (O) leaf on the intercellular CO_2 concentration. In the inset the ratio of the rate of photosynthesis (P_x) at a certain intercellular CO_2 concentration to the maximum rate of photosynthesis (P_{max}) at saturating CO_2 concentrations is plotted for a leaf before and after chilling. Other conditions are the same as given for Fig. 1.

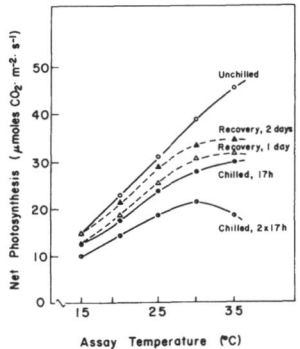 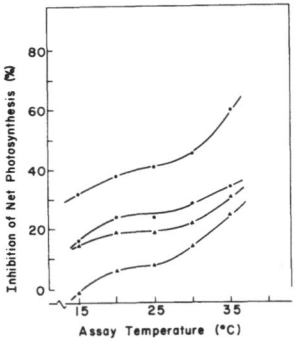

FIGURE 3. The dependence of the rate of photosynthesis on the assay temperature in a tomato plant before and after chilling. Photosynthesis was measured prior to chilling (O) after one night (17h) of chilling at 1°C (O) and after two nights (2 × 17) of chilling (O). The degree of recovery of photosynthesis was measured during the two days subsequent to the second night of chilling. Measurements were made at 1500 $\mu l/l$ ambient CO_2 and at an irradiance of 1800 $\mu E \cdot m^{-2} \cdot s^{-1}$.

FIGURE 4. The dependence of the percent of inhibition of photosynthesis by dark chilling on the assay temperature. O 17h chilling; O 2X17h chilling, $\triangle$ 1 day of recovery, $\triangle$ 2 days of recovery.

The degree of inhibition of light-and-CO_2-saturated photosynthesis brought on by dark chilling at 1°C depends both upon the temperature at which photosynthesis is assayed subsequent to the chilling period and upon the number of chilling periods. Figures 3 and 4 show that for plants grown at 32°C (day temperature) the chilling impairment of photosynthesis increased with increasing assay temperature but in a non-linear fashion. A second night of chilling approximately doubled the severity of the inhibition at all assay temperatures. The persistence of chilling impairment of photosynthesis is demonstrated by the fact that even after two days of recovery, at growth temperatures, light- and CO_2-saturated photosynthesis remained depressed by 20% when assayed at 35°C.

To summarize, the correlation between declining photosynthetic rates and declining transpiration rates shown in Fig. 1B has led numerous investigators (1,2,4) to suggest that increased stomatal resistance to CO_2 flux is the cause of the chilling-impairment of photosynthesis. Since photosynthesis in chilled tomato plants remained depressed by about 35% at saturating levels of CO_2 or when the intercellular CO_2 level was readjusted to the unchilled value, it is clear that inhibition at the level of the chloroplast also occured. It is likely that this is a general pattern of the effect of prechilling on photosynthesis in sensitive plants since there is evidence for both stomatal and nonstomatal causes for chilling impairment of photosynthesis in attached leaves of *Xanthium strumarium* (6) and *Zea mays* (17).

Figure 5A shows that maximum inhibition by chilling of CO_2-saturated photosynthesis is not observed at limiting light intensities. In fact, at incident irradiances below 200 $\mu E \cdot m^{-2} \cdot s^{-1}$, net photosynthesis in the plant before

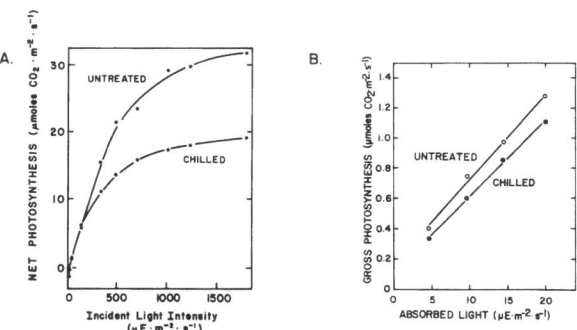

FIGURE 5. A. The dependence of CO_2-saturated photosynthesis of a chilled and unchilled attached leaf on incident irradiation. B. The effect of chilling in the dark on the quantum yield for photosynthetic CO_2 fixation in an attached tomato leaf. The ambient CO_2 concentration was 1500 $\mu l \cdot l^{-1}$ and the ambient temperature was 25°C.

and after 16 h of dark chilling was identical. That is, the apparent quantum yield for CO_2 reduction was not significantly altered by chilling even though the maximum rate was depressed nearly 40%. This observation is not in accord with the notion that impaired water oxidation capability is a significant element of chilling injury to photosynthesis.

The insensitivity of the quantum yield of CO_2 reduction to chilling was examined in greater detail (17). The absolute quantum yield of CO_2 reduction was calculated from the slope of the dependence of the CO_2 reduction rate on the amount of light absorbed (Fig. 5B). All measurements were made at saturating intercellular CO_2 levels and irradiances near or below the light compensation point thus circumventing entirely any involvement of changing stomatal conductance. The data show a decrease in quantum yield of only 10% (from 0.056 prior to chilling to 0.051 subsequent to chilling) while the maximum rate (both CO_2- and light-saturated) was inhibited nearly 25% in the same leaf. Photosystem II electron transport from the oxidation of water to the reduction of the secondary quinone acceptor B is carried out by an integral membrane polypeptide complex (18). Each of these operate independently with no electron shuttling between complexes but cooperate to reduce a common plastoquinone acceptor pool (19). The association between photosystem II reaction centers and water oxidizing enzyme complexes is such that the failure of one water-oxidizing complex leads directly to the loss of photochemical activity of one photosystem II reaction center (19). The effect that the loss of photochemical activity of a reaction center will have on the quantum yield will depend on the extent to which energy transfer between photosystem II centers can occur. Joliot et. al. (20) demonstrated that an inactive photosystem II center will not allow escape of exciton energy if the center is in a "quenching form". The inhibition of water oxidation would cause the reaction center of the photosystem II complex to be in a quenching state (20). Consequently, loss of photochemical activity due to loss of water oxidation capability should have a direct effect on the quantum yield of photosystem II turnover. On the other hand, the effect of impaired water oxidation capacity of the maximum rate of CO_2 reduction would be substantially less severe than the effect on the quantum yield of CO_2 reduction. The rate limiting step(s) in light- and CO_2-saturated photosynthesis is clearly not associated with photosystem II activity. Consequently, damage to a small fraction of the photosystem II centers would immediately be reflected in quantum yield measurements (where light is limiting) while any effect on light- and CO_2-saturated photosynthesis would be absent or greatly reduced. This prediction is born out in experiments with both attached leaves (Fig. 6) and isolated chloroplasts (data not shown, see reference 16) in which photosystem II was partially inhibited by DCMU. Figure 6 shows, that for an attached tomato leaf in which the light- and CO_2-saturated rate of photosynthesis has been inhibited by 23% by DCMU, the quantum yield of CO_2 reduction is decreased by 40%. Our data show clearly that photosynthesis impaired by chilling does not conform to what is expected of impaired water-oxidation. Instead, any effect of chilling observed on the quantum yield is small compared to the effects on maximum rates.

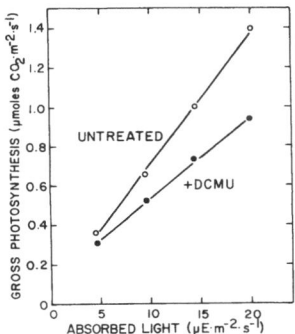

FIGURE 6. The effect of DCMU on the quantum yield for photosynthetic CO_2 fixation in an attached tomato leaf. After completion of the control measurements the attached leaf was immersed in 15 μM DCMU for 5 min at 25°C. It was then stored overnight at 25°C in the dark for measurement the following day. Experimental conditional were identical to those given for Figure 2. The quantum yield was reduced from 0.066 to 0.041 by the DCMU treatment. The rate of light- and CO_2-saturated photosynthesis was 25 μ mol $CO_2/m^2/s$ prior to treatment and 19 μmol $CO_2/m^2/s$ after the treatment with DCMU.

A second line of evidence, maximum rates of photosystem II turnover in chloroplasts isolated from unchilled or chilled leaves, demonstrates two points. First, the rate of photosystem II turnover is only marginally less in chloroplasts from chilled leaves compared to unchilled leaves (Table 1). Secondly, in all cases the rate of photosystem II turnover far exceeds the electron flux required to support the rate of light- and CO_2-saturated CO_2 reduction measured in attached leaves (Table 1).

Table 1. Comparison of the effect of prechilling on CO_2 reduction versus photosystem II activity.

The experimental conditions for the measurement of CO_2 reduction are given in Figure 1. Rates of photosystem II electron flow were determined from oxygen consumption with a Clark-type oxygen electrode due to the aeorobic oxidation of dimethyl-methylenedioxy-p-benzohydroquinone (0.25 mM) in a reaction mixture containing 50 mM HEPES-KOH (pH 8.0), 0.1 M sorbitol, 5 mM MgCl$_2$, 10 mM KCl, 0.1 mM MnCl$_2$, 1 μM DBMIB and chloroplasts containing 10 μg chlorophyll/ml.

	Ambient CO_2	Rate of CO_2 reduction	Rate of electron flux required to support CO_2 reduction	Rate of PS II electron transport	Conc of active active PS II centers
	(ul/l)	(umol CO_2/ m^2 s^1)	(mmol elec/ mol chl s)	(mmol elec/ mol chl s)	chl molecules/ active reaction center
Untreated	300	19 + 3, n = 6	115		
	1500	28 + 4, n = 10	170	245 + 15, n = 12	620 + 90, n = 46
Chilled	300	10 + 2, n = 8	60		
	1500	19 + 2, n = 7	115	215 + 25, n = 12	680 + 40, n = 6

The third line of support, measurement of the concentration of active photosystem II centers, shows there is vanishingly little difference between chloroplasts isolated from chilled versus unchilled leaves (Table I). In this experiment, proton release from the oxidation of water was measured and a reduction in the amount of protons released, calculated on the basis of total chlorophyll content, is a necessary consequence of reduced water oxidation capacity. The fact that we observed very similar values for proton release in the two sorts of chloroplasts indicates that prechilling has a correspondingly small effect on the water oxidation capacity.

CONCLUSION

We have studied photosynthesis in tomatoes subsequent to a chilling night and showed that there is a major contribution to the inhibition by a direct effect of chilling at the level of the chloroplast in addition to a smaller contribution by reduced stomatal conductance to CO_2. It has been a widely held belief that the reduced capacity for water oxidation often observed in chloroplasts isolated from sensitive plants is a major cause of the reduced photosynthesis measured in attached prechilled leaves. From our data we can conclude significant reduction in the capacity of water oxidation does not occur in tomato due to prechilling and is not a primary element of the·inhibition of CO_2 fixation observed in attached leaves. This conclusion is based on three lines of evidence, 1) measurement of quantum yields of CO_2 reduction in attached leaves and quantum yields of diiminodurene reduction by photosystem II of isolated chloroplasts, 2) measurement of maximum rates of photosystem II turnover in isolated chloroplasts, and 3) measurement of the concentration of active photosystem II centers in isolated chloroplasts.

ACKNOWLEDGEMENTS

This research was supported by USDA/CRGO Grant 79-59-2171-1-1-309-1 to D. R. Ort.

REFERENCES

1. Crookston, R.K., J. O'Toole, R. Lee, J. L. Ozbun and D. H. Wallace (1974) Crop Sci. 14, 457-464.
2. Izhar, S. and D. H. Wallace (1967) Crop Sci. 7, 546-547.
3. Pasternak, D. and G. L. Wilson (1972) New Phytol. 71, 683-689.
4. Kishitani, S. and S. Tsunoda (1974) Photosynthetica 8, 161-167.
5. Chatterton, N. J., G. E. Carlson, W. E. Hungerford and D. R. Lee (1972) Crop Sci. 12, 206-208.
6. Drake, B. G. and K. Raschke (1974) Plant Physiol. 53, 808-812.
7. Drake, B. G. and F. Salisbury (1972) Plant Physiol. 50, 572-575.
8. Peoples, T. R. and D. W. Koch (1978) Crop Sci. 18, 255-258.
9. Kaniuga, Z. and W. Michalski (1978) Planta 140, 129-136.
10. Kaniuga, Z., B. Sochanowicz, J. Zabek, and K. Krzystyniak (1978) Planta 140, 121-128.
11. Margulies, M. M. and A. T. Jagendorf (1960) Arch. Biochem. Biophys. 90, 176-183.
12. Margulies, M. M. (1972) Biochim. Biophys. Acta 267, 96-103.
13. Smillie, R. M. and R. Nott (1979) Plant Physiol. 63, 796-801.
14. Smillie, R. M. (1979) In Low temperature stress in crop plants. The role of the membrane (J. Lyons, D. Graham and J. Raison, eds.) Academic Press, New York, pp. 187-202.
15. Martin, B., D. R. Ort and J. S. Boyer (1981) Plant Physiol. 68, 329-334.
16. Martin, B. and D. R. Ort (1982) Plant Physiol. In Press.
17. Raschke, K. (1970) Planta 91, 336-363.
18. Diner, B. A. and F. A. Wollman (1980) Eur. J. Biochem. 110, 521-526.
19. Joliot, P. and B. Kok (1975) In Bioenergetics of photosynthesis (Govindjee, ed.) Academic Press, New York, pp. 387-412.
20. Joliot, P., P. Bennoun and A. Joliot (1973) Biochim. Biophys. Acta 305, 317-328.

PHOTOSYNTHESIS IN C_4 PLANTS AT LOW TEMPERATURES

S. P. LONG

1. ABSTRACT

The possible causes of the apparently characteristic low photosynthetic capacity of C_4 species at low temperatures are examined through a computer simulation of metabolite fluxes. Predictions for *Zea mays* are compared with measured steady-state rates of CO_2 uptake and O_2 evolution, and with the length of photosynthetic induction following dark-light transitions. Incorporation of the known *in vitro* changes in Pyruvate P_i dikinase activity into the simulation suggests temperature effects on photosynthetic gas exchange similar to those observed *in vivo*.

2. INTRODUCTION

In contrast to C_3 species, C_4 species show a sharp decline in photosynthetic rates of CO_2 assimilation (F_c) per unit of leaf area with decrease in temperature below 15°C. Reduction in leaf temperature from 15°C to 5°C reduced F_c by 84% in *Zea mays* L. (11) and by 76% in the cool temperate C_4 species, *Spartina anglica* Hubbard (13). *In vitro* studies have suggested three potential causes:

1. Pyruvate P_i dikinase, an enzyme unique to the photosynthetic C_4 cycle, is cold-labile showing a sharp increase in activation energy (E_a) below 11.7°C (15).

2. Phosphoenolpyruvate carboxylase (PEPc), the primary carboxylase of C_4 photosynthesis is cold-labile showing an increase in E_a below 10.8°C (16).

3. Translocation of photosynthetic intermediates between the mesophyll and bundle sheath, essential to C_4 photosynthesis, may be inhibited. Cytoplasmic streaming has been shown to decrease with temperature in a number of tropical plants (14).

This study explores the significance of these effects to steady and non-steady state photosynthetic fluxes of CO_2 and O_2 through a computer simulation of C_4 photosynthetic metabolism.

3. PROCEDURE

3.1. Materials and methods

Fully emerged second leaves of *Z. mays* cv. LG11 seedlings and mature leaves from various positions on *S. anglica* plants were used. Both species were grown at $20^{\circ}C$ in a 14h photoperiod of 0.4 mmol (photons) $m^{-2} s^{-1}$. Photosynthetic CO_2 assimilation and O_2 evolution were measured in temperature controlled chambers in an open gas-exchange system (12) and a leaf-disc electrode. (3) respectively.

3.2. Model

Fluxes between pools of metabolic intermediates as indicated in Fig. 1 were simulated. Where fluxes between two intermediates are catalysed by a single enzyme Michaelis-Menten kinetics were assumed accounting for simultaneous limitation by two substrates following the general equation: (4)

$$F \ 1 \ (VAB)/(k_{as} \ k_b + Bk_a + Ak_b + AB) \ - - - - - - - - - \ (1)$$

Where
- F = flux (mmol s^{-1})
- V = maximal velocity (mmol s^{-1})
- A = concentration of first substrate (mmol m^{-3})
- B = concentration of second substrate (mmol m^{-3})
- k_{as} = dissociation constant for enzyme-substrate complex
- k_a = Michaelis constant for substrate A
- k_b = Michaelis constant for substrate B

To correct for competitive inhibition of RubP carboxylase by oxygen, k_b of equation (1) was obtained as follows:

$$k_b = k_b^C \ (1 + O_2 /k_1^O) \ - - - - - - - - - - \ (2)$$

Where k_b^C is the k_m for CO_2 in the absence of O_2 and k_1^O is the concentration of O_2 giving half-maximal inhibition (1, 4).

Inhibition of RubP oxygenase activity was similarly accounted for. Estimates of maximal velocities and kinetic parameters for each enzyme were taken from previous publications (1, 5, 8, 9, 15, 17).

Only enzyme catalysed fluxes F_2, F_3, F_8 and F_9 were considered reversible. Other fluxes were considered irreversible either because of a large free energy change (e.g. carboxylases) or because of the rapid removal of end-products of the forward reaction (e.g. Pyruvate P_i dikinase). An Activation half-time of 60s for enzymes during dark-light transition was assumed.

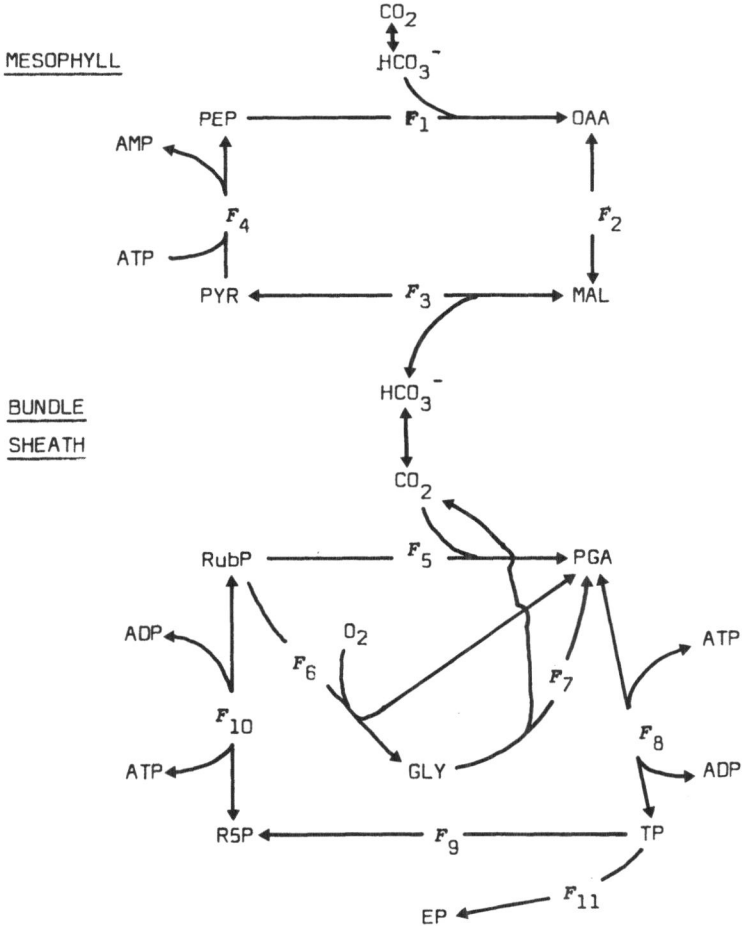

Fig. 1 The photosynthetic C_4 pathway in *Zea mays*, illustrating the metabolite pools and exchanges represented in the computer simulation. Abbreviations: PEP, Phosphoenolpyruvate; OAA, Oxaloacetate; MAL, Malate; PYR, Pyruvate; RubP, Ribulose-1:5-biphosphate; PGA, Glycerate-3-phosphate; TP, Triose phosphates; R5P, Ribulose-5-phosphate; EP, end products.

F_1	-	Phosphoenolpyruvate carboxylase (PEP_c)
F_2	-	Malate dehydrogenase
F_3	-	'Malic' enzyme, decarboxylating
F_4	-	Pyruvate P_i dikinase (PDK)
F_5	-	Ribulose biphosphate carboxylase (RubPc)
F_6	-	Ribulose biphosphate oxygenase (RubPo)
F_7	-	C_2 pathway of glycollate metabolism
F_8	-	PGA kinase/glyceraldehyde 3P dehydrogenase
F_9	-	Pathway of R5P regeneration
F_{10}	-	Ribulose-5-phosphate kinase
F_{11}	-	Starch/Sucrose synthesis

Rate constants for conversion of TP to R5P, Glycollate to Serine, and PGA to Pyruvate were estimated from half-times of label transfer in $^{14}C/^{12}C$ pulse and chase feeding experiments (5, 8, 16). Removal of intermediates into storage carbohydrates was predicted from kinetic data for ADP glucose pyrophosphorylase accounting for the increase in maximal velocity and substrate affinity effected by increase in PGA concentration (5).

The effects of temperature on maximal velocities of individual enzymes were predicted from the Arrhenius equation. Where activation energies (E_a) were unknown, 50 000 J mol^{-1} was assumed. Solubilities of O_2 and CO_2 were corrected for temperature and HCO_3^-/CO_2 ratios determined from the Henderson-Hasselbalch equation, assuming pH 7.5 for the mesophyll cytoplasm and pH 8.3 for the chloroplasts.

The simulation assumed NADPH.H to be saturating and a fixed rate of ATP regeneration, assuming a total adenylate nucleotide concentration of 1 mmol m^{-3} (1, 7). Initial concentrations for metabolites in chloroplasts were, in the absence of data for C_4 species, based on those reported (1) for *Spinacia oleracea*.

Changes in intermediate pool sizes were determined by integration of the rate equations using Gear's variable-order variable-step method (6). The program for this simulation was written in Fortran-10 and run on a DEC-10 (Digital Equipment Corp.) system. Predicted changes in pool sizes and photosynthetic O_2 and CO_2 fluxes were made at 100 ms intervals.

4. RESULTS AND DISCUSSION

The marked dependence of rates of photosynthetic gaseous exchanges on temperature in *Zea mays* is illustrated in Table 1. A reduction in rate of 84% is seen both for CO_2 assimilation (F_c) of attached leaves in normal atmospheric conditions and for O_2 evolution (F_0) of cut leaf discs in an enriched CO_2 atmosphere of 8 650 mg m^{-3}. Reduction of leaf temperature from 15°C to 5°C lengthens the induction roughly five times to about 40 minutes (Table 2).

TABLE 1

Steady-state rates of CO_2 assimilation (F_c) and O_2 evolution (F_o) per unit of leaf area at different leaf temperatures. Both rates are expressed in $\mu mol\ m^{-2}\ s^{-1}$. Measurements of F_c and F_o were made at photon flux densities of 1.80 and 1.15 $mmol\ m^{-2}\ s^{-1}$, respectively.

Leaf Temperature	Zea mays F_o	Zea mays F_c	Spartina anglica F_c
20°C	21.7 ± 2.8	18.2 ± 2.0	21.6 ± 2.1
15°C	13.6 ± 2.4	11.1 ± 1.6	15.9 ± 1.4
10°C	10.8 ± 3.4	8.2 ± 0.9	8.0 ± 0.7
5°C	2.8 ± 1.0	1.8 ± 0.6	3.9 ± 0.3

The computer simulation predicts a reduction in rate of 86-88%. The length of the induction will, beyond the 45-120s required to activate the enzymes, depend not only on the activity of the enzymes but the initial pool sizes. Since these are not accurately known for leaves of *Z. mays* realistic predictions of the actual lengths of induction cannot be made. However, given a fixed initial pool size, periods of induction at one temperature relative to another should be nearly constant. All leaves were similarly pre-treated and thus initial levels of metabolites should have been similar in all leaves. The simulation predicted a four-fold increase in the length of the induction at 5°C relative to 15°C, in close agreement to the relative change observed (Table 2).

TABLE 2

Measured and predicted rates of CO_2 assimilation (F_c), O_2 evolution (F_o), and the time taken to reach maximum rate (t) at 5°C. Results are expressed relative to their respectivee rates at 15°C.

Measured	F_c^5/F_c^{15}	F_o^5/F_o^{15}	t^5/t^{15}
Spartina anglica	0.24 ± 0.02	-	4.4 ± 1.0
Zea mays	0.16 ± 0.05	0.16 ± 0.07	5.1 ± 0.4
Predicted			
Zea mays[1]	0.12	0.14	4.0
Zea mays[2]	0.75	0.76	1.3
Zea mays[3]	0.23	0.25	3.7
Zea mays[4]	0.12	0.14	>100

[1] Assumes loss of activity of both Pyruvate P_i dikinase (PDK) and PEP carboxylase (PEPc) as predicted from *in vitro* studies.
[2] Assumes stable PDK but not PEPc.
[3] Assumes stable PEPc but not PDK.
[4] Assumes 100-fold reduction in translocation of metabolites between mesophyll and bundle-sheath.

If it is assumed in the simulation that Pyruvate P_i dikinase (PDK) is stable *in vivo* below $10^{\circ}C$ then the predicted reduction in F_c from $15^{\circ}C$ to $5^{\circ}C$ is only 25% and the increase in the length of induction, 30%. Both changes are significantly smaller than those actually observed. Isolated activities of PDK are only just sufficient to allow the *in vivo* rates of CO_2 assimilation. In the simulation loss of PDK activity below $11^{\circ}C$ results in a strong limitation to CO_2 assimilation through the slow supply of PEP whilst pyruvate accumulates, in turn leading to a large accumulation of malate through the reversibility of 'malic' enzyme activity. This is in accordance with previous ^{14}C studies of pool turnover in C_4 plants at low temperature. These have shown a marked decrease in transfer of assimilated ^{14}C from C_4 dicarboxylates to PGA (2).

Inhibition of translocation of dicarboxylates to the bundle sheath and pyruvate/alanine to the mesophyll could also account for the results obtained by ^{14}C labelling. If a 100-fold reduction in conductance of metabolite flux between the two cell types is assumed in the simulation the final steady-state photosynthetic rate is unchanged, but the length of induction is greatly increased to over 100x that at $15^{\circ}C$ (Table 2). This is because much larger concentration gradients would be built up in the pathway. In practice no evidence of a significant lengthening of the lag period beyond that predicted from known temperature effects on the enzymes was found.

If it is assumed in the simulation that PEP carboxylase remains stable *in vivo* below $10.4^{\circ}C$ then a reduction in photosynthetic rate of 75-77% is suggested. This reduction is much greater than that suggested if it is assumed that PDK is stable at low temperatures since PEP carboxylase is strongly limited by the supply of PEP.

In conclusion, the hypothesis based on *in vitro* enzymological studies that the sharp reduction in the capacity of *Zea mays* to fix CO_2 at low temperatures results from the cold lability of Pyruvate P_i dikinase is compatible with *in vivo* observations of photosynthetic CO_2 and O_2 fluxes.

5. REFERENCES

1. Barber, J. (ed.), 1976. The Intact Chloroplast. Elsevier, Amsterdam.
2. Brooking, I. R., Taylor, A. O., 1973. Plants under climatic stress V. Chilling and light effects on radio-carbon exchange between photosynthetic intermediates of *Sorghum*. Plant Physiology 52, 180-2.
3. Delieu, T., Walker, T. A., 1981. Polarographic measurement of photosynthetic O_2 evolution by leaf discs. New Phytologist 89. 165-78.
4. Fromm, H. J., 1975. Initial Rate Enzyme Kinetics. Springer-Verlag, Berlin.
5. Gibbs, M., Latzko, E., 1979. Photosynthesis II. Photosynthetic Carbon Metabolism and Related Processes. Encyclopaedia of Plant Physiology 6, Springer-Verlag, Berlin.
6. Hall, G., Watt, J. M. (eds.), 1976. Modern Numerical Methods for Ordinary Differential Equations. Clarendon Press, Oxford.
7. Hampp, R., Goller, M., Ziegler, H., 1982. Adenylate levels, Energy charge, and Phosphorylation potential during Dark-Light and Light-Dark transition in Chloroplasts, Mitochondria and Cytosol. Plant Physiology 69, 448-55.
8. Hatch, M. D., Slack, C. R., 1970. The C_4-Dicarboxylic Acid Pathway of Photosynthesis. Progress in Phytochemistry, Interscience, New York. pp. 35-106.
9. Hatch, M. D., 1981. Regulation of C_4 photosynthesis and the mechanism of light/dark modulation of Pyruvate P_i Dikinase activity. Photosynthesis IV, Balaban International, pp. 227-236.
10. Heldt, H. W., Laing, W., Lorimer, G. H., Stitt, M., Wirtz, W., 1981. On the regulation of CO_2 fixation by light. Photosynthesis IV, Balaban International, pp. 213-226.
11. Long, S. P., East, T. M., Baker, N. R., 1982. Chilling damage to Photosynthesis in Young *Zea mays* I. Effects of light and temperature variation on photosynthetic CO_2 assimilation. Journal of Experimental Botany (In press).
12. Long, S. P., Woolhouse, H. W., 1978. The responses of net photo-synthesis to vapour pressure deficit and CO_2 concentration in *Spartina townsendii (sensu lato)*, a C_4 species from a cool temperate climate. Journal of Experimental Botany 29, 567-77.
13. Long, S. P., Woolhouse, H. W., 1978. The responses of net photo-synthesis to light and temperature in *Spartina townsendii (sensu lato)*, a C_4 species from a cool temperate climate. Journal of Experimental Botany 29, 803-14.
14. Patterson, B. D., Graham, D., 1977. Effect of Chilling Temperatures on the Protoplasmic Streaming of Plants from Different Climates. Journal of Experimental Botany 28, 736-43.
15. Shirahashi, K., Hayakawa, S., Sugiyama, T., 1978. Cold Lability of Pyruvate, Orthophosphate Dikinase in the Maize Leaf. Plant Physiology, 62, 826-30.
16. Uedan, K. T., Sugiyama, T., 1976. Purification and characterization of phosphenol-pyruvate carboxylase from maize leaves. Plant Physiology 7, 906-10.

EFFECTS OF FREEZING STRESS ON PHOTOSYNTHETIC REACTIONS IN COLD ACCLIMATED AND UNHARDENED PLANT LEAVES

G.H. KRAUSE and R.J. KLOSSON

Botanisches Institut der Universität Düsseldorf,
Universitätsstraße 1, D-4000 Düsseldorf 1, Germany

ABSTRACT

Leaves of cold acclimated and unhardened plants of *Triticum aestivum*, *Viola wittrockiana*, *Tetragonia tetragonioides*, and *Hedera helix* were subjected to controlled frost treatments. After thawing, rates of photosynthetic CO_2 uptake and respiratory CO_2 release by the leaves were measured at room temperature. Simultaneously, physical parameters related to photosynthetic activities of the thylakoid membranes, such as chlorophyll fluorescence, light scattering and membrane potential-dependent absorbance changes were monitored.

According to measured rates of photosynthesis, cold acclimation led to significant frost tolerance of *Triticum*, *Viola*, and *Hedera*. The temperature inflicting 50% inhibition, T_{50}, was decreased by acclimation from about $-6^{\circ}C$ to $-15^{\circ}C$ in *Triticum* and *Viola*; T_{50} in hardened leaves of *Hedera* was about $-23^{\circ}C$. In *Tetragonia* acclimation lowered the T_{50} by maximally $2^{\circ}C$ from about -5.5 to $-7.5^{\circ}C$.

Except for *Tetragonia*, respiration was distinctly less sensitive to a freeze-thaw cycle than photosynthetic CO_2 uptake; T_{50} of respiration frequently was 1-3 degrees lower than T_{50} of photosynthesis.

In leaves of *Triticum*, *Viola*, and unhardened *Tetragonia*, the physical signals indicating light-driven processes in the thylakoids were affected at significantly lower freezing temperatures than CO_2 assimilation.

From these results, a decrease in photosynthetic CO_2 fixation presently appears as the earliest detectable sign of freezing injury in plant leaves.

1. INTRODUCTION

Former studies with hardened and unhardened spinach leaves (7,8,12) have shown that macroscopically visible freezing damage was well correlated to inhibition of photosynthetic CO_2 assimilation. To elucidate the mechanism of freeze-thaw injury to the photosynthetic apparatus, thylakoid membranes were isolated from frost-treated leaf tissue and their photosynthetic activities tested (7). In parallel, physical parameters of the whole leaves, namely chlorophyll fluorescence, light scattering at 535 nm and potential-dependent absorbance changes were measured (8). From work with isolated chloroplasts it is known that light-induced quenching of chlorophyll fluorescence measured in the absence of substrate levels of CO_2 is linearly related to the intrathylakoid proton concentration (1,13) and thus can be used as an indicator of the light-dependent proton gradient across the thylakoid membrane. Similarly, the relationship between slow photoinduced light scattering changes and the proton gradient is well established (9-11). As a further indicator of thylakoid energization served the fast absorbance change (maximum at about 520 nm) that is caused by an absorbance shift of the carotenoids due to the build-up of a membrane potential (4). A close correlation was found between changes in these physical signals of whole leaves and inactivation of thylakoids, as revealed after isolation (8,12). Moreover, it was shown that inhibition of photosynthetic CO_2 assimilation occurred at slightly higher freezing temperatures than inactivation of the energy-conserving system of the thylakoids.

In the present study, effects of freezing stress on photosynthetic reactions is investigated using several further plant species that can acquire cold hardiness to varying degrees. Since it is difficult to isolate active thylakoids routinely from those plants, the physical parameters mentioned above are used to characterize damage to thylakoids *in situ*.

2. MATERIAL AND METHODS

Plants of *Triticum aestivum* L., *Viola wittrockiana* Gams, and *Tetragonia tetragonioides* (Pall.) O. Ktze. were cultured in the greenhouse (temperatures not below 12°C). 4-6 week old plants of *Tetragonia* and *Triticum* or 6-8 week old *Viola* plants were used for the experiments. Hardening was achieved by exposing plant batches in winter (Nov. to Febr.) for 4 weeks to field conditions (*Triticum* and *Viola*) or in the greenhouse to temperatures of +3°C in the night and +10°C during daytime (*Tetragonia*). Hardy

Hedera leaves were harvested in the winter months from field-grown plants.

For frost treatment detached leaves were cooled slowly ($6^{\circ}C/h$) from $+5^{\circ}C$ to different minimum temperatures at which they remained for 2 h before warmed up at the same rate to $+5^{\circ}C$. Thawed leaves were stored for maximally 24 h at $+4^{\circ}C$ before physiological measurements were made at room temperature. Controls were kept for the whole time under these conditions. No significant decline in activities was observed during the storage.

Rates of photosynthesis in a limiting intensity of red light and of dark respiration were determined as described before (7). Photoinduced absorbance changes were monitored at 535 nm (10). Slow changes in these signals denote altered light scattering; the fast component of the light/ dark transition served as a relative measure of the steady-state membrane potential. The general experimental conditions of these measurements were as previously described for spinach leaves (8).

3. RESULTS

The temperature range of frost treatment that lies between incipient and full inhibition of photosynthetic CO_2 uptake varied with the species and usually was widened by cold acclimation (Table 1). The narrowest range was observed in *Tetragonia*, which was also the least hardy species. Like in spinach, inhibition of photosynthesis was well correlated to irrevers- ible damage of the leaves, as apparent by water infiltration of the tissue. Table 1 also gives temperatures of spontaneously starting ice formation. It can be seen that in the non-acclimated state the leaves tolerated ice formation to some degree. Acclimation to low temperatures markedly in- creased frost tolerance.

The general pattern of freezing damage in leaves of the four species, as revealed by inactivation of photosynthetic activities, was similar to that observed with spinach (7,8,12). However, except for *Tetragonia* (see below), respiratory CO_2 release was inactivated by markedly lower tempera- tures than photosynthetic CO_2 assimilation. This is shown in Fig. 1 for un- hardened and hardened wheat leaves. When the rates of photosynthesis had declined by more than 50%, respiration was still unaffected. In general, the delay in the decline of respiration was more pronounced in hardened leaves.

Measurements of physical parameters indicating energization of the thyla koid membranes *in vivo* revealed that - like in spinach but with exception

Table 1. Temperatures of cold treatment (in °C) leading to incipient and full inhibition of photosynthesis and to spontaneously beginning ice formation (determined visually). The data are from 2 experiments. In *Triticum, Viola* and *Tetragonia,* the temperature of ice formation was studied in unhardened leaves only. In spinach, hardening did not lower this temperature (7).

Leaf material		Inhibition of photosynthesis		Beginning of ice formation
		incipient	full	
Triticum	unhardened	-5.5 / -5.5	-7 / -8	-5 to -5.5
aestivum	hardened	-13 / -14.5	-18 / -19	
Viola	unhardened	-4 / -6	-8 / -12	-3 to -4
wittrockiana	hardened	-13 / -14.5	-18 / -24	
Tetragonia	unhardened	-5.5 / -5.5	-6.5/ -6.5	-3 to -4
tetragonioides	hardened	-7.5 / -7.5	-9 / -9	
Hedera helix	hardened	-20 / -24	-27 / -30	-7 to -8

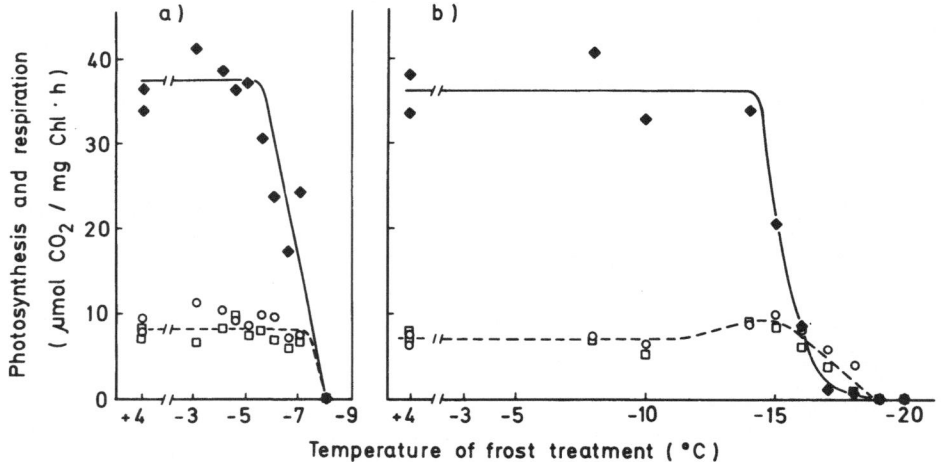

FIGURE 1. Photosynthesis and respiration of unhardened (a) and cold acclimated (b) wheat leaves. Photosynthetic CO_2 uptake (◆) was measured in air (300 µl·l⁻¹ CO_2); the actinic light (half band width 630-680 nm, intensity 45 W·m⁻²) was about half-saturating. CO_2 evolution in the dark was determined in air in the presence (□) and absence (O) of CO_2. The rates are plotted as functions of the minimum temperature of frost treatment.

of *Hedera* (see below) - the photosynthetic energy-conserving system is damaged by more severe freezing than CO_2 assimilation. Fig. 2 shows that in the steady-state of illumination the light scattering signal of the controls is large in the absence of CO_2 in the gas phase, i.e., when only limited utilization of photosynthetic energy is possible; the signal is minimal in

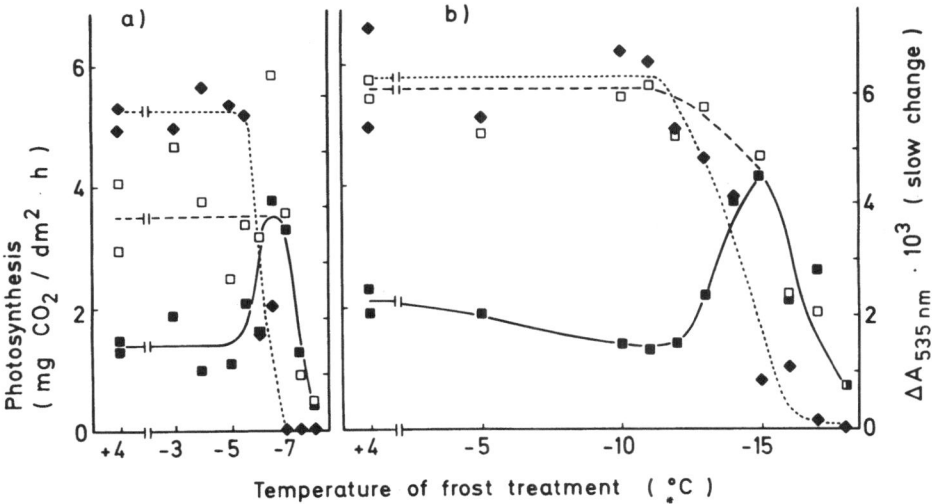

FIGURE 2. Effects of frost treatment on photoinduced light scattering changes in unhardened (a) and hardened (b) wheat leaves. The data denote the slow apparent absorbance decrease at 535 nm observed upon darkening after 2 min in the light. Signals were measured in the presence (—□—) and absence (--□--) of CO_2 in air. For comparison, steady-state rates of photo synthesis (···◆···) are given. Experimental conditions as for Fig. 1.

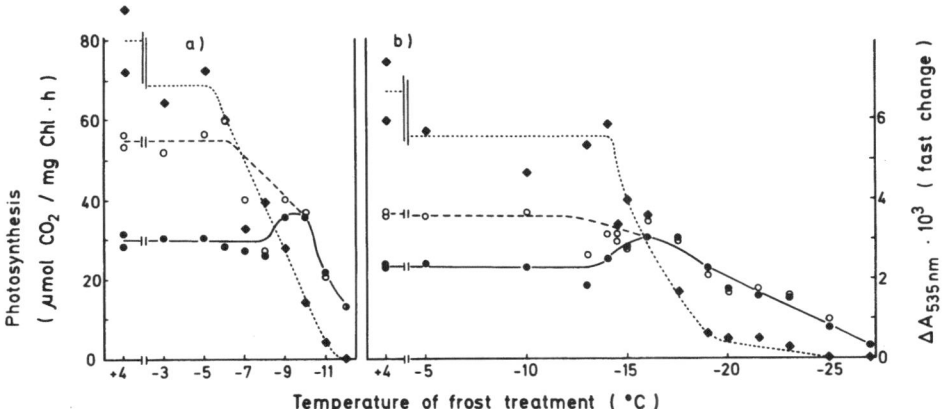

FIGURE 3. Membrane potential-dependent absorbance changes (measured at 535 nm) and rates of photosynthesis of unhardened (a) and hardened (b) leaves of *Viola* as functions of the minimum temperature of frost treatment. —●—, ΔA in presence of CO_2; --O--, ΔA in absence of CO_2. The data represent the fast component of the absorbance signal at 535 nm seen when the actinic light was switched off after 2 min of illumination. ···◆···, photosynthesis; the light intensity, 100 $W \cdot m^{-2}$, was about 80% saturating. Other conditions as for Fig. 1.

the presence of CO_2, when the energy is utilized by carbon reduction (cf. ref. 8). With progressing inhibition of CO_2 uptake by frost treatment, the size of the signal seen in the presence of CO_2 approaches that observed in the absence of CO_2. This indicates that in partially damaged leaves the utilization of photosynthetic energy is restricted by carbon metabolism. In the experiment of Fig. 2 this effect is more clearly seen in hardened leaves where there is less scattering of the data. In studies with spinach (12) it was found that in such intermediate stages of injury the ATP level in the leaves is still high. At lower freezing temperatures, the parallel decline of light scattering signals measured in the presence and absence of CO_2 denotes inactivation of the thylakoids. A very similar course of injury is seen when membrane potential-dependent absorbance changes are considered, as depicted in Fig. 3 for unhardened and hardened *Viola* leaves.

In Fig. 4, obtained with hardened *Viola*, all parameters that indicate the capacity of thylakoid energization (measured in the absence of CO_2) are affected at lower minimum temperatures of frost treatment than photo-synthesis. A delay in inhibition of thylakoid activities, however smaller, was also visible in unhardened leaves of *Tetragonia*.

Hedera helix, an evergreen woody plant, behaved differently from the herbaceous species. Acclimated *Hedera* leaves exhibited rates of photo-synthesis only slightly above respiration rates (Fig. 5). Strongly lowered rates of photosynthesis have been observed before in hardened spruce needles (15). Also, respiration was low in *Hedera*, as compared to the herbaceous plants tested. An effect of CO_2 on the physical parameters of thylakoids was not observed. Apparently, acclimation leads to a reduction in the capacity of carbon metabolism. Frost treatment affected photo-synthesis at considerably higher temperatures than respiration. As shown in Fig. 6, the physical signals that indicate energy-conserving reactions are lowered over a wide temperature range of about $15^{\circ}C$.

To summarize the results obtained with the four species investigated, the minimum temperatures inflicting 50% inhibition, T_{50}, of the various parameters are presented in Fig. 7. It can be seen that in *Triticum*, *Viola* and *Hedera* the T_{50} values of photosynthesis were usually 1-3 degrees higher than T_{50} values of respiration. In *Tetragonia*, as well as in spinach (7), both activities declined closely together; in some experiments (unpublished) a slightly earlier inhibition of photosynthesis than of respiration was

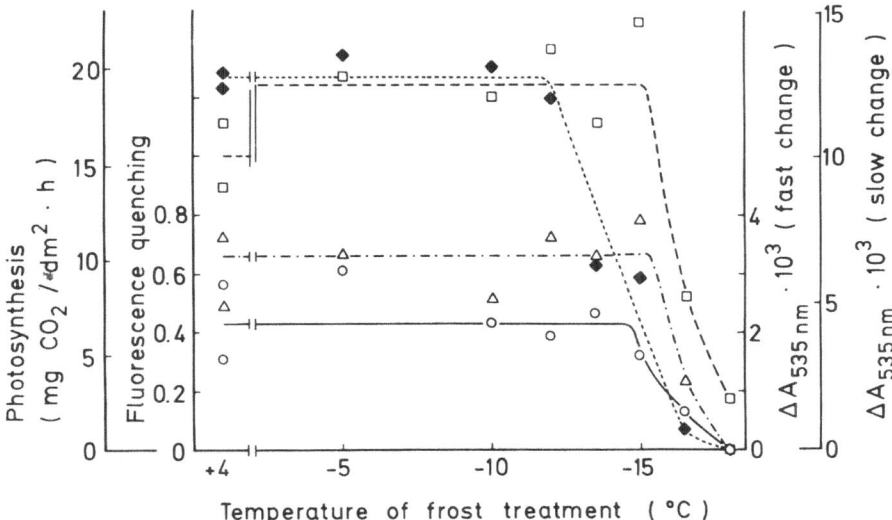

FIGURE 4. Effects of frost treatment of hardened *Viola* leaves on photosynthesis (···◆···), photoinduced chlorophyll fluorescence quenching (−·△−·),
light scattering changes ($\Delta A_{535\ slow}$, −−□−−), and membrane potential-dependent
absorbance changes ($\Delta A_{535\ fast}$, —O—). The physical parameters were measured
in the absence of CO_2. Fluorescence quenching is defined as (P−S)/S, where P
is the peak of fluorescence induction, S the level of stationary fluorescence
emission. $\Delta A_{535\ slow}$ represents the light scattering increase seen upon
illumination after a 3 min dark period. Other conditions as for Fig. 3.

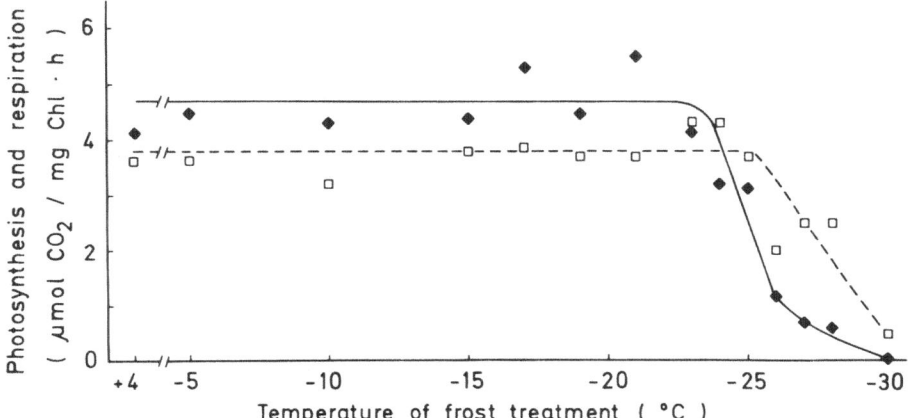

FIGURE 5. Photosynthesis and respiration of cold acclimated *Hedera* leaves
as functions of the minimum temperature of frost treatment.
◆, photosynthesis; □, respiration in the presence of CO_2. Conditions
as for Fig. 1. Photosynthesis was about 75% light-saturated.

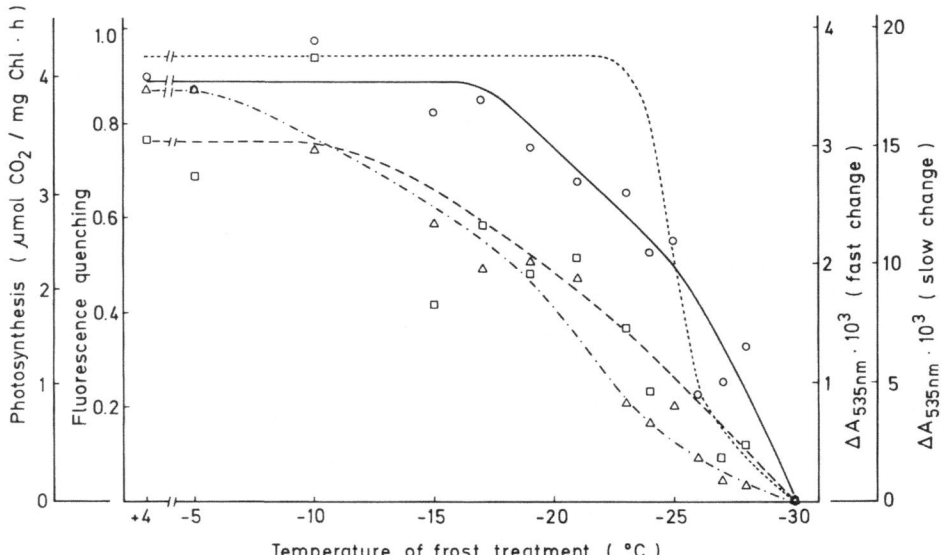

FIGURE 6. Effects of frost treatment of cold acclimated *Hedera* leaves on photoinduced chlorophyll fluorescence quenching (—·Δ—·), light scattering changes ($\Delta A_{535\ slow}$, — — □ — —) and membrane potential-dependent absorbance changes ($\Delta A_{535\ fast}$, — O —). The signals were measured as for Fig. 4, but in the presence of 300 $\mu l \cdot l^{-1}$ CO_2 in air. (Effects of CO_2 on these paramters were absent in hardened *Hedera* leaves). For comparison, the inactivation curve of photosynthesis (······, see Fig. 5) is given.

also observed in spinach leaves. In *Triticum, Viola* (and *Spinacia*; ref. 8) T_{50} values of signals of light scattering, potential-dependent absorbance changes and quenching of chlorophyll fluorescence were usually considerably lower than those of photosynthesis. This difference is particularly pronounced in *Viola* leaves. In *Tetragonia,* which acquires only little frost tolerance, the differences between T_{50} values of photosynthesis and physical signals are small (unhardened leaves) or within statistical deviation not seen at all (hardened leaves). In contrast, in hardened leaves of *Hedera*, T_{50} values of the physical signals are in part higher than T_{50} of photosynthesis. As already shown in Fig. 6, inactivation of thylakoids seems to start before the low rates of photosynthesis are affected.

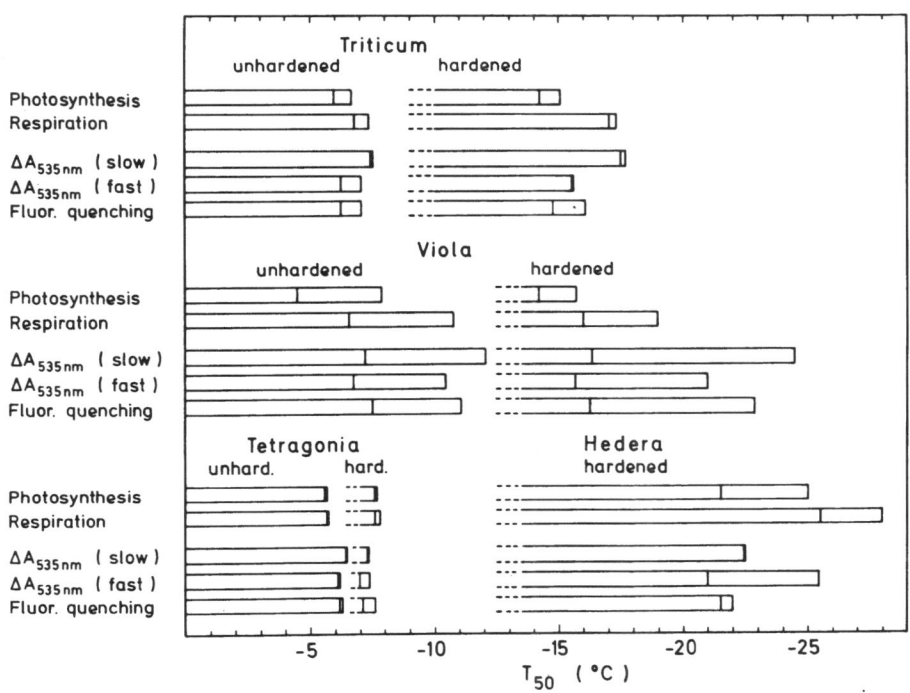

FIGURE 7. T$_{50}$ values (in ^{O}C) of photosynthesis, respiration and of physical parameters indicating thylakoid membrane energization, obtained with un-hardened and hardened leaves. The physical paramters were measured in the absence of CO_2 in air; for definitions see legends to Figs. 3 and 4. T$_{50}$ denotes the minimum temperature of the freeze-thaw treatment resulting in 50% inhibition of the respective activities. Values from 2 experiments with each unhardened and hardened species are given.
Note: *Viola* was cultured in summer and hardened artificially in a phytotron (higher T$_{50}$ values of unhardened and hardened leaves) or cultured in the greenhouse in winter and hardened under field conditions (lower T$_{50}$ values).

4. DISCUSSION

In hardened and unhardened leaves of the species tested in this and former (7,8,12) studies, freezing stress either affected photosynthetic CO_2 assimilation at higher temperatures than respiration (*Triticum, Viola, Hedera*), or both processes were inhibited almost in parallel (*Tetragonia, Spinacia*). In spinach, the decrease in photosynthesis appeared irreversible (7). Thus photosynthesis, measured after freezing and thawing can be viewed as a sensitive indicator of first stages of freezing injury of the leaf tissue. One should note that photosynthesis also appears to be the most sensitive physiological parameter to reveal leaf damage by other adverse factors such as heat (16) and osmotic stress (5,6).

The course of inactivation of the energy-conserving apparatus of the thylakoids was elucidated by monitoring slow and fast components of photo-induced absorbance changes at 535 nm, and by chlorophyll fluorescence quenching. In spinach leaves, the decline of these signals upon freezing/thawing coincided with inactivation of electron transport and concomitant proton pumping and photophosphorylation by the thylakoids (7,8,12). In the presently studied herbaceous plants, the pattern of breakdown of the physical signals was very similar to that observed in spinach. Thus the mechanism of freezing injury appears to be alike in these species.

With the exception of hardened *Tetragonia*, where the range between beginning and full leaf damage is only $1.5^{o}C$, leaves of the herbaceous plants showed a marked delay of breakdown of thylakoid energization as compared to photosynthetic CO_2 uptake. Furthermore, stimulation of the physical signals, frequently observed in the presence of CO_2 in intermediary stages of leaf injury (Figs. 2 and 3), as earlier reported for spinach (8), denoted declining energy utilization by photosynthetic carbon metabolism. From this follows that inactivation of the energy-conserving system of the thylakoids is not the primary cause of inhibition of photosynthesis in the herbaceous plants. Presently, the mechanism of primary inactivation of photosynthetic CO_2 assimilation is not clear. Ribulose-bisphosphate carboxylase might be denatured by freeze-dehydration (3) or be inactivated secondarily due to disorder of compartmentation in the cell. The plasmalemma and tonoplast have been claimed to be most sensitive to freezing stress (for. ref. see 2,14). Possibly, a primary loss of integrity of those membranes and the resulting breakdown of compartmentation might specifically inhibit CO_2 fixation in the Calvin cycle, while thylakoid reactions would be less affected. Indeed, isolated spinach thylakoids were relatively insensitive to freezing and thawing in the presence of crude cell sap (R.J. Klosson, unpublished). The above hypothesis is supported by preliminary electron microscopic studies (V. Ahrer-Steller, unpublished): In the stage of beginning decline of photosynthesis, single damaged cells, surrounded by unaltered tissue, can be detected in spinach leaves. In these cells, the plasmalemma and tonoplast were broken and the turgor lost. The chloroplast stroma was swollen, but the thylakoid membranes appeared in part unchanged. Strongly enlarged intrathylakoid spaces seen upon more severe freezing might denote thylakoid inactivation.

Although cold acclimation leads to a significantly increased freezing tolerance of *Spinacia* (7), *Viola* and *Triticum* (cf. ref. 14), the pattern of freezing inactivation was not different from that in unhardened leaves, indicating that the mechanism of injury is not changed by hardening. The widened temperature range of partial freezing damage often observed with hardened leaves (Table 1) probably denotes differential capability of individual cells to acquire hardiness.

In the hardened leaves of the evergreen *Hedera helix*, the course of freeze-inactivation might indicate a mechanism different from that pre-vailing in the herbaceous plants. However, the rates of photosynthesis in these leaves are extremely low. This may be the reason for the lower frost sensitivity of CO_2 assimilation than of the energy-conserving system of the thylakoids.

ACKNOWLEDGEMENTS

The authors thank Professor K.A. Santarius for helpful discussions. The study was supported by the Deutsche Forschungsgemeinschaft.

REFERENCES

1. Briantais J-M, Vernotte C, Picaud M, Krause G H (1979) A quantitative study of the slow decline of chlorophyll a fluorescence in isolated chloroplasts. Biochim Biophys Acta 548: 128-138.
2. Heber U, Schmitt J M, Krause G H, Klosson R J, Santarius K A (1981) Freezing damage to thylakoid membranes in vitro and in vivo. In Morris G J, Clarke A, eds. Effects of low temperature on biological membranes, pp. 263-283. London: Academic Press.
3. Huner N P A, Carter J V (1982) Differential subunit aggregation of a purified protein from cold-hardened and unhardened Puma rye. Z Pflan-zenphysiol 106: 179-184.
4. Junge W (1977) Membrane potentials in photosynthesis. Annu Rev Plant Physiol 28: 503-536.
5. Kaiser W M, Kaiser G, Prachuab P K, Wildman S G, Heber U (1981) Photosynthesis under osmotic stress. Inhibition of photosynthesis of intact chloroplasts, protoplasts, and leaf slices at high osmotic potentials. Planta 153: 416-422.
6. Kaiser W M, Heber U (1981) Photosynthesis under osmotic stress. Effect of high solute concentrations on the permeability properties of the chloroplast envelope and activity of stroma enzymes. Planta 153: 423-429.
7. Klosson R J, Krause G H (1981) Freezing injury in cold-acclimated and unhardened spinach leaves. I. Photosynthetic reactions of thylakoids isolated from frost-damaged leaves. Planta 151: 339-346.
8. Klosson R J, Krause G H (1981) Freezing injury in cold-acclimated and unhardened spinach leaves. II. Effects of freezing on chlorophyll fluorescence and light scattering reactions. Planta 151: 347-352.

9. Köster S, Heber U (1982) Light scattering and quenching of 9-amino-acridine fluorescence as indicators of the phosphorylation state of the adenylate system in intact spinach chloroplasts. Biochim. Biophys Acta 680: 88-94.

10. Krause G H (1973) The high-energy state of the thylakoid system as indicated by chlorophyll fluorescence and chloroplast shrinkage. Biochim Biophys Acta 292: 715-728.

11. Krause G H (1974) Changes in chlorophyll fluorescence in relation to light-dependent cation transfer across thylakoid membranes. Biochim Biophys Acta 333: 301-313.

12. Krause G H, Klosson R J, Tröster U (1982) On the mechanism of freezing injury and cold acclimation of spinach leaves. In Li P H, Sakai A, eds. Plant cold hardiness and freezing, Vol. 2. New York: Academic Press, in press.

13. Krause G H, Vernotte C, Briantais J-M (1982) Photoinduced quenching of chlorophyll fluorescence in intact chloroplasts and algae. Resolution into two components. Biochim Biophys Acta 679: 116-124.

14. Levitt J (1980) Response of plants to environmental stresses. Vol. 1. New York: Academic Press.

15. Senser M, Beck E (1977) On the mechanism of frost injury and frost hardening in spruce chloroplasts. Planta 137: 195-201.

16. Thebud R, Santarius K A (1982) Effects of high temperature stress on various biomembranes of leaf cells in situ and in vitro. Plant Physiol, in press.

THE EFFECT OF TEMPERATURE ON CHLOROPHYLL FLUORESCENCE INDUCTION OF
CUCUMBER LINES DIFFERING IN GROWTH CAPACITY AT SUBOPTIMAL CONDITIONS

P.R. VAN HASSELT, J. WOLTJES AND F. DE JONG
Department of Plant Physiology, University of Groningen, Biological Centre,
P.O. Box 14, 9750 AA Haren (Gn), The Netherlands.

ABSTRACT

Temperature dependent chlorophyll a fluorescence induction transients
in leaf discs of two cucumber lines were measured between 27°C and 0°C.
Cucumber cv Farbio was compared with line 45 with a better growth capacity
at suboptimal temperature (20°C D/12°C N). With decreasing temperature
breaks were evident in the rate of change of the fast fluorescence maximum
P. After growth at normal temperature (25°C D/20°C N) the temperature
dependent yield of P of Farbio showed one break at 13°C. The curve of Farbio
grown at suboptimal temperature (20°C D/12°C N) showed two breaks at 19°C
and 4°C. A similar curve with breaks at 21°C and 7°C was observed in line
45 grown at normal conditions. After growth at suboptimal conditions line
45 showed breaks at 16°C and 8°C. Temperature dependent changes of the
fluorescence yield of the slow maximum M minus the yield of the minimum S
showed optimum curves. Young leaves (10 day-old) of both lines showed an
optimal value of M-S at 3°C. In older leaves (35 day-old) the optimum
shifted to 12°C with Farbio and 18°C with line 45 suggesting an influence
of plant age on fluorescence properties. Plants grown at suboptimal conditions retained their optimum at 6°C (Farbio) or 3°C (line 45).

1. INTRODUCTION

Many important horticultural crops, e.g. tomato and cucumber, are of a
tropical or subtropical origin. In temperate climates such thermophilic
plants are generally grown in greenhouses in order to maintain optimal
temperature conditions for growth and production. During the last years
costs of heating of greenhouses have been risen dramatically due to the
increase of fuel prices and breeding programs have been started to select
varieties which grow and produce well at suboptimal temperatures (1).

Chlorophyll a fluorescence has been used by several authors as an
intrinsic probe to monitor temperature dependent chloroplast functioning

e.g. in algae (2) and isolated chloroplasts of higher plants (3,4), and in intact leaves to study heat- (5) and chilling (6) sensitivity.

Because monitoring chlorophyll fluorescence is a fast and nondestructive method for measuring temperature dependent changes in chloroplast functioning, and temperature dependent fluorescence was found to increase faster in chill sensitive than in chill resistant species (7), it was decided to study the possibility of using temperature dependent chlorophyll a fluorescence as a criterium for breeding plants with a higher growth capacity at suboptimal temperature. In this paper results of preliminary experiments on the temperature dependence of chlorophyll a fluorescence induction transients of two cucumber lines differing in growth capacity at suboptimal temperature are reported.

2. PROCEDURE

2.1. Material and methods

2.1.1. Plant material. Plants (*Cucumis sativus* L.) of cold sensitive cv Farbio and cold tolerant line 45 were grown in pots in two Conviron E7 growth cabinets. Seeds were obtained from Dr. A.P.M. den Nijs of the Institute of Horticultural Plant Breeding (I.V.T.) at Wageningen. One growth cabinet was kept at normal conditions: a daylength of 16 h and a temperature of 25°C during the day and 20°C during the night (25°C D/20°C N). The other cabinet was controlled at suboptimal conditions: a daylength of 12 h and temperature of 20°C D/12°C N. Light intensity between 400–700 nm was measured with a Lambda LI 185 meter and was 230 $\mu E\ m^{-2}s^{-1}$ at growth level in both cabinets.

2.1.2. Fluorescence measurements. Chlorophyll fluorescence (> 710 nm) of leaf discs (diameter 7 mm) was measured with a portable fluorometer (Model SF10, Brancker Research Ltd., Ontario, Canada) and registered on a recorder. Intensity of the exciting light (LED MV 5220) was 10 $\mu E\ m^{-2}s^{-1}$. A leaf disc was placed in humid filter paper in a hole (9 mm diameter, 2 mm depth) of the aluminium upper part of a metal cuvet. The temperature of the cuvet was controlled by circulating ethanol of appropriate temperature through it, measured with a thermistor and registered on a recorder. The rate of temperature decline (0.3°C/min) and the time of fluorescence measurements were controlled by a selfbuilt microcomputer. Experiments started at 30°C. The temperature declined linearly during 10 min to 27°C and remained constant during the following 20 min (equilibration period). After these 30 min dark preincubation fluorescence was measured during 10 min.

Subsequently the temperature again decreased 3°C in 10 min and remained constant during 20 min before fluorescence was measured at 24°C. In this way chlorophyll fluorescence induction was measured every 40 min from 27°C to 0°C.

3. RESULTS AND DISCUSSION

On illumination of dark adapted green cells the intensity of chlorophyll a fluorescence shows characteristic transients called fluorescence induction (8). The changes in fluorescence yield were attributed to variations in the redox state of Q, the primary acceptor of photosystem II which quenches chlorophyll fluorescence in its oxidized state (9). The fluorescence induction curve is generally described as consisting of three peaks (I, P and M) with dips (I and S) in between and the final steady state level is often termed T (8).

Figure 1 shows typical changes in the shape of the fluorescence induction curve with decreasing temperature. Fluorescence yield increased rapidly to a peak value (P). The preceding faster transients 0, I, D (8) could not be resolved at the recording time used. After P, fluorescence decreased to a minimum S followed by a slow peak M. Remarkably, in most experiments around 24°C a small second slow peak (M_1) with a second minimum (S_1) could be distinguished from the first slow peak (M_2). The minimum in between M_1 and M_2 is called S_2. At lower temperature all transients, in particular M_2, occurred after a longer illumination time. Hence the level of M_2 and T could not be established at temperatures below 18°C within the recording time used. Therefore only temperature dependent changes of P, S_1 and M_1 were used. Two slow peaks M_1 and M_2 were also observed during fluorescence induction in *Bryopsis* chloroplasts and it was shown that the peak M_1 was linked to a proton gradient across the thylakoid membrane (10). All peaks, with the exception of M_2, generally increased with a decrease of the temperature (Fig. 1). It should be noted that temperature lowering affected P and M differently. In several experiments M_1 increased faster than P and sometimes P temporally decreased as a result of lower temperature (Fig. 3A). These observations could be explained by the observations that P is mainly affected by changes in photosystem II functioning (8) while M_1 reflects in addition structural changes of the thylakoid membrane e.g. due to protonaton (11).

Figure 2A shows that the intensity of the fast induction maximum P in leaf discs of Farbio grown at normal conditions slowly increased with

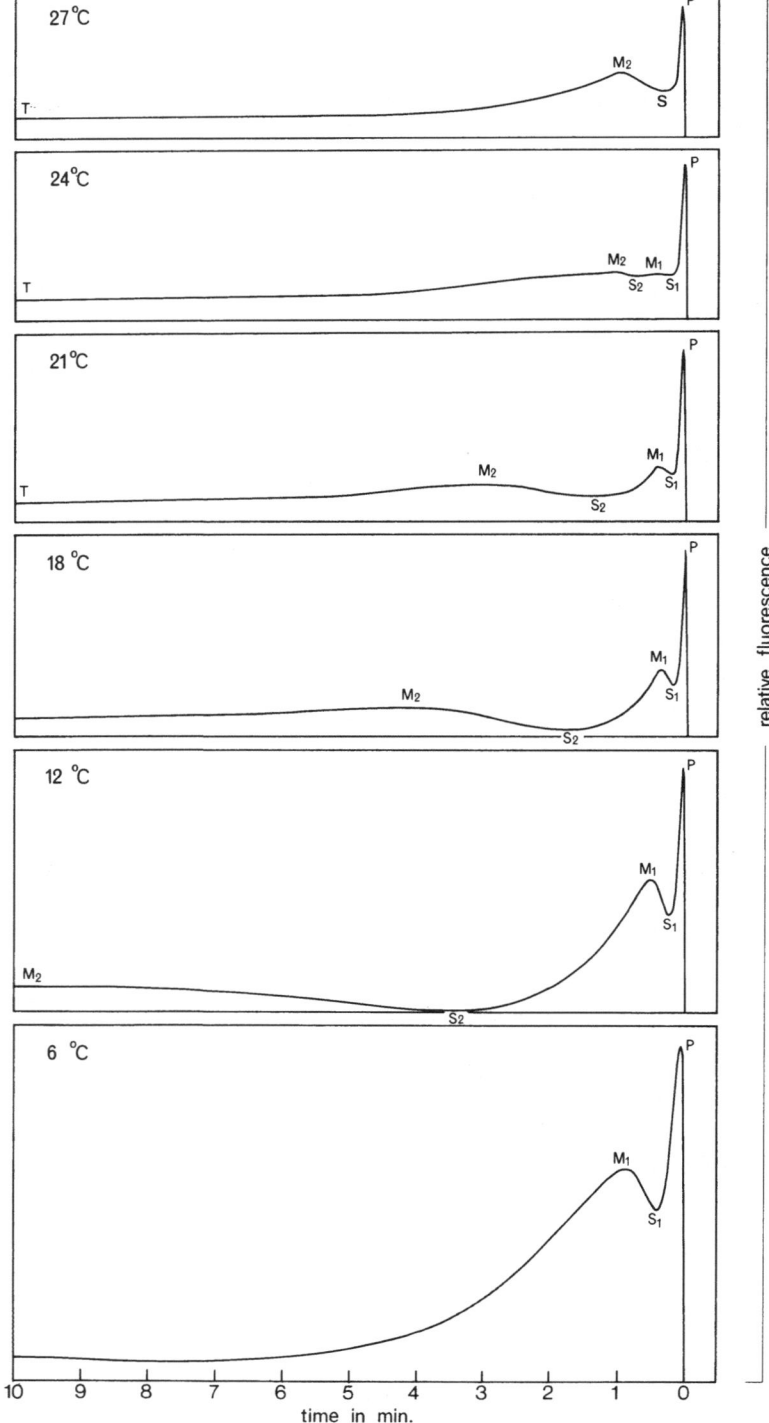

FIGURE 1. The effect of temperature on chlorophyll fluorescence induction transients of a cucumber (cv. Farbio) leaf disc. Recording tracings of changes in fluorescence yield at the indicated temperatures are shown. Terminology of the transients is according to reference (8).

decreasing temperature. At 13°C a break was evident. Below 13°C a faster change was observed. The value of P of leaves grown at suboptimal conditions (Fig. 2B) was affected differently by decreasing temperature. Two breaks were observed at 19°C and 4°C. Between the breakpoints no decrease of P occurred. A similar curve with breaks at 21°C and 7°C showed the temperature dependent intensity of P of line 45 grown at normal conditions (Fig. 3A). Remarkably, the value of P between the breakpoints showed a small decrease. Below 7°C there was a fast increase of fluorescence. Two breaks, at 16°C and 8°C, were also observed in line 45 grown at suboptimal conditions (Fig. 3B). Evidently suboptimal conditions caused with line 45 a shift of the first break from 21°C to 16°C. The second break around 6°C was hardly affected.

From these results it can be concluded that the rate of change of the fast fluorescence induction maximum P of cucumber leaves shows abrupt temperature dependent alterations. At low temperature between 4°C and 8°C the alterations of the rate were generally most evident. It was observed that exposition of plants to temperatures which induce chilling injury caused a fast decrease of chlorophyll fluorescence (6). Possibly the break observed in cucumber leaves at low temperature above the freezing point indicates the induction of chilling injury. Interestingly, the less growing cultivar Farbio showed no break at low temperature at normal conditions (Fig. 3A). At present, however, it is not known whether there is a correlation between a better growth capacity at suboptimal temperature and a higher chilling sensitivity.

When line 45 plants were grown at suboptimal instead of normal temperature, the first breakpoint in the curve of the temperature dependence of P shifted to a 5°C lower temperature (Fig. 3B). This could be ascribed to an adaptation of chloroplast membrane properties of line 45 to the suboptimal growth conditions. The origin of the observed breaks in the curves of temperature dependent chlorophyll fluorescence induction cannot be explained presently. At temperatures above 0°C lipid phase changes are unlikely to occur in chloroplast membranes because the main lipid components have phase transitions far below 0°C (4,7). Recently, however, it was observed that suboptimal conditions caused a higher degree of unsaturation of cucumber leaf phospholipids (12). Hence phase transition of specific lipids e.g. phospholipids affecting chloroplast membrane functions cannot be ruled out as a cause of temperature dependent chlorophyll fluorescence changes.

262

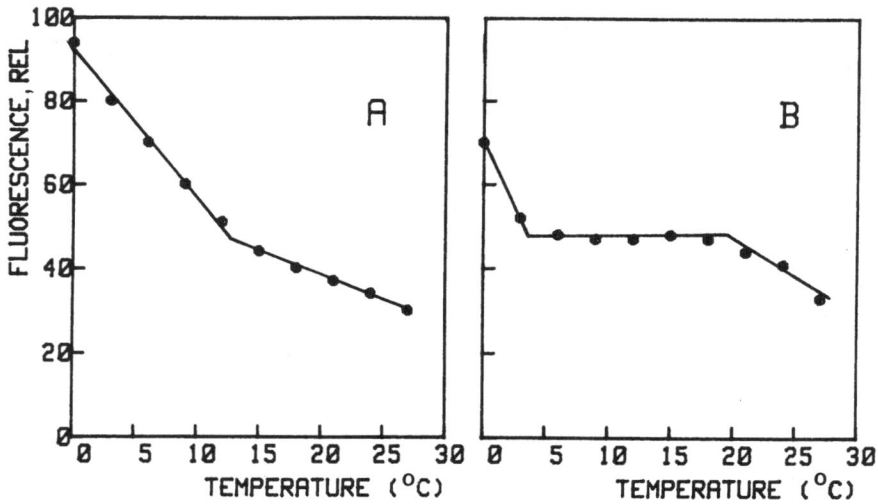

FIGURE 2. The effect of temperature on the chlorophyll fluorescence yield of the fast peak P of Farbio. Leaf discs of a 9 days old 5th leaf grown at 25°C D / 20°C N (A) and of a 9 days old 4th leaf grown at 20°C D / 12°C N (B) were used.

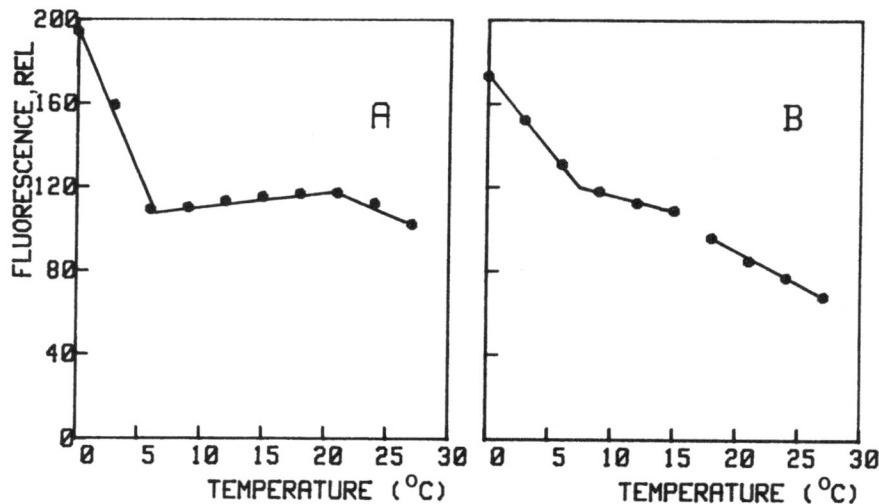

FIGURE 3. The effect of temperature on the chlorophyll fluorescence yield of the fast peak P of line 45. Leaf discs of a 16 days old 2th leaf grown at 25°C D / 20°C N (A) and of a 31 days old 2th leaf grown at 20°C D / 12°C N (B) were used.

Several authors who studied temperature dependence of chlorophyll fluorescence in higher plants found no breaks in the rate of fluorescence
increase (3,7). These authors have measured steady state fluorescence with
rather high light intensities and in addition they often used DCMU. High
light intensity as well as DCMU cause saturation of the fluorescence
response (Q mostly reduced). Hence small temperature induced changes
affecting the rate of oxidation of Q will probably not result in changes
of chlorophyll a fluorescence induction as were observed in this study.

The slow fluorescence change SMT reflects processes in which the ultrastructure of the thylakoid membrane plays a fundamental role (8). With
decreasing temperature a fast increase of the slow fluorescence maximum M
was observed in the present experiments. Therefore the effect of temperature on the empirical value $M_{(1)} - S_{(1)}$ was investigated (Fig. 4). In both
varieties M-S values plotted against temperature showed an optimum around
3°C in young leaves (A). In older leaves (B) the optimum shifted to higher
values (Farbio 12°C; line 45 18°C). This observation suggests that plant
age is influencing chlorophyll fluorescence properties. It should be noted
that leaves of plants grown at suboptimal conditions retained their optimal
value at a low temperature (C). The observed lower optimal temperature of
line 45 (3°C as compared to 6°C for Farbio) might be related to its better
growing capacity at suboptimal conditions. Further experiments to get more
insight in the origin of temperature dependent changes of chlorophyll
fluorescence induction transients and its possible use in the selection of
plants which grow well at lower temperatures are in progress.

REFERENCES

1) Den Nijs,A.P.M.(1980) Adaptation of the glasshouse cucumber to lower temperatures in winter by breeding. Acta Hortic. 118:65-72.
2) Fork, D.C. and Murata, N. (1979) Effect of growth temperature on the lipid
and fatty acid composition, and the dependence on temperature of light
induced redox reactions of cytochrome f and light energy redistribution
in the thermophilic blue-green alga Syncchococcus lividus. Plant Physiol.
63:524-530.
3) Murata, N. and Fork, D.C. (1975) Temperature depedence of chlorphyll a
fluorescence in relation to the physical phase of membrane lipids in Algae and higher plants. Plant Physiol. 56:791-796.
4) Murata, N., Throughton, J.H. and Fork, D.C. (1975) Relationships between
the transition of the physical phase of membrane lipids and photosynthetic
parameters in Anacystis nidulans and lettuce and spinach chloroplasts.
Plant Physiol. 56:508-517.

264

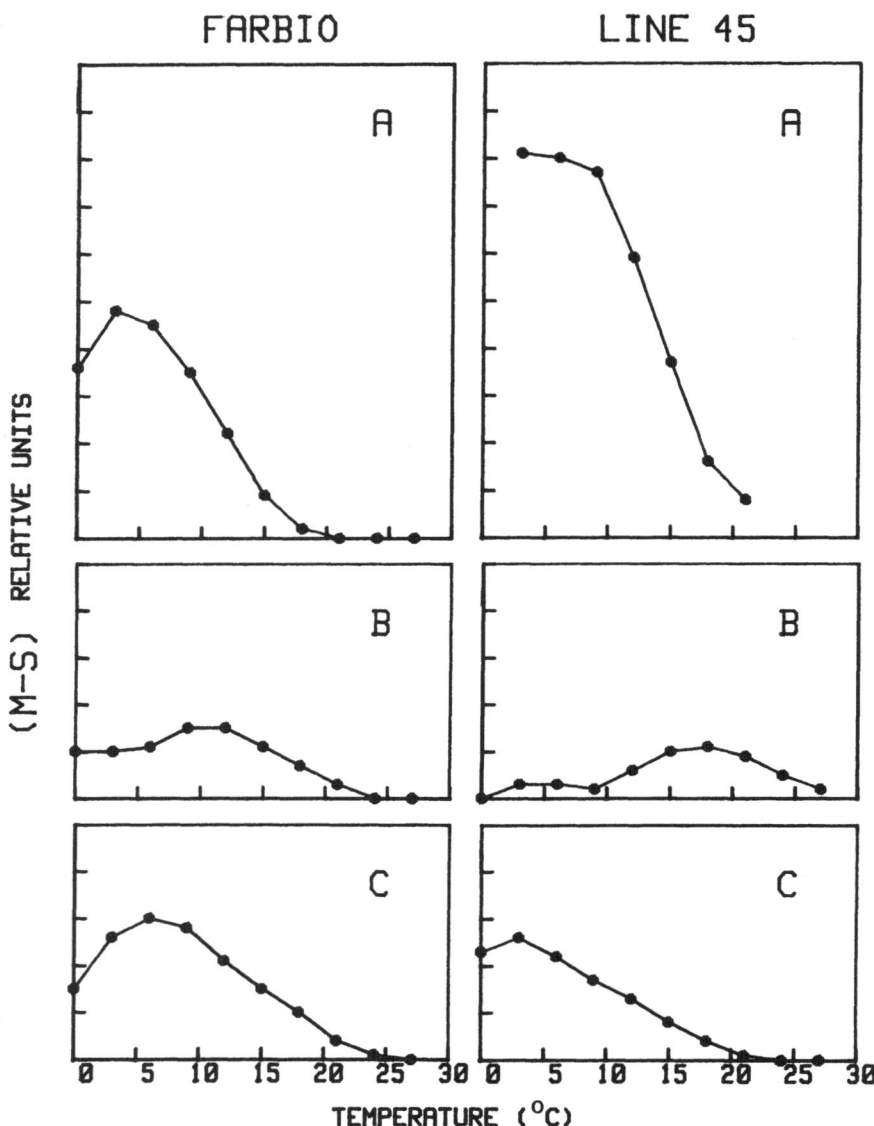

FIGURE 4. The effect of temperature on the difference between the fluorescence yield of the slow maximum M_1 and its preceeding dip S_1 of leaf discs from the first leaf of Farbio and line 45 plants. Plants grown for 10 days (A) or 35 days (B) at 25°C D / 20°C N followed by 25 days at 20°C D / 12°C N.

5) Schreiber, U. and Berry, J.A. (1977) Heat induced changes of chlorophyll fluorescence in intact leaves correlated with damage of the photosynthetic apparatus. Planta 136:233-238.

6) Smillie, R.M. (1979) In Lyons, J.M., Graham, D. and Raison, J.K., eds. Low temperature stress in crop plants, pp. 187-202. New York: Academic Press.

7) Melcarec, P.K. and Brown, G.N. (1977) The effect of chilling stress on the chlorophyll fluorescence of leaves. Plant and Cell Physiol. 18:1099-1107.

8) Govindjee and Papageorgiou, G. (1971) In Giese, ed. Photophysiology Vol. VI, pp. 1-46. New Yorj: Academic Press.

9) Duysens, L.N.M. and Sweers, H.E. (1963) Mechanism of two photochemical reactions in algae as studied by means of fluorescence. In Ashida, J., ed. Studies on micro algae and photosynthetic bacteria, pp. 353-372. Tokyo: Univ. of Tokyo Press.

10) Yamagishi, A., Satoh, K. and Katoh, S. (1978) Fluorescence induction in chloroplasts isolated from the green alga Byopsis maxima. Plant and Cell Physiol. 19:17-25.

11) Yamamoto, Y. and Nishimura, M. (1976) Characteristics of light induced H^+ transport in spinach chloroplasts at lower temperatures. Plant and Cell Physiol. 17:17-21.

12) Horvath, I., Vigh, L., Woltjes, J., Van Hasselt, Ph.R. and Kuiper, P.J.C. (1982) In Wintermans, J.F.G.M. and Kuiper, P.J.C., eds. Proceedings of the 5th International Symposium on the Biochemistry and Metabolism of Plant Lipids (in press). Amsterdam: Elsevier.

EFFECTS OF LOW TEMPERATURES ON BARLEY CHLOROPLAST MEMBRANES

MICHEL HAVAUX [+], CONSTANTIN CLEANIS and ROBERT LANNOYE.

Laboratoire de Physiologie végétale, Université Libre de Bruxelles, 28 av. Paul Heger, 1050 Bruxelles, Belgium.

ABSTRACT

We investigated the effects of cold treatment (3 weeks at 2°C) on the photosynthetic apparatus of winter barley and maize. The ratio of the maximum (P level) to the minimum (0 level) chlorophyll fluorescence induction signal (Kautsky phenomenon), measured in vivo at 25°C, was not modified during cold adaptation of barley. On the other hand, this P/0 ratio, which is a good indicator of the activity of PSII, decreased rapidly in cold treated maize seedlings. In both species, the slow fluorescence quenching (from P to T) was not suppressed by the low temperature treatment. However, a noticeable modification of the slow fluorescence transients (P-S-M-T) was observed in cold adapted barley leaves. The M peak was strongly reduced after 3 weeks at 2°C. In non-hardened barley plants, the slow fluorescence changes were markedly inhibited at 2°C. At this temperature, chlorophyll fluorescence was blocked at P level. During cold acclimation of barley seedlings, a progressive restoration of those slow changes was observed at 2°C. This observation indicates that cold hardening of barley induces adaptation of the photophysiological activity to low temperatures. This chloroplast membrane adaptation was examined at the protein level. Electrophoretic analysis showed that cold had no influence on the chlorophyll-protein complexes of barley thylakoids. Likewise, one and two- dimensional polyacrylamide gel electrophoresis of chloroplastic membrane proteins showed no marked difference between plants grown at 25°C or hardened 3 weeks at 2°C. These results suggest that cold adaptation of barley thylakoids occurs mainly at the lipid level.

[+] Michel HAVAUX is "Aspirant F.N.R.S.".

INTRODUCTION

As environment becomes colder, winter cereals rapidly adapt to chilling and proceed to harden by developing systems that allow them to survive subzero temperatures (12,17). Despite the intensive research of the last years, the physiological and biochemical changes which lead to increased freezing tolerance of plants are poorly understood. Alteration of the membrane permeability properties is a commonly observed manifestation of freezing in biological systems and it is generally accepted that cell membranes are the primary site of freezing injury, due to extracellular ice formation (7,12,13). In consequence, cold acclimation must involve cellular alterations which protect cell membranes from freezing stresses. One component of this adaptation involves accumulation of cryoprotective compounds acting as membrane stabilizers (12,17,27). Moreover, growth at low temperature is generally accompagnied by changes in the chemical composition of the cell membranes (8, 12,22,28). These changes would enable the membranes to resist dehydration stresses caused by extracellular ice formation. More it has been shown that resistant plants are able to control the fluidity of their membranes while exposed to decreasing temperatures (15,25,26). This phenomenon has been interpreted as an adaptation to cold which permits continued metabolic activity necessary for hardening.

In this paper, the effect of cold hardening on the photosynthetic membranes of winter barley is investigated and the low temperature responses of barley and of a chill-sensitive species (maize) are compared.

MATERIAL AND METHODS

Growth conditions

Maize plants (*Zea mays* L., var. LG9) were grown in a compost-sand mixture (1:1) and winter barley plants (*Hordeum vulgare* L., convar. distichon, var. Sonja) were grown hydroponically with the following nutrient solution :

N	S	P	K	Ca	Mg		
40	5	5	20	15	15	(%)	(10 meq/l)

The plants were grown at 25°C in a 16h photoperiod of 2800 lux during 10 days (barley), or 2 weeks (maize). Cold stress was imposed by transferring the plants to a cold room (2°C; 3 weeks) where they were exposed to the same photoperiod.

Chlorophyll fluorescence

Induction kinetics of chlorophyll fluorescence from intact leaves (dark adapted for at least 20 min) were measured using bifurcated fiberoptics to transmit excitation light to and chlorophyll fluorescence emission from the upper surface of the leaves. A plexiglass rod was used as light guide. Fluorescence was excited with $17.5 \ 10^{-4}$ W/m^2 blue-green light supplied by a tungsten filament source filtered through a Schott BG 18 filter. Chlorophyll fluorescence was detected at wavelenghts longer than 660 nm (Kodak Wratten filter 70) using a silicon photodiode with integrated amplifier (Centronic OSI 5). The signal from the photosensor was recorded on a dual-beam storage oscilloscope (fast fluorescence transients) or a chart recorder (slow transients).

Separation of chlorophyll-protein complexes

Chloroplasts were extracted by grinding the leaves in a medium containing 0.33M mannitol, 10 mM KH_2PO_4, 3 mM NaCl, 10 mM $MgCl_2$ and 0.1 % BSA (pH 7.8) in a Waring Blendor. The homogenate was filtered through 8 layers of cheesecloth and the chloroplasts were sedimented at 2200g for 5 min. The chloroplast pellet was washed and resuspended in 50 mM K_2HPO_4, 10 mM KCl (pH 7.8). After 15 min centrifugation at 12000g, the thylakoids were finally suspended in a few milliliters of the same medium and stored in liquid N_2. Total chlorophyll was determined in 80% acetone (2).

Chloroplast thylakoids were solubilized at 4°C in 0.3M Tris-HCl (pH 8.8) containing 10% glycerol and 1% SDS to give a final SDS/chlorophyll ratio 10/1 .

Electrophoresis was performed immediately (at 4°C in the dark) with samples containing 75-90 µg of chlorophyll on a 15% acrylamide gel with a 5% acryl-amide stacking gel according to Laemnli (11). A current of 1.25 mA/gel for 15 min followed by 5.5 mA/gel for 75 min was supplied.

Chlorophyll-protein complexes were characterized by their visible wavelenght absorption spectra as in (5). Proteins were stained by placing the gels in 0.2% Coomassie blue in methanol-acetic acid-water (5-1-5) for 1h and then destained in the methanol-acetic acid-water.

Two-dimensional electrophoresis of thylakoid proteins

Thylakoid membranes were isolated according to Santarius (21) and delipi-dated by extracting twice with 90% acetone. The pellet was freed from acetone under vacuum. Amounts of acetone extracted thylakoids containing about 300 µg proteins were solubilized as in (4).

Thylakoid proteins were separated by two-dimensional electrophoresis according to O'Farrel (16). Slab gels were stained as described above.

RESULTS AND DISCUSSION

The chlorophyll fluorescence induction phenomenon (Kautsky effect) provides a rapid and sensitive test for monitoring the primary processes of photosynthe sis and for estimating the potential photosynthetic activity (3,18).
In particular, the ratio of the maximum (P level) to the minimum (O level) fluorescence signal is a good indicator of the electron flow through PSII. This ratio was measured in vivo at 25°C in maize and barley plants grown at 2°C (Figure 1).

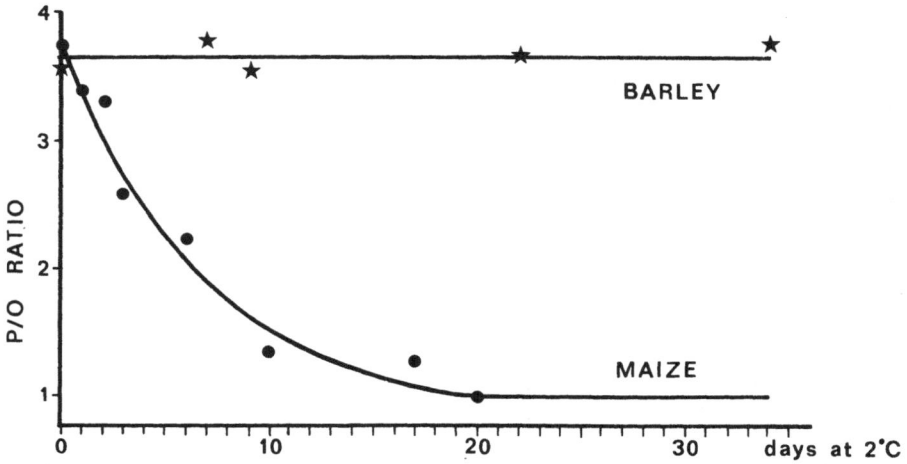

Fig. 1 - Maximum (P) / Minimum (O) chlorophyll fluorescence ratio (measured at 25°C) in maize and barley leaves during growth at 2°C.

Cold treatment had no effect on the P/O value in barley leaves, indicating a stable primary photochemical efficiency. On the other hand, P/O decreased rapidly in stressed maize leaves.
After 20 days at 2°C, no more variable fluorescence (P - O) was detected. The loss of the variable component of chlorophyll fluorescence reflects damage on the electron transport capacity. Indeed, the fast rise of fluores- cence from O (which remained unchanged as chilling injury developed) to P is attributed to Q, the primary electron acceptor of PSII, which quenches chlorophyll fluorescence in the oxidized state. During the first few seconds

of illumination, Q (present in the oxidized form in dark-adapted leaves) is converted in the non quenching form by reductants generated by PSII. In consequence, the variable fluorescence is determined by the ratio of Q reduced to Q oxidized. A low P/O ratio provides evidence that exposure of sensitive plants to chilling temperatures causes injury to the thylakoid structure affecting the electron flow mediated by PSII.

In both species, the slow fluorescence quenching from P to the steady-state T was not suppressed by the low temperature treatment (Figure 2).

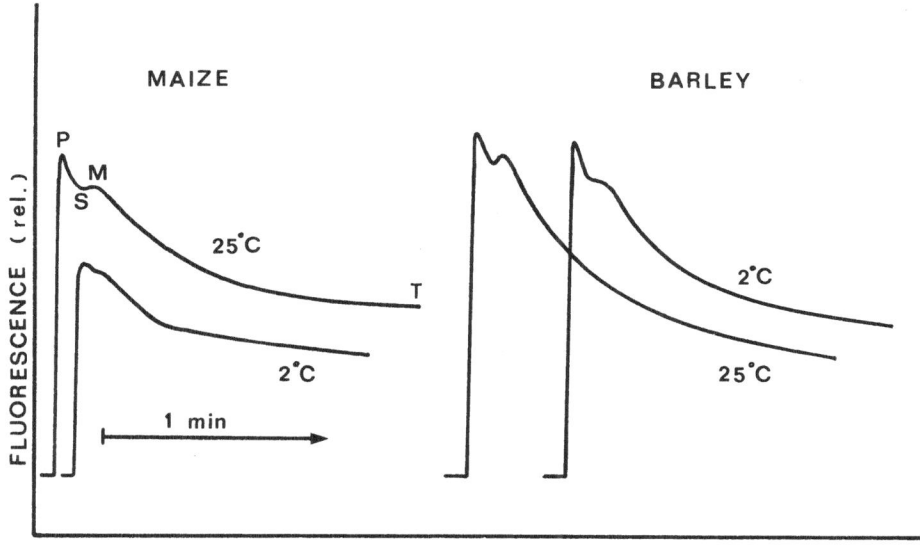

Fig. 2 - Slow chlorophyll fluorescence transients (at 25°C) in leaves of maize and barley plants grown at 2°C (10 days - maize or 3 weeks - barley) or kept at 25°C.

The slow chlorophyll fluorescence changes are a complex phenomenon which has not been explained satisfactorily. In addition to the reoxidation of Q, they result from changes in the distribution of excitation energy between the weakly fluorescent PSI and the more fluorescent PSII. This increased energy transfert from PSII to PSI appears to be under the control of energy dependent conformational changes in the thylakoids.

Whereas this "spillover" between the 2 photosystems was not affected by the cold treatment, a noticeable modification of the slow fluorescence transients (P-S-M-T) was observed in cold adapted barley leaves. The M peak was strongly reduced after 3 weeks at 2°C, suggesting structural modifications in the

272

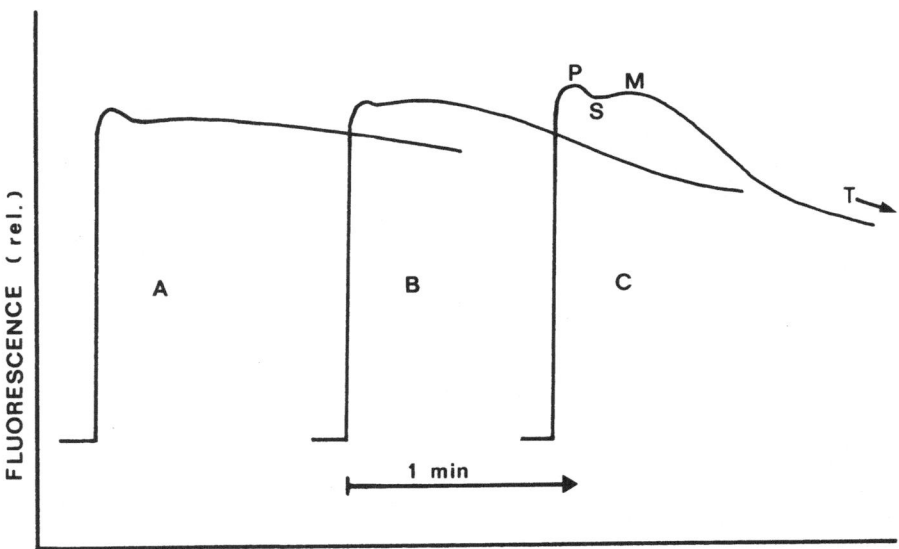

Fig. 3 - Slow chlorophyll fluorescence transients in barley leaves
at 2°C : (A) Control plants grown at 25°C, (B) and (C) plants
hardened at 2°C during 10 days and 3 weeks respectively .

thylakoids. The slow chlorophyll fluorescence transients were also monitored
at 2°C in barley leaves (Figure 3).

Low temperature inhibited markedly the slow fluorescence changes in non hardened leaves. At this temperature, fluorescence decreased very slowly from the
P level. After cold hardening of barley seedlings, the typical P-S-M-T
sequence was partly recovered. The progressive restoration of the slow fluorescence transients shows that cold acclimation of barley induced adaptation
of the photosynthetic apparatus to low temperature.

Cold adaptation of barley thylakoid membranes was examined at the protein
level. Membranes were solubilized in SDS and subjected to discontinuous SDS-
polyacrylamide gel electrophoresis at 4°C. 7 chlorophyll containing bands
were resolved (Figure 4) and characterized by their absorption spectra (not
shown here). Using Anderson's nomenclature (1), they were designated CP1a,
and CP1 (P-700 chlorophyll a protein complexes); LHCP[1] , LHCP[2] and LHCP[3]
(light-harvesting chlorophyll a/b protein complexes); CPa (possible reaction
center of PSII) and FC (free chlorophyll). Two small green bands (X and Y),
seen between CP1 and LHCP[1] in both hardened and unhardened material, could not

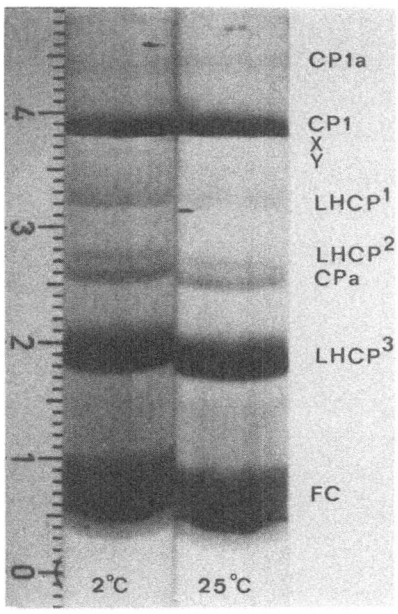

Fig.4 - Unstained polyacrylamide gels of the separated chlorophyll-protein complexes of barley plants hardened 3 weeks at 2°C or kept at 25°C.

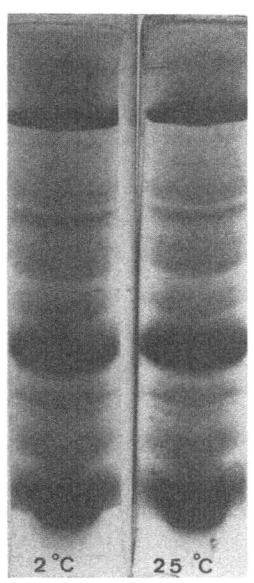

Fig.5 - Coomassie blue stained polyacrylamide gels of the separated chloroplast membrane proteins of barley plants hardened 3 weeks at 2°C or kept at 25°C.

1st D

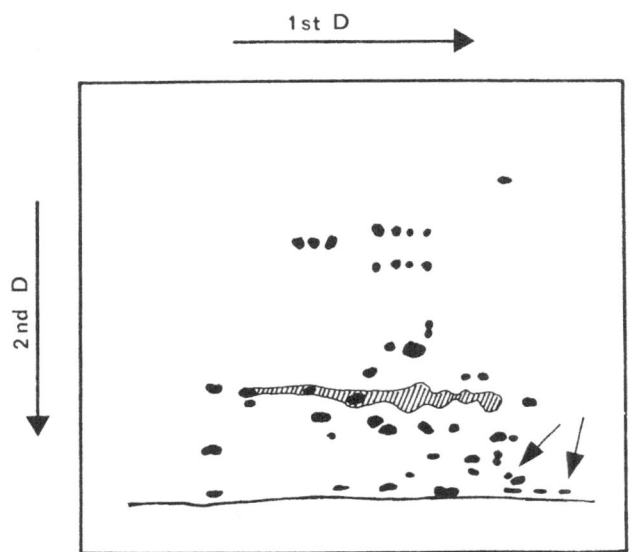

2nd D

Fig.6 - Schematic map of the chloroplast membrane polypeptides of barley plants grown at 25°C or hardened 3 weeks at 2°C. The 2 polypeptides not found in plants grown at 2°C are marked by arrows.

been characterized so far. In addition to those chlorophyll-protein complexes, Coomassie blue staining of the gels allowed us to detect 7 additionnal membrane proteins (Figure 5). No clear effect of growth at low temperature on the thylakoid proteins was observed. Figure 6 shows thylakoid proteins map obtained by means of two-dimensional electrophoresis. Although two minor polypeptides were detected only in membranes isolated from controls grown at 25°C, no marked changes were observed between hardened and non hardened chloroplasts. Our results indicate a great stability of the chloroplastic membrane proteins, which contrasts with the cold induced qualitative changes generally observed in the soluble proteins fraction (9,14,20). The adaptation of the photophysiological activity to low temperature is apparently caused by changes in the membrane lipids rather than in the membrane proteins.

Cold hardening of plants is generally accompagnied by augmentation of membranes, as deduced from total phospholipid levels (8,23), and by changes in membrane lipid composition. Most plants show an increase in phosphatidylcholine and phosphatidylethanolamine during frost hardening (22,28).

A pronounced shift from satured to polyunsatured fatty acids has been observed at low temperature (6,10). Those changes indicate alteration of the physical properties of membranes. It has been reported that fluidity of chloroplast membrane lipids, measured by spin labelling, increased during cold hardening and the magnitude of the changes was related to the frost resistance of the plants (25,26). However, Pomeroy and Raison (19) showed recently that change in membrane fluidity was not a prerequisite for the development of frost hardiness in wheat. Further experiments are under way to examine the role of membrane lipid changes in cold adaptation of winter barley.

Acknowledgements

We wish to thank "Le Fonds National Belge de la Recherche Scientifique" for financial support.

REFERENCES

1. Anderson JM, Waldron JC and Thorne SW (1978) Chlorophyll-protein complexes of spinach and barley thylakoids. FEBS Lett 92 : 227-233.
2. Arnon DI (1949) Copper enzymes in isolated chloroplasts. Polyphenol oxidase in Beta vulgaris. Plant Physiol 24 : 1-15.
3. Barber J (1976) Ionic regulation in intact chloroplasts and its effect on primary photosynthetic processes. In Barber J, ed. Topics in photosynthesis. Vol I, pp 89-134. Amsterdam : Elsevier.

4. Boschetti A, Sauton-Heiniger E, Schaffner JC and Eichenberger W (1978) A two-dimensional separation of proteins from chloroplast thylakoids and other membranes. Physiol Plant 44 : 134-140.

5. Bricker TM and Newman DW (1981) The chlorophyll-proteins of soybean (Glycine max L. var Wayne) cotyledons. Z Pflanzenphysiol Bd 104 : 91-96.

6. Chapman DJ and Barber J (1981) Adaptation in degree of fatty acid unsaturation of chloroplast membrane lipids to growth temperature. In Akoyunoglou G, ed Photosynthesis, vol VI, 359-368. Philadelphia : Balaban International Science Services.

7. Heber U, Tyankova L and Santarius KA (1973) Effects of freezing on biological membranes in vivo and in vitro. Biochim Biophys Acta 291 : 23-27.

8. Horvath I, Vigh L, Belea A and Farkas T (1980) Hardiness dependent accumulation phospholipids in leaves of wheat cultivars. Physiol Plant 49 : 117-120.

9. Huner NPA and Macdowall FDH (1976) Chloroplastic proteins of wheat and rye grown at warm and cold-hardening temperatures. Can J Biochem 54 : 848-853.

10. Kabata K, Sadakane H, Kurose M, Kobayakawa A, Watanabe T and Hatano S (1980) Changes in fatty acid composition of membrane fractions during hardening of Chlorella ellipsoidea. J Fac Agr Kyushu Univ 25 : 91-97.

11. Laemmli UK (1970) Cleavage of structural proteins during the assembly of the head of bacteriophage T4. Nature (London) 227 : 680-685.

12. Levitt J (1980) Responses of plants to environmental stresses, vol I. New York : Academic Press.

13. Lyons JM, Raison JK and Steponkus PL (1979) The plant membrane in response to low temperature : an overview. In Lyons JM et al, eds. Low temperature stress in crop plants, pp 1-24. New York : Academic Press.

14. Mäkinen A and Stegemann H (1981) Effects of low temperature on wheat leaf proteins. Phytochem 20 : 379-382.

15. Mazliak P (1981) Regulation à court terme et à long terme de l'activité des enzymes membranaires par la température. Physiol vég 19 : 543-563.

16. O'Farrell PH (1975) High resolution two-dimensional electrophoresis of proteins. J Biol Chem 250 : 4007-4021.

17. Olien CR and Smith MN (1981) Protective systems that have evolved in plants. In Olien CR and Smith MN, eds. Analysis and improvement of plant cold hardiness, pp 61-87. Boca Raton : CRC Press.

18. Papageorgiou G (1975) Chlorophyll fluorescence : an intrinsic probe of photosynthesis. In Godvinjee, ed. Bioenergetics of photosynthesis, pp 319-371. New York : Academic Press.

19. Pomeroy MK and Raison JK (1981) Maintenance of membrane fluidity during development of freezing tolerance of winter wheat seedlings. Plant Physiol 68 : 382-385.

20. Rochat E et Therrien HP (1975) Etude des protéines des blés résistant, Kharkov, et sensible, Selkirk, au cours de l'endurcissement au froid. I. Protéines solubles. Can J Bot 53 : 2411-2416.

21. Santarius KA (1980) Membrane lipids in heat injury of spinach chloroplasts. Physiol Plant 49 : 1-6 .

22. Senser M and Beck E (1982) Frost resistance in spruce (Picea abies (L.) Karst.) IV. The lipid composition of frost resistance and frost sensitive spruce chloroplasts. Z Pflanzenphysiol Bd 105 : 241-253.

23. Singh J, de la Roche IA and Siminovitch D (1975) Membrane augmentation in freezing tolerance of plant cells. Nature 257 : 669-670.

24. Steponkus PL (1978) Cold hardiness and freezing injury of agronomic crops. Adv Agron 30 : 51-98.

276

25. Vigh L, Horvath I, Farkas T, Horvath LI and Belea A (1979) Adaptation of membrane fluidity of rye and wheat seedlings according to temperature. Phytochem 18 : 787-789.
26. Vigh L, Horvath I, Horvath LI, Dudits D and Farkas T (1979) Protoplast plasmalemma fluidity of hardened wheats correlates with frost resistance. FEBS Lett 107 : 291-294.
27. Volger HG and Heber U (1975) Cryoprotective leaf proteins. Biochem Biophys Acta 412 : 335-349.
28. Willemot C (1979) Chemical modification of lipids during frost hardening of herbaceous species. In Lyons JM et al, eds. Low temperature stress in crop plants, pp 411-430. New York : Academic Press.

EFFECTS OF COLD ON CO_2 EXCHANGE IN WINTER RAPE LEAVES[*]

U. MACIEJEWSKA, J. TOMCZYK, A. KACPERSKA
Institute of Botany, University of Warsaw, Warsaw, Poland

ABSTRACT

Changes in the rates of photosynthesis and dark respiration induced by cold pretreatment ($2°C$) of winter rape plants were investigated.

At the begining of cold pretreatment (one to four days) photosynthesis as well as dark respiration readily responded to changes in the ambient temperature. Prolongation of plant exposure to cold brought about not only depression of photosynthesis but also reduction of the photosynthetic response to changes in temperature and light conditions. On the contrary, prolonged cold pretreatment slightly affected the sensitivity of dark respiration to temperature. In cold-pretreated plants the rate of dark respiration was higher than in the untreated ones at every temperature at which the measurements were carried out.

To obtain full reversion of cold-induced changes in photosynthesis and dark respiration rates caused by prolonged (6, 8 or 14 days) cold pretreatment it was necessary to expose plants for 3-7 days to higher ($25°C$) temperature.

It seems that low temperature brings about two types of effects concerning photosynthesis and respiration : a direct response of photosynthesis and respiration and an indirect one, which may depend on the cold-induced modifications in physical and metabolic properties of chloroplasts and mitochondria.

INTRODUCTION

In spite of the abundant data concerning the effects of low temperature on the photosynthetic and respiratory activities in so-called chilling-resistant species (Sycheva & Vasiukova 1972, Barta & Hodges 1970, Svec &

---

[*]This work was supported by Grant No. FG-Po/310 from the USDA, Agricultural Research Program, Public Law 480.

Hodges 1973, Sawada & Miyachi 1974 a, b, c, Steponkus et al. 1977) the information on the possible modification of CO_2 exchange pattern during the so-
called cold hardening of herbaceous plants is rather scarce. Our previous
experiments showed that CO_2 assimilation decreased and photosynthetic carbon metabolism was modified in winter rape plants, the final effect depending
on the duration of plant exposure to cold (Maciejewska et al. 1974, Sosinska
et al. 1977, Sosinska & Kacperska-Palacz 1979). On the other hand, Sycheva &
Vasiukova (1972) reported that mitochondrial activity increased in relation
to cold hardening.

Since availability of adenine- and nicotinamide-nucleotides seems to be
the key factor for plant hardening capability (Levitt 1972, Kacperska-Palacz
1978) mutual relationship between processes engaged in synthesis of these
products, i.e. photosynthesis and dark respiration, might be of great importance in cold hardening process. Therefore, comparative studies were undertaken on the effects of low temperature on the rates of CO_2 uptake in light
and CO_2 evolution in darkness during prolonged exposure of winter rape
plants to 2° C.

MATERIAL AND METHODS

Winter rape plants (*Brassica napus* L., var. *oleifera* L.) were grown, as
previously described, on aqueous nutrient solution (Kacperska-Palacz et al.
1977) or in soil (Obłój & Kacepreska 1981) in growth cabinet at a day/night
temperature of 25/20° C, under a 16 h day, at light intensity of about
15 W m^{-2} from day light fluorescent tubes (Polam, Poland). After 5 weeks
of growth, half of plants were subjected to continuous (day/night) low
temperature (2° C), the other conditions remaining unchanged. The other
continued growth at the 25/20° C temperature. The rates of CO_2 uptake in
light (APS) and CO_2 evolution in darkness (DR) were followed at various
temperatures after different periods of plant cold pretreatment.

In other experiments the rates of CO_2 exchange were measured after
transferring the cold-pretreated plants to warm temperature (25/20° C) for
one, two or seven days.

The CO_2 exchange was measured by using Infrared Gas Analyzer (Junkalor).
Above-ground portions of plants sealed in a plexiglass chamber were included
into a closed circuit system with CO_2 analyzer and kept at the desired
temperature (temperature of measurement). The roots of the plants were
maintained outside the chamber, in water at the room temperature. The volume

of the system was 21 l, air flow rate 20 l per min. The light, supplied with white photolamps Narva 500 W (230 W m^{-2}), was passing throught a water filter to revent heating. The measurements of CO_2 exchange were carried out at 25°, 10° or 1° (± 1° C) inside the assimilation chamber. The rate of the temperature change was 4° C per 10 min.

In some experiments the light intensity during measurements of CO_2 uptake at 25° C and 1° C was differentiated into 15, 30, 55, 95, 170 and 230 W m^{-2} in order to determine the light response curves.

Intensity of CO_2 exchange was calculated on dry weight basis (leaf dry weight or leaf, petiole and stem dry weight for photosynthesis or respiration, respectively).

At least two separate series of experiments were carried out. The results of each series showed the same trends in CO_2 exchange pattern, although they differed in the maximal rate of CO_2 uptake. Thus, the results of only one representative series for each variant of the experiment were taken into consideration in the presented paper. The presented curves were calculated from experimental data by the least squares method.

RESULTS

The data shown in Fig. 1 indicate that photosynthesis in the control and in the cold-pretreated plants responded differently to the change of the ambient temperature. In the control plants (cf. thin lines) CO_2 assimilation readily decreased with the reduction of the ambient temperature, independently of the day of experiment. In the cold-pretreated ones (cr. thick lines) a marked decrease of photosynthesis rate was observed after two days of exposure to cold, even when measured at 25° C. Prolongation of the cold pretreatment caused further depression of photosynthesis and also reduced its response to the change in the ambient temperature. After four days of cold pretreatment the rate of photosynthesis in hardened plants, measured both at high (25° C) and at low (1° C) temperature remained even lower than that in unharedened plants, when measured at 1° C.

In contrast to photosynthesis, respiration rate depended more on the ambient temperature than on the period of pretreatment with cold : it decreased with reduction of the ambient temperature, independently of the day of experiment (Fig. 2). It should, however, be emphasized, that in hardened plants the respiration rate was always higher than that in unhardened plants, independently of the temperature of measurement. Furthermore, a marked increase

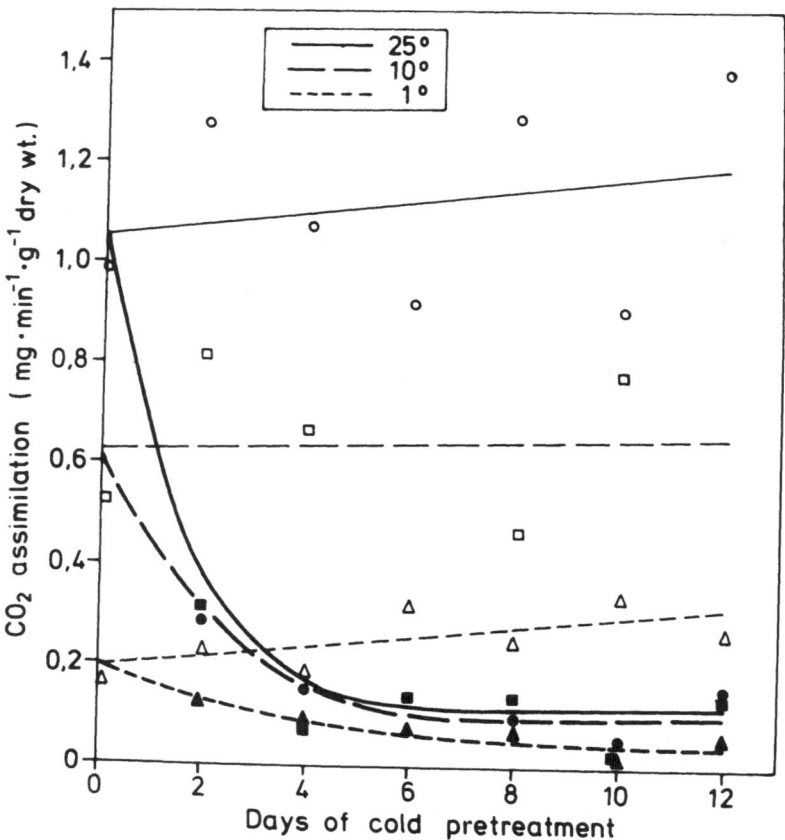

Fig. 1. Effects of cold pretreatment on CO_2 assimilation rate in rape
plants : (O,□,△). control, (●,■,▲) cold pretreated plants.

of dark respiration was observed in plants exposed to cold for two or four days
if measurements were performed at 25° C.

As a result of the temperature-induced alterations in CO_2 exchange, APS:DR
ratio in plants was changed. In unhardened plants lowering of the ambient
temperature increased the ratio from 11 (at 25° C) to about 14 (at 1° C). The
effect was due to the higher inhibition of dark respiration than that of photo-

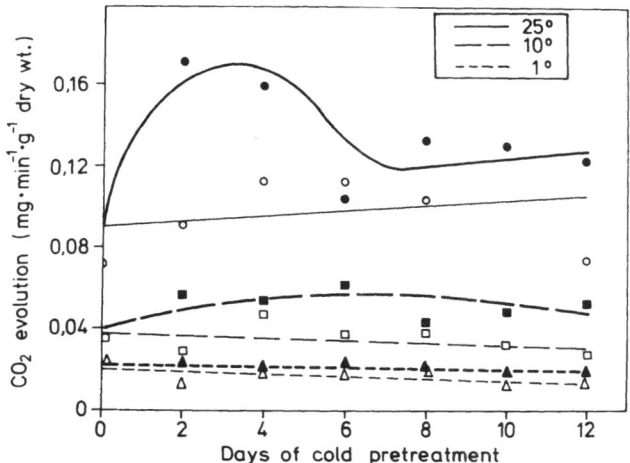

Fig. 2.   Effects of cold pretreatment on dark respiration rate in rape
plants : (O ,□ ,Δ) control, (●,■,▲) cold-pretreated plants.

synthesis. The opposite effect was found in the cold-pretreated plants. The
ratio markedly decreased after two days of plant exposure to cold and prolor
gation of cold pretreatment caused a further decrease of the ratio down to
values close to 1.0.

We also observed that low temperature pretreatment affected photosynthesis response to light, too. At 25° C, photosynthesis in unhardened
plants continuously increased with light intensity and did not reach the
saturation point even at 230 W m^{-2}, whereas photosynthesis in hardened
plants slowly responded to an increase in light intensity and reached
plateau at 170 W m^{-2}. At 1° C the effect of cold pretreatment on photosynthesis response to the change in light intensity was less visible
(data not shown).

In order to check the reversibility of the cold-induced changes in
CO_2 exchange, the rates of photosynthesis and dark respiration were investigated in cold-pretreated plants transferred to warm temperature for
1-7 days. In his experiment plants were grown in soil (data not shown). It
was found that the cold-induced changes in CO_2 exchange were gradually
reversed during dehardening : the rate of CO_2 assimilation increased,
whereas the rate of dark respiration decreased. More hardy plants needed
a longer period of dehardening to achieve photosynthesis and respiration

rates similar to those noted in the control. During dehardening the increase of APS : DR ratio was also observed. The values noted at 25° C for fully dehardened plants exceeded those noted in the control.

DISCUSSION

The above presented data indicate that temperature affects CO_2 exchange in the studied plants in two ways : directly and indirectly. The direct effect of temperature, consisting in a marked but easily reversible decrease of both photosynthesis and dark respiration rates at lowered ambient temperature does not seem to depend on physical and stomatal factors only, since CO_2 output is reduced more than CO_2 uptake. The more pronounced effect of low temperature on dark respiration than on photosynthesis might suggest that mitochandria are more sensitive to sudden temperature drop than chloroplasts. Another, indirect effect of low temperature on CO_2 exchange in winter rape plants seems to differ for photosynthesis and for respiration. In the case of photosynthesis the effect consists in a long lasting depression of CO_2 assimilation and a decreased responsiveness of the process to temperature increase and higher light intensity. Such a diminished photosynthetic response to temperature and light conditions was already observed by Barta & Hodges (1970) for cold-hardened wheat plants.

It seems that it is due mostly to modified properties of photosynthetic apparatus than to the cold-induced disturbance in sink/source relationship (Neales & Incoll 1968). The inhibition of long distance transport of photosynthates can not be taken into account because the roots of the studied plants were kept at room temperature. Thus, the acceptor activity was not reduced and possibly did not inhibit the photosynthesis rate in the upper part of the plants. Short distance transport of photosynthates can not be taken into account either because the inhibitory effects of the prolonged cold pretreatment were sustained for several days during dehardening.

The structural alterations of chloroplast components upon hardening may be expected both from the results of our previous studies on the cold-induced changes in photosynthetic carbon metabolism, and from the data reported by other authors. As a matter of fact, Steponkus et al. (1977) demonstrated the cold-Induced modifications in thylakoid membranes isolated from hardy spinach plants.

The higher rates of respiration in the cold-pretreated plants than in the untreated ones at any temperature of measurement seem to indicate that the effectiveness of cellular respiration increases during cold hardening. According to Miller et al. (1974), Pomeroy & Andrews (1975), cold hardening slightly affects the mitochondrial respiratory activities. Thus, the observed at high (25° C) temperature increased rates or even "burst" of CO_2 output in the cold-pretreated plants may indicate that some respiratory intermediates (possibly glycolysis products) which accumulated at low temperature were rapidly used up when the direct inhibitory effect of cold on the mitochondrial activity dissapeared. Similar enhancement of respiration by higher temperature in cold-hardened pea and potato plants was previously described by Sycheva & Vasiukova (1972).

The modifyng effect of cold pretreatment on photosynthesis and respiration concedes relatively easily during the first few days of the experiment, but it took about a week to disappear in the cold-adapted tissue. This may point to a new steady state of metabolism achieved in the latter tissue. It seems that the previously described (Sobczyk & Kacperska-Palacz 1980) biphasic response of winter rape tissue to cold can also be interpreted in terms of photosynthetic and respiration activities : after the reaction phase, during which the tissue readily responds to temperature change, plants pass to the restitution phase, when the cold-induced disturbance are overcome and a new metabolic balance is established.

The observation, that during prolonged exposure of plants to cold the respiratory metabolism is not inhibited to such extent as photosynthesis may have an important biological implication. The maintenance of a high level of ATP (Sobczyk & Kacperska-Palacz 1978) as well as of a high level of reducing power (Kuraishi 1968) in cold adapted plants may depend on respiratory activities more than on photosynthesis. However, these problems necessitate further studies.

REFERENCES

Barta, A.L. & Hodges, H.F. Characterization of photosynthesis in cold hardening in cold hardening winter wheat. Crop Sci. 10 : 535-238 (1970).
Kacperska-Palacz, A. Mechanism of cold acclimation in herbaceous plants. In Li, P.H. & Sakai, A., eds. Plant cold hardiness and freezing stress, pp. 139-152. Academic Press, New York (1978).
Kacperska-Palacz, A., Długokecka, E., Breitenwald, J. & Wciślińska, B. Physiological mechanisms of frost tolerance : Possible role of protein in plant adaptation to cold. Biol. Plant. 19 : 10-17 (1977).
Kuraishi, S., Arai, N., Ushijima, T. & Tazaki, T. Oxidized and reduced nicotinamide adenine dinucleotide phosphate levels of plants hardened and unhardened against chilling injury. Plant Physiol. 43 : 238-242 (1968).

Levitt, J. Responses of plants to environmental stresses. pp. 110-228. Academic Press, New York (1972).

Maciejewska, U., Maleszewski, S., Kacperska-Palacz, A. Effect of cold hardening on the carbon photosynthetic metabolism in rape leaves. Bull. Acad. Pol. Sci. 22 :   513-517 (1974).

Miller, R.W., de la Roche, I. & Pomeroy, M.K. Structural and functional responses of wheat mitochandrial membranes to growth at low temperatures. Plant Physiol. 53 :   426-433 (1974).

Neales, T.F. & Incoll, L.D. The control of leaf photosynthetisis rate by the level of assimilate concentration in the leaf : a review of the hypothesis. Bot. Rev. 34 :   107-125 (1968).

Ob/lój, H. & Kacperska, A. Dessication tolerance changes in winter rape leaves grown under different environmental conditions. Biol. Plant. 23 : 209-213 (1981).

Pomeroy, M.K. & Andrews, Ch. J. Effect of temperature on respiration of mitochondria and shoot segments from cold-hardened and nonhardened wheat and rye seedlings. Plant Physiol. 56 :   703-706 (1975).

Sawada, S. & Miyachi, S. Effects of growth temperature on photosynthetic carbon metabolism in green plants I. Photosynthetic activities of various plants acclimatized to varied temperatures. Plant and Cell Physiol. 15 : 111-120 (1974 a).

------- II. Photosynthetic $^{14}CO_2$-incorporation in plants acclimatized to varied temperatures. Plant and Cell Physiol. 15 : 225-238 (1974 b).

------- III. Differences in structure, photosynthetic activities and acitivities of ribulose diphosphate carboxylase and glycolate oxidase in leaves of wheat grown under varied temperatures. Plant and Cell Physiol. 15 : 239-248 (1974 c).

Sobczyk, E.A. & Kacperska-Palacz, A. Adenine nucleotide changes during cold acclimation of winter rape plants. Plant Physiol. 62 :   875-878 (1978).

Sobczyk, E.A. & Kacperska-Palacz, A. Changes in some enzyme activities during cold acclimation of winter rape plants. Acta Physiol. Plant. 2 :   123-131 (1980).

Sosińska, A. & Kacperska-Palacz, A. Ribulosediphosphate and phosphoenolpyruvate carboxylase activities in winter rape as related to cold acclimation. Z. Pflanzenphysiol. 92 : 455-458/1979).

Sosińska, A. Maleszewski, S. & Kacperska-Palacz, A. Carbon photosynthetic metabolism in leaves of cold-treated rape plants. Z. Pflanzenphysiol. 83 : 285-292 (1977).

Steponkus, P.L., Garber, M.P., Myers, S.P. & Lineberger, R.D. Effects of cold acclimation and freezing on structure and function of chloroplast thylakoids. Cryobiology 14 : 303-321 (1977).

Svec, L.V. & Hodges, H.F. Respiratory activity in barley seedlings during cold hardening in controlled and natural environments. Can. J. Plant. Sci. 53 : 457-463 (1973).

Sycheva, Z.F., Vasiukova, V.A. Effect of frost hardening on respiratory metabolism of actively vegetating plants. Fizol. Rast. 19 : 824-830 (1972).

MOLECULAR SPECIES OF PHOSPHATIDYLGLYCEROLS ASSOCIATED WITH THE CHILLING
SENSITIVITY OF HIGHER PLANTS

NORIO MURATA
Department of Biology, University of Tokyo, Komaba, Meguro-ku, Tokyo 153,
Japan

ABSTRACT

Chilling-sensitive and chilling-resistant plants were compared with respect
to the compositions of fatty acids and molecular sepcies of a leaf-phospho-
lipid, phosphatidylglycerol. The sum of palmitic plus $trans$-Δ^3-hexadecenoic
acid contents ranged from 50 to 57% of the total fatty acids of this lipid
in chilling-resistant plants, and from 60 to 78% in chilling-sensitive plants.
The sum of the contents of dipalmitoyl plus 1-palmitoyl-2-($trans$-Δ^3-
hexadecenoyl) species ranged from 3 to 19% of the total species of phosphatidylglycerol in chilling-resistant plants, and from 26 to 65% in chilling-
sensitive plants. These findings suggest that these two molecular species
of phosphatidylglycerol are closely associated with the chilling sensitivity
of the plants.

INTRODUCTION

In a current concept for the mechanism of chilling injury (1,2), the
primary event is the formation of a gel phase of lipids in cellular membranes
at low temperatures. In cells of the blue-green alga, *Anacystis nidulans*,
which contain simple membrane systems, the phase separation of lipids in the
cytoplasmic membranes is directly related to chilling injury (3-5). The
membrane systems of higher plant cells are much more complex than those of
the blue-green algae; it is still uncertain which membranes in the cells
undergo the phase transition at low temperatures to induce the chilling
injury.

Murata and Fork (6) studied the temperature dependence of the kinetics of

---

Abbreviations: 16:0, palmitic acid; 16:1^t, $trans$-Δ^3-hexadecenoic acid;
18:0, stearic acid; 18:1, oleic acid; 18:2, linoleic acid; 18:3, α-linolenic
acid; PG, phosphatidylglycerol; GC-MS, gas chromatography-mass spectrometry;
GLC, gas-chromatography; TLC, thin-layer chromatography.

the electrochromic shift of carotenoid absorption which was produced by the
membrane potential across the thylakoid membranes. Characteristic break points
in the Arrhenius plot of the dark decay of the spectral change at about 10°C
were observed in the chilling-sensitive but not in the chilling-resistant
plants; this phenomenon was explained in terms of the leakiness of the
thylakoid membranes in the phase separation state.

Pike and Berry (7), using parinaric acid as a fluorescent probe, observed
that phospholipid preparations derived from desert plant leaves underwent the
thermotrophic phase transition at about room temperature. Fork et al. (8),
using the same technique, reported similar phase behaviour in the phospholipid
preparations from chloroplasts of chilling-sensitive plants. These findings
suggest the possibility that the chilling-sensitive but not the chilling-
resistant plants contain phospholipid molecular species which undergo the
phase transition at about room temperature and are therefore associated with
the chilling sensitivity of the plants.

The constituent lipids of the thylakoid membranes of the chloroplasts are
monogalactosyl diacylglycerol, digalactosyl diacylglycerol, sulfoquinovosyl
diacylglycerol, phosphatidylglycerol and phosphatidylcholine. The major lipid
components among them, the galactolipids, contain a high proportion of poly-
unsaturated fatty acids (9), suggesting that the phase transition temperatures
of all the molecular species of these lipids should be far below room temperature. On the other hand, phosphatidylglycerol (PG) contains a relatively
high proportion of saturated fatty acids (9), and therefore may contain
molecular species which undergo the phase transition at about room temperature.

In the present study, we isolated PG from leaves of chilling-sensitive and
chilling-resistant plants, and compared the compositions of fatty acids and
molecular species (determined by combination of two fatty acid molecules) in
this lipid.

MATERIALS AND METHODS

Leaves were collected from plants cultivated in a green-house or in a growth
chamber, or growing on the Komaba campus of the University of Tokyo or in farm
fields. Some leaves were purchased from a local market.

Lipids were extracted from the leaves according to Bligh and Dyer (10),
and the lipid extract was subjected to the column chromatography with DEAE-
Sepharose CL-6B to separate the acidic lipids (11). From the acidic lipid
fraction, PG was isolated by TLC on silica gel.

To investigate the fatty acid compositions of PG, this lipid was subjected
to methanolysis, and the resultant methyl esters were analyzed by GLC. The
distributions of fatty acids in the C-1 and C-2 positions of PG were investigated by means of enzymic hydrolysis with the lipase from *Rhizopus delemar*
(12).

To investigate the molecular species composition, PG was treated with
phospholipase C, and the resultant diacylglycerols were acetylated with acetic
anhydride in pyridine (13). The monoacetyldiacylglycerols thus prepared were
subjected to GLC and GC-MS. Details of the analytical procedures will be
described elsewhere (14).

RESULTS

In Table 1. 10 species of chilling-sensitive plants and 11 species of
chilling-resistant plants are compared with respect to the fatty acid
compositions of PG from their leaves. This lipid contained 16:0, 18:0, 18:1,
18:2, 18:3 and a unique fatty acid, *trans*-Δ^3-hexadecenoic acid ($16:1^t$); the
latter exists exclusively in PG (15). The relative content of 16:0 varied
from 14% (spinach) to 45% (cabbage) in the chilling-resistant plants, and
from 22% (cyclamen) to 55% (kalanchoe) in the chilling-sensitive plants.
The relative content of $16:1^t$ ranged from 5% (cabbage) to 40% (red clover)
in the chilling-resistant plants, and from 6% (kalanchoe) to 39% (cyclamen)
in the chilling-sensitive plants. The variation in the relative contents
of 16:0 and $16:1^t$ were very large in both the chilling-sensitive and the
chilling-resistant plants. No relationship seems to exist between the chilling
sensitivity of plants and the relative content of either of these fatty acids
alone. Wide variations in the relative contents of C_{18} acids were also found
in both groups of plants (Table 1). Nevertheless, the sum of 16:0 plus $16:1^t$
contents tended to be higher in the chilling-sensitive than in the chilling-
resistant plants. It ranged from 50 to 57% in the chilling-resistant plants,
and from 60 to 78% in the chilling-sensitive plants (Table 1).

The distribution of fatty acids in the C-1 and C-2 positions of *sn*-glycerol
of PG from the leaves was studied. The results indicate that $16:1^t$ was
exclusively localized at the C-2 position, whereas 16:0 was found at both
the C-1 and C-2 positions. The C_{18} acids were more abundant at the C-1 than
the C-2 position.

A comparison of the molecular species compositions of PG in chilling-
resistant and chilling-sensitive plants is presented in Table 2. The

TABLE 1.   Fatty acid compositions of phosphatidylglycerol from leaves of chilling-sensitive and chilling-resistat plants.

Plant	Fatty acid composition (mol %)						
	16:0	16:1^t	18:0	18:1	18:2	18:3	16:0 + 16:1^t
Chilling-sensitive plants							
Sweet potato	41	37	4	6	5	7	78
Taro	42	35	3	3	9	8	77
Rice (Indica)[*]	35	36	2	2	8	13	71
Rice (Japonica)[**]	34	32	1	2	7	18	68
Castor bean	37	30	5	20	3	5	67
Tobacco	31	32	2	6	10	19	63
Squash	36	26	2	3	8	25	62
Kalanchoe	55	6	2	8	15	14	61
Cyclamen	22	39	1	20	9	9	61
Corn	32	28	1	1	9	29	60
Chilling-resistant plants							
Pea	37	20	1	12	13	17	57
Cluster amaryllis	35	20	1	4	16	24	55
Dandelion	20	33	0	3	17	27	55
Japanese radish	35	19	1	2	6	37	54
Red clover	14	40	1	7	13	25	54
Chinese cabbage	33	20	1	1	7	38	53
Spinach	14	39	0	2	4	41	53
Welsh onion	35	17	1	3	18	26	52
Cabbage	45	5	1	3	21	25	50
Oat	28	22	1	2	8	39	50
Wheat	22	28	1	2	9	38	50

[*] contained 4% octadecatetraenoic acid

[**] contained 6% octadecatetraenoic acid

TABLE 2. Molecular species composition of phosphatidylglycerol from leaves of chilling-sensitive and chilling-resistant plants.

Plant	$\begin{bmatrix}16:0\\16:0\end{bmatrix}$	$\begin{bmatrix}16:0\\16:1\end{bmatrix}^t$	$\begin{bmatrix}18:0\\16:0\end{bmatrix}$	$\begin{bmatrix}18:0\\16:1\end{bmatrix}^t$	$\begin{bmatrix}18:1\\16:0\end{bmatrix}$	$\begin{bmatrix}18:1\\16:1\end{bmatrix}^t$	$\begin{bmatrix}18:2\\16:0\end{bmatrix}$	$\begin{bmatrix}18:2\\16:1\end{bmatrix}^t$	$\begin{bmatrix}18:3\\16:0\end{bmatrix}$	$\begin{bmatrix}18:3\\16:1\end{bmatrix}^t$	16:0/16:0 + 16:0/16:1^t
Chilling-sensitive plants											
Sweet potato	26	39	3	3	7	8	4	4	2	4	65
Taro	20	42	1	3	3	3	11	4	7	6	62
Castor bean	20	22	7	1	15	29	3	1	1	1	42
Tobacco	19	20	0	5	1	15	11	7	7	16	39
Sponge cucumber	16	23	2	13	4	19	5	3	4	11	39
Corn	20	17	0	0	0	4	5	11	17	26	37
Kalanchoe	32	2	3	0	19	0	26	1	11	6	34
Squash	17	14	0	2	0	5	13	3	23	23	31
Cyclamen	5	21	0	3	11	34	8	8	4	5	26

Molecular species composition (mol %)

TABLE 2 continued.

Plant	Molecular species composition (mol %)										
	[16:0/16:0]	[16:0/16:1]t	[18:0/16:0]	[18:0/16:1]t	[18:1/16:0]	[18:1/16:1]t	[18:2/16:0]	[18:2/16:1]t	[18:3/16:0]	[18:3/16:1]t	16:0/16:0 + 16:0/16:1]t
Chilling-resistant plants											
Pea	12	7	1	0	27	6	15	10	15	7	19
Lettuce	13	4	0	1	1	1	26	4	26	24	17
Cluster amaryllis	6	5	0	2	6	3	28	7	23	20	11
Red clover	4	6	0	0	5	23	11	14	12	25	10
Japanese radish	5	3	1	0	3	1	5	4	51	26	8
Welsh onion	5	2	1	0	4	1	27	11	26	24	7
Chinese cabbage	4	3	1	0	2	0	9	3	47	32	7
Spinach	1	5	0	0	1	0	5	0	16	72	6
Oat	3	2	0	0	2	0	11	3	33	45	5
Cabbage	3	1	0	0	4	0	45	2	43	2	4
Dandelion	2	2	0	0	3	12	16	29	15	20	4
Wheat	1	2	0	0	2	0	11	5	32	47	3

16:0/16:0 species accounted for 1% (spinach and wheat) to 13% (lettuce) of
the total PG in the chilling-resistant plants, and 5% (cyclamen) to 32%
(kalanchoe) in the chilling-sensitive plants. The proportion of 16:0/16:1^t
species ranged from 1% (cabbage) to 7% (pea) in the chilling-resistant plants,
and from 2% (kalanchoe) to 39% (sweet potato) in the chilling-sensitive plants.
Thus, the relative content of either of these molecular species alone was not
correlated with the chilling sensitivity of the plants. The sum of the relative contents of 16:0/16:0 plus 16:0/16:1^t species, however, did correlate
well with the chilling sensitivity. It varied from 3% (wheat) to 19% (pea)
in the chilling-resistant plants, and from 26% (cyclamen) to 65% (sweet potato)
in the chilling-sensitive plants. Among the chilling-sensitive plants, the
sensitivity is greatent in sweet potato, taro, rice and castor bean and least
in cyclamen and kalanchoe. Among the chilling-resistant plants, wheat, oat
and spinach are the most hardy and pea and lettuce are the least. It may be
seen in Table 2 that the order of the 16:0/16:0 plus 16:0/16:1^t contents seems
to correspond to the degree of chilling sensitivity of the plants.

DISCUSSION

The compositions of fatty acids and molecular species of PG indicate that
the sum of the 16:0 plus 16:1^t contents relative to the total fatty acids of
PG and the sum of the 16:0/16:0 plus 16:0/16:1^t species contents relative to
the total species of PG correlate well with the chilling sensitivity of the
plants. The results are summarized in Fig. 1. Clear separations of the
distributions of these parameters between the chilling-sensitive and the
chilling-resistant plants are seen. Therefore, the parameters may be used
as a marker of chilling sensitivity of the plants.

The 16:0/16:0 species of PG undergoes the gel to liquid crystalline phase
transition at 42°C (16). Although the phase behaviour of the 16:0/16:1^t
species of PG is not known, the phase transition temperature of this species
may be estimated. Trans unsaturation is much less effective than cis unsaturation in decreasing the phase transition temperature (17), and an unsaturation bond near the end of a hydrocarbon chain is much less effective
than one at the center of the chain (18). Thus the introduction of a trans
unsaturation bond at the Δ^3 position may decrease the phase transition temperature only slightly. It is reasonable, therefore, that these two molecular
species (16:0/16:0 and 16:0/16:1^t) of PG would form the gel phase domains in
the membranes at the chilling temperature and cause chilling injury in the

292

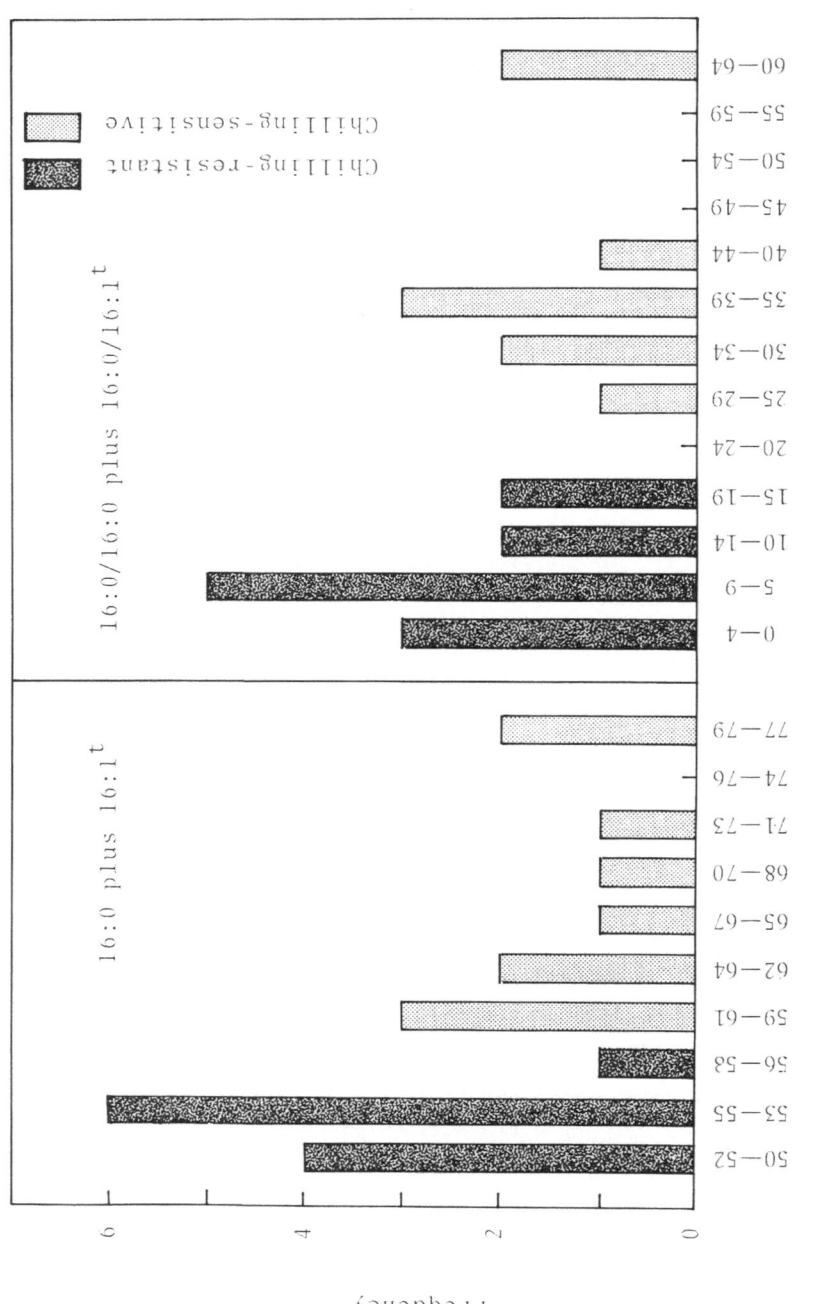

FIGURE 1. Histograms of the sum of the 16:0 plus 16:1^t contents relative to the total fatty acids in PG, and of the sum of the 16:0/16:0 plus 16:0/16:1^t contents relative to the total species of PG.

plants. This consideration explains why we can regard the sum of the $16:0/16:0$ and $16:0/16:1^t$ contents in PG as a measure of the chilling sensitivity of higher plant leaves.

The molecular species $18:0/16:0$ and possibly $18:0/16:1^t$ should also undergo the phase transition at about room temperature. The contents of these molecular species were very low in the chilling-resistant plants, and amounted to a considerable level in some chilling-sensitive plants. The other molecular species containing either 18:1, 18:2 or 18:3 would be expected to undergo the phase transition below room temperature, and therefore would not take part in chilling injury (11).

The $16:0/16:1^t$ species concerned in this study should be 1-palmitoyl-2-($trans$-Δ^3-hexadecenoyl)-PG, since $16:1^t$ was esterified only to the C-2 position of snglycerol in this lipid.

REFERENCES

1. Raison JK. 1973. *J. Bioenerg.* 4: 258-309.

2. Lyons JM. 1973. *Annu. Rev. Plant Physiol.* 24: 445-466.

3. Ono T, Murata N. 1981. *Plant Physiol.* 67: 176-181.

4. Ono T, Murata N. 1981. *Plant Physiol.* 67: 182-187.

5. Ono T, Murata N. 1982. *Plant Physiol.* 69: 125-129.

6. Muratan, Fork DC. 1977. *Biochim. Biophys. Acta* 461: 365-378.

7. Pike CB, Berry JH. 1980. *Plant Physiol.* 66: 238-241.

8. Fork DC, van Ginkel G, Harvey G. 1981. *Plant & Cell Physiol.* 22: 1035-1042.

9. Harwood JL. 1980. *In* the Biochemistry of Plants. Edited by Stumpf PK. Vol. 4. pp. 1-55. Academic Press, New York.

10. Bligh EG, Dyer WJ. 1959. *Can. J. Biochem. Physiol.* 37: 911-917.

11. Murata N, Sato N, Takahashi N, Hamazaki T. 1982. *Plant & Cell Physiol.* 23: in press.

12. Fischer W, Heinz E, Zeus M. 1973. *Hoppe-Seyler's Z. Physiol. Chem.* 354: S. 1115-1123.

13. Kito M, Ishinaga M, Nishihara M, Kato M, Sawada S, Hata T. 1975. *Eur. J. Biochem.* 54: 55-63.

14. Murata N. 1982. *Plant & Cell Physiol.* 23: in press.

15. Weenink RO, Shorland FB. 1964. *Biochem. Biophys. Acta* 84: 613-614.

16. Jacobson EK, Papahadjopoulos D. 1975. *Biochemistry* 14: 152-161.

17. Phillips MC, Hauser H, Paltauf F. 1972. *Chem. Phys. Lipids* 8: 127-133.

18. Barton PG, Gunstone FD. 1975. *J. Biol. Chem.* 250: 4470-4476.

INVESTIGATIONS ON THE HEAT-SENSITIVITY OF THYLAKOID MEMBRANES IN SPINACH
LEAVES: THE INFLUENCE OF LIGHT AND SHORT-TIME ACCLIMATIZATION TO HIGH
TEMPERATURES

E. WEIS

Botanisches Institut der Universität Düsseldorf,
Universitätsstraße 1, D-4000 Düsseldorf 1, Germany

ABSTRACT

The electrochromic pigment absorption change at 518 nm was measured in
intact spinach leaves after heat-stress to test the thermal sensitivity
of the thylakoid membrane in vivo. The temperature which induced 50 %
inhibition of the absorption change increased by 3° - 5°C when the leaves
were illuminated before and during heat-stress or when they were preincubated for at least 1 h at 35°C (short-time acclimatization). In 35°C -
acclimated leaves light had no further protecting effect.

Chlorophyll fluorescence studies showed that both the light-induced
and the temperature-induced increase in thermal stability of the thylakoid
membrane are closely correlated with an interconversion of the photosynthetic pigment apparatus; photosystem II was converted to a low fluorescent state. The results are interpreted in terms of a close mechanistic
relationship between light-adaptation and short-time temperature acclimatization of the photosynthetic apparatus. The occurrence of a temperature-
induced change in the distribution of excitation energy between photosystem I and photosystem II is suggested to be a primary event in temperature acclimatization.

1. INTRODUCTION

The ability of a given genotype to adjust the heat-sensitivity of the
photosynthetic apparatus within certain limits to the thermal environment
is well documented; it has been suggested that high-temperature acclimatization results in an increased stability of the pigment-protein complexes
in the thylakoid membrane (for a review see ref. 2). This suggestion well
agrees with the observation that the lipid composition of cellular membranes
varies with the growth temperature (e.g. refs. 4 and 8). However, at least
part of the high-temperature acclimatization has been shown to occur within

a few hours. Therefore, changes in the chemical composition of the chloroplast membranes cannot play a significant role in this short-time acclimatization; rather, rapid changes occurring in the stroma could modulate the membrane structure (9). For instance, alterations in the ionic milieu can change the physical phase of the thylakoid membrane and, hence, can influence the thermal stability of membrane-bound protein complexes, as has been discussed recently (13; see also ref. 3).

Under certain conditions, the thermal stability of the photosynthetic apparatus is increased by light as well (5, 14). Most likely, the mechanism of this light-induced stabilization involves the light-induced transport of protons from the stroma to the thylakoid space and accompanying ion fluxes, i.e., the light-induced energization of thylakoids (14). As is widely accepted, light affects the structure of thylakoids and changes the distribution of the available excitation energy between PS I and PS II; this regulation has been discussed in terms of light-adaptation of the photosynthetic apparatus (for a review see ref. 7).

In the present paper, the relation between light-adaptation and short-time temperature acclimatization of thylakoid membranes in intact spinach leaves was studied, and it is discussed whether there could be a mechanistic linkage between these two regulation mechanisms. The light-induced electrochromic pigment absorption change at 518 nm which is known to indicate the light-induced electrical field gradient across the thylakoid membrane (15) was used for testing the sensitivity of the thylakoid membrane to mild heat stress. Structural and functional modulations of the thylakoid pigment apparatus were demonstrated by chlorophyll fluorescence studies.

2. MATERIAL AND METHODS

Spinach (*Spinacia oleracea* L.) was grown in a growth chamber under the following conditions: 10 h light at $20^{\circ}C$ / 14 h dark at $15^{\circ}C$, light-intensity about 15,000 lux; 70 % rel. humidity. For short-time temperature acclimatization plants or freshly detached leaves were incubated for various times at $35^{\circ}C$ in the dark in air saturated with water vapour. Control leaves were kept under the same conditions at $20^{\circ}C$.

For heat-treatment control leaves or $35^{\circ}C$ - acclimated leaves were kept for 6 min in a temperature-controlled water bath which was equipped with a fiber optic to allow illumination (red light, RG 630 filter from Schott) during the heat-treatment. If the leaves were heated in the light, they were first pre-illuminated for 3 min at 20^OC and then rapidly transferred into the illuminated water bath.

Light-induced absorption changes of leaves at 518 nm were measured under conditions described earlier (14). The illumination with red actinic light (RG 630; 200 W/m^2) was short (4 s) in order to avoid substantial light-induced light-scattering changes.

Corrected chlorophyll fluorescence spectra were measured at 77^OK in a Farrand spectrofluorimeter (MK 1) as follows: discs from control leaves or 35^OC - acclimated leaves were placed for 10 min either in the dark or in the light in a mortar and then rapidly frozen in liquid N_2. The leaf-discs were carefully ground at low temperature and the resulting frozen leaf-powder (about 25 µg chlorophyll) was mixed with ice-powder obtained from 4 ml water. For fluorescence measurement this mixture was transferred into a low-temperature cuvette. Actinic light: 465 - 485 nm.

Induction kinetics of chlorophyll fluorescence were measured at room temperature with leaves that were predarkened for 3 h.

3. RESULTS

Mild heat stress causes an increase in the ionic conductivity of the thylakoid membrane and, hence, a break-down of the light-induced electrical field gradient across the membrane before affecting photosynthetic electron transport (10,11); the decrease in the electrical gradient is accompanied by an inhibition of the stromal carboxylation reaction and the overall photosynthetic CO_2-fixation (10,12,14). However, this mild heat effect is reversible and, therefore, is not related to lethal damage of leaves. Nevertheless, a transient decrease in the photosynthetic productivity occurring during peak temperatures most likely will diminish the ecological resistance of a given species and, thus reflects a primary response to heat stress. In the present study the electrochromic absorption change (ΔA at 518 nm) of thylakoid pigments is used for testing this primary heat effect. Fig. 1 shows the light-induced 518 nm change of intact leaves as a function of the pretreatment temperature (6 min heat stress) to which the leaves were exposed in the dark or in the light. If control leaves (20^OC - acclimated) were heated in the light, the temperature which induced 50 % inhibition ($T_{0.5}$) increased by 3° - 5^OC. Essentially the same shift in $T_{0.5}$ was observed when leaves were preincubated for 2 h at 35^OC in the

298

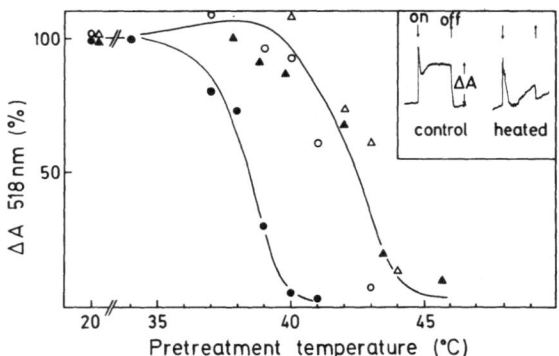

FIGURE 1. Light-induced absorption change (ΔA) at 518 nm of leaves as a
function of the temperature of a 6 min heat-treatment. Δ A was determined
from the light-off signal after 4 s of illumination, as indicated in the
insert. ● = leaves stored at 20°C in the dark and then heat-treated in
the dark; O = leaves stored at 20°C in the dark and heat-treated in the
light; ▲ = leaves stored for 2 h at 35°C in the dark and then heat-
treated in the dark; Δ = leaves stored for 2 h at 35°C in the dark and
then heat-treated in the light. Insert: traces of light-induced absorption
change at 518 nm of leaves stored at 20°C and then treated in the dark
for 6 min at 20°C ("control") and at 39°C ("heated"). Measurements at 20°C.

dark (35°C - acclimatization). In this case, illumination during the heat
stress treatment had no additional stabilizing effect. Therefore, it might
be concluded that in leaves both illumination and short-time acclimatization
to high temperature could alter the thylakoid membranes in such a way that
they become less sensitive to heat stress.

Chlorophyll fluorescence has been suggested to be a probe of changes
in the structural organization of PS I[1] and PS II complexes and related
changes in excitation-energy distribution between the two photosystems
(for a review see ref. 7). Fig. 2 shows fluorescence emission spectra at
77°K which indicate the distribution of fluorescence yield between PS I
and PS II/LHC. The fluorescence spectra were measured with particles prepared from frozen leaf-discs. In the present experiment the ratio between
PS I fluorescence at 735 nm and PS II fluorescence at 685 nm (F_I/F_{II}) was

---

[1] Abbreviations: PS I or II, photosystem I or II; LHC, light-harvesting
complex; DCMU, 3-(3´, 4´-dichlorophenyl)1,1-dimethylurea.

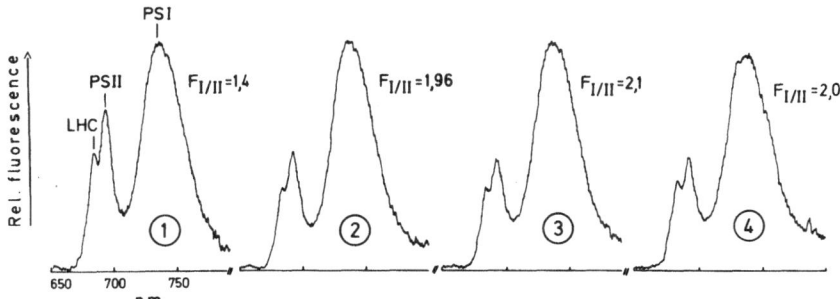

FIGURE 2. Chlorophyll fluorescence spectra measured at 77^OK. Preparation of frozen leaf-discs as described in the text. (1) leaf stored at 20^OC in the dark; (2) leaf stored at 20^OC in the dark and then illuminated for 10 min; (3) leaf stored for 2 h at 35^OC in the dark; (4) leaf stored for 2 h at 35^OC in the dark and then illuminated for 10 min. The spectra were normalized on the peak of PS I fluorescence.

1.4 in control leaves (curve 1) but increased up to about 2.0 if the leaves were illuminated before freezing in liquid N_2 (curve 2) indicating that the relative fluorescence yield of PS II decreased in light-adapted leaves. Essentially the same increase in F_I/F_{II} occurred if the leaves becomes acclimated to high temperatures (2 h at 35^OC in the dark; curve 3). Sometimes, this temperature-induced increase in F_I/F_{II} was even somewhat larger than that observed after illumination at 20^OC. However, illumination of 35^OC - acclimated leaves had no further effect on the fluorescence spectrum (curve 4). These results seem to indicate that either illumination or short-time acclimatization to high temperatures shift the pigment system to a state in which the yield of PS II fluorescence is low in relation to that of PS I fluorescence.

Induction curves of fluorescence measured at room temperature in the presence of DCMU showed that the maximum yield of the variable fluorescence decreased in 35^OC - adapted leaves by about 30 % (Fig. 3) whereas the F_o-fluorescence remained almost unchanged. In the presence of DCMU the electron acceptors of PS II become completely reduced and the emitted fluorescence represents the maximum yield of the fluorescence of PS II pigments (7). Hence, this experiment support the result shown in Fig. 2, that the fluorescence yield of pigments around PS II centers decreases during the high-temperature acclimatization.

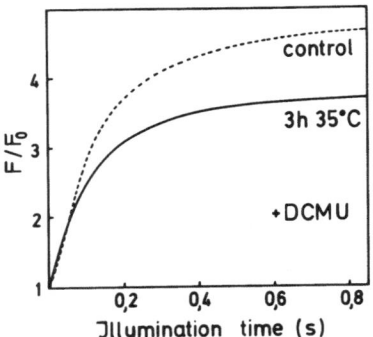

FIGURE 3. The time courses of fluorescence increase (F/F_0) of leaves which were stored for 3 h at 20°C ("control") or at 35°C in the dark. Before measurements, the leaves were rapidly infiltrated with a solution of 50 μm DCMU. The measurements were carried out at room temperature. The intensity of the blue actinic light (corning filters 5030, 9782) was 2.2 W/m².

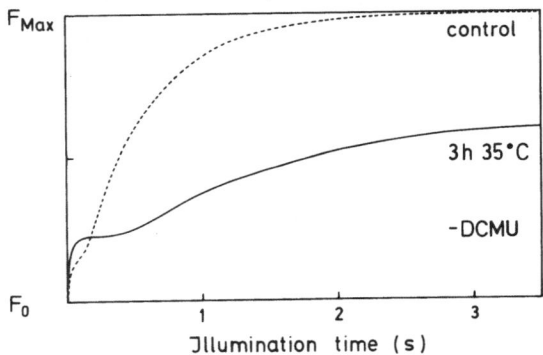

FIGURE 4. The time courses of fluorescence increase of leaves which were stored for 3 h at 20°C ("control") or at 35°C in the dark. The measurements were carried out at room temperature. The fluorescence curves were normalized on the maximum fluorescence yield (F_{Max}) determined in a separate measurement in presence of DCMU. Fluorescence was measured at 740 nm; the intensity of the actinic light (630-680 nm) was 8 W/m².

Fig. 4 shows fluorescence induction curves of predarkened leaves measured in the absence of DCMU. The curves were normalized on the maximum
fluorescence yield (F_{Max}), determined in a separate measurement in the
presence of DCMU. In the absence of DCMU, the variable fluorescence is
controlled by the redox state of the plastoquinone pool. In $20^{o}C$ - acclimated leaves the fluorescence reached F_{Max} after about 3 s. This most likely
indicates that during the first 3 s the available excitation-energy preferentially is channeled into PS II, i.e., the pigment system is in the so-
called state 1 (7). In this case, the plastoquinone pool becomes almost
completely reduced, presumably because the rate of reoxidation by PS I is
low. However, in predarkened $35^{o}C$ - acclimated leaves the fluorescence
increased during the first 3 s up to a steady state level but did not reach
the maximum yield, indicating that the plastoquinone pool remained in a
more oxidized state. This suggests that a larger share af the available
energy was channeled into PS I and, hence, the reoxidation of plastoquinone
by PS I was accelerated. Such situation would be expected when the pigment
system is in state 2. Essentially the same results were obtained with leaves
which were gased with CO_2-free air (not shown) suggesting that Calvin cycle
induction phenomena are not involved during the first 3 seconds of illumination.

Usually, it is suggested that in intact cells prolonged illumination
transforms the photosynthetic pigment apparatus to state 2 in which less
excitation-energy is delivered to PS II; in predarkened cells, a larger
proportion of the absorbed light-energy is transferred to PS II (state 1;
for discussion see refs. 1 and 7). The similarity between light-induced
and temperature-induced change in the fluorescence spectra (Fig. 2), the
temperature-induced decrease in the maximum yield of the variable fluorescence (Fig. 3) and the experiment shown in Fig. 4 might indicate that a
similar interconversion of the pigment apparatus can also occur in the dark
by increasing the temperature up to $35^{o}C$.

Fig. 5 shows the time course of the decrease in the maximum yield of
variable fluorescence and of the increase in the thermal stability of the
thylakoid membrane. 1 - 2 hours were sufficient to complete the acclimatization ($t_{0.5}$ = 20 - 40 min). The close correlation between the change in
the thermal stability and the change in fluorescence yield could indicate
a mechanistic relation between these two effects. As shown earlier (9) the
short-time acclimatization was completely reversible, although the time

302

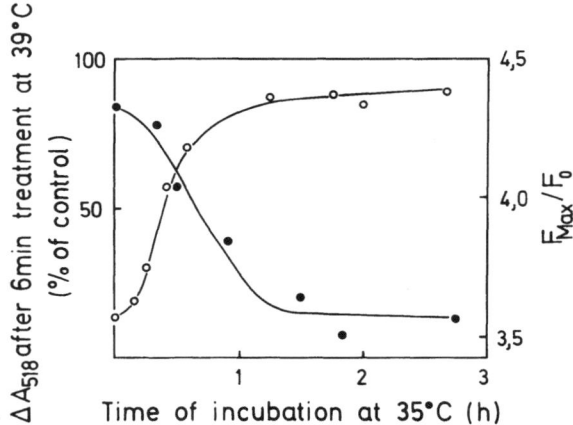

FIGURE 5. The time course of high-temperature acclimatization.
O = ΔA_{518} absorption signal after a heat-treatment (6 min at 39°C in the dark) as a function of the time of incubation of leaves at 35°C in the dark. 100 % = ΔA of a non-treated leaf. ● = Maximum yield of variable fluorescence (F_{Max}/F_0) as a function of the time of the incubation of leaves at 35°C in the dark. F_{Max} was determined in the presence of DCMU.

necessary for retransition was considerably longer. 12 - 24 h after decreasing the temperature from 35°C to 20°C all fluorescence properties were similar to that observed with control leaves (not shown).

4. DISCUSSION

State 1 - state 2 transition of the photosynthetic pigment system has been widely accepted as a regulatory mechanism which ensures optimal light energy distribution between the two photosystems. Essentially, two mechanisms of control of energy distribution have been discussed: a metal cation dependent mechanism and a light-dependent mechanism which includes the phosphorylation of polypeptides of the light-harvesting complex by a membrane kinase (for discussion see 1 and 7). It was found that the kinetics of the state 1 - state 2 transition is highly sensitive to temperature (6), possibly due to the relatively high activation energy of redistribution of pigment-protein complexes within the thylakoid system. The present results seem to indicate that changes in temperature can even trigger such interconversion. In intact cells, the energy distribution usually is controlled by the light-condition, but other environmental conditions such as variations in temperature could also affect this mechanism, possibly due to changes in the intracellular activity of metal cations occurring during temperature acclimatization. This would mean that short-time acclimatization of the photosynthetic apparatus mainly is a regulatory phenomenon in the sense that the photochemical activity is optimized during large temperature variations and during exposure to peak temperatures, respectively.

Obviously, light protect the thylakoid membranes in $20^{o}C$ -acclimated leaves against mild heat stress. However, in high-temperature acclimated leaves which are less sensitive anyhow light has no stabilizing effect, so that in the light there is no difference between the sensitivity of low- and high-temperature acclimated leaves to mild heat stress. Since in nature heat stress usually occurs in the light, the question arises whether there is an ecological significance of the observed protecting effect of light. Possibly, this change in the thermal stability is a secondary effect related to structural and functional changes of the membrane: thylakoids which are shifted to state 2 by illumination or by temperature acclimatization are less sensitive to mild heat stress. Further studies might clarify the physiological significance of the observed structural and functional changes of the photosynthetic apparatus occurring during temperature acclimatization.

ACKNOWLEDGEMENT

The author is grateful to Miss A. Stauffenberg for excellent technical assistance.

REFERENCES

1. Anderson JM (1981) Consequences of spatial separation of photosystem 1 and 2 in thylakoid membranes of higher plant chloroplasts. FEBS letters 124: 1-10.
2. Berry JA, Björkman O (1980) Photosynthetic response and adaptation to temperature in higher plants. Ann Rev. Plant Physiol 31: 491-543.
3. Fork DC, Murata N, Avron M (1977) The effect of temperature on the physical phase of chloroplast membrane lipids and photosynthesis. Carnegie Inst Year Book 76: 220-234.
4. Holton RW, Blecker HH, Onore M (1964) Effect of growth temperature on the fatty acid composition of a blue-green alga. Phytochemistry 3: 595-602.
5. Kislyuk M (1979) Protecting and injurious effects of light on photosynthetic apparatus during and after heat-treatment of leaves. Photosynthetica 13: 386-391.
6. Murata N, Troughton JH, Fork DC (1975) Relationship between the transition of the physical phase of membrane lipids and photosynthetic parameters in *Anacystis nidulans* and lettuce and spinach chloroplasts. Plant Physiol 56: 508-517.

7. Papageorgiou G (1975) Chlorophyll fluorescence: An intrinsic probe of photosynthesis. In Gowindjee ed. Bioenergetics of Photosynthesis, pp. 319-371. New York: Academic Press.
8. Pearcy RW (1978) Effect of growth temperature on the fatty acid composition of the leaf lipids in *Atriplex lentiformis* (Torr.) Wats. Plant Physiol 61: 484-486.
9. Santarius KA, Müller M (1979) Investigations on heat resistance of spinach leaves. Planta 146: 529-538.
10. Weis E (1981) Reversible heat-inactivation of the Calvin cycle: A possible mechanism of the temperature regulation of photosynthesis. Planta 151: 33-39.
11. Weis E (1981) Reversible effects of high, sublethal temperatures on the light-induced light scattering changes and electrochromic pigment absorption shift in spinach leaves. Z Pflanzenphysiol 101: 169-178.
12. Weis E (1981) The temperature sensitivity of dark-inactivation and light-activation of the ribulose-1,5-bisphosphate carboxylase in spinach chloroplasts. FEBS letters 129: 197-200.
13. Weis E (1982) The influence of metal cations and pH on the heat-sensitivity of photosynthetic oxygen evolution and chlorophyll fluorescence in spinach chloroplasts. Planta 154: 41-47.
14. Weis E (1982) The influence of light on the heat-sensitivity of the photosynthetic apparatus in isolated spinach chloroplasts. Plant Physiol (in press).
15. Witt HT (1971) Coupling of quanta, electrons, fields, ions and phosphorylation in the functional membrane of photosynthesis. Rev Biophys 4: 365-477.

THE EFFECT OF SALT SPECIES AND CONCENTRATION ON PHOTOSYNTHESIS AND GROWTH
OF PEA PLANTS (*Pisum sativum* L. cv. Alaska)

E. Hasson, A. Poljakoff-Mayber and J. Gale, Department of Botany, The Hebrew
University of Jerusalem, Jerusalem, 91904, Israel

ABSTRACT

Osmotic adjustment ($\Delta\pi$), stomatal resistance (r_s), net photosynthesis
(P_N) and three growth parameters (fresh weight-FW; dry weight-DW and length-L)
were measured in pea plants grown in different concentrations (85-390 mosmol)
of either NaCl, KCl, Na_2SO_4 or K_2SO_4. Complete osmotic adjustment was not
attained in any of the treatments, and $\Delta\pi$ decreased with increasing external
salinity, indicating a possible reduction in turgor. FW and L of the
plants decreased accordingly.

Stomatal resistance in plants grown in NaCl rose with increasing salinity and in parallel a decrease in DW and in P_N was observed. Stomatal
resistance generally rose with increasing concentrations of potassium
salts, but at a KCl concentration of $\sim$ 100 mosmol r_s was lower and P_N
was double than that of the controls. However, the dry weight of these
plants was lower than of plants exposed to K_2SO_4. The latter had the
highest dry weight of the four treatments which was even higher than that
of the controls. At $\sim$ 100 mosmol K_2SO_4, P_N was not affected although r_s
increased $\sim$ 40% above that of the controls, indicating increased capacity
of the photosynthetic mechanism.

High levels of potassium seem to benefit photosynthesis in Alaska peas
exposed to salinity. The osmotic component does not appear to be the main
factor affecting P_N. The nature of the specific ions appear to be of
greater importance, although the mechanism of their effect is as yet unclear.

(*Key words:* Ion specific effects, salinity and photosynthesis.)

INTRODUCTION

Dry matter production and maintenance of turgor as a result of a fav-
orable water balance, are two of the main factors necessary to enable growth
of plants. Salinity in the growth substrate interferes with both. Any

increase in external osmolarity, if not accompanied by an equivalent change in the vacuolar sap will result in a decrease in turgor of plant tissues and commensurate physiological functions. Extension growth has been shown to be the most sensitive function affected [5].

Many angiosperm halophytes maintain their turgor mainly by uptake of external solutes [2, 4] while in glycophytes osmotic adaptation by accumulation of organic solutes is more common [4, 9]. In glycophytes, adaptation is usually incomplete.

Plant growth may be evaluated by: fresh weight (FW) or dry weight (DW) increment, or by extension growth (length). The first and the last may depend more directly on osmotic adaptation and maintenance of a positive water balance. Dry weight, which depends on photosynthetic activity and allocation of photosynthate, is also affected by salinity [3, 6]. These effects may be ion specific affecting the photosynthetic apparatus, the ratio between photosynthesis and respiration or the synthesis of organic osmotica for the vacuole and cytoplasm. The specific effects of ions may depend on their internal concentration, and on cation-anion composition [1, 9]. Studies of the osmotic and specific effects of various types of salinity on osmotic adaptation, photosynthesis and growth of peas are presented in the following.

METHODS

Peas (*Pisum sativum* L. cv. Alaska) were exposed during growth to various levels of salinity induced by one of four salts — NaCl, Na_2SO_4, KCl or K_2SO_4 — added to half strength Hoagland solution. Media of all treatments always contained the minimal amounts of potassium necessary for normal growth (6 mM). The plants were grown in vermiculite for 10 days, in a growth room at 21/18°C. They were exposed to the same level of salinity from imbibition to harvest. The osmolarity of the substrate (π_e) was checked periodically during growth. Osmolarity of expressed sap (π_i) was measured using a Knauer osmometer. Potassium and sodium contents were measured by atomic absorption spectrometry. Stomatal resistance and rate of photosynthesis were measured as described by Sésták et al. [7]. Measurements were carried out at 25±1C, $[CO_2]$ of 300-350 µl l^{-1}, light intensity of $\sim$ 350 µmol m^{-2} s^{-1} 400-700 nm, and humidity of Δ_e (leaf to air) = $\sim$ 13 mbar.

RESULTS AND DISCUSSION

In none of the treatments was there a complete osmotic adjustment. The osmolarity of expressed sap (π_i) from roots and shoots exposed to the four salts increased linearly with increasing external concentrations. Plots of π_i as a function of π_e for roots of all treatments and for shoots of plants exposed to KCl, K_2SO_4 and NaCl salinities gave slopes of ~ 0.5. The most complete adjustment was observed in shoots of plants exposed to Na_2SO_4 (~ 0.8), which maintained a more or less equal $\Delta\pi$ at most levels of salinity (Table 1). In all other treatments $\Delta\pi$ decreased with increasing salinity, with one exception — plants grown at 93 mosmol K_2SO_4 (~ 0.2 MPa). The order of maintenance of $\Delta\pi$ being NaCl < KCl < K_2SO_4 < Na_2SO_4.

Table 1. Osmotic adaptation of shoots of pea plants grown in different concentrations of various salts.
Results are given as $\Delta\pi = \pi_i - \pi_e$ in mosmol and $\Delta\pi$ as percent of $\Delta\pi$ for control plants grown in Hoagland solution. π_e and π_i — osmotic concentrations of culture solution and expressed sap respectively in mosmol.

π_e	π_i	$\Delta\pi$	%	π_e	π_i	$\Delta\pi$	%
	KCl				K_2SO_4		
89	347	258	93	93	380	287	103
178	377	199	72	185	395	210	76
267	445	178	64	278	445	167	60
355	437	82	29	370	520	150	54
	NaCl				Na_2SO_4		
95	321	226	81	91	350	259	93
190	351	161	58	183	375	192	69
285	421	136	49	274	485	211	76
380	432	52	19	365	580	215	77
Hoagland control: π_e = 27; π_i = 305; $\Delta\pi$ = 278							

Potassium and sodium content of plants also differed in the four treatments (Table 2). The presence of SO_4^{--} apparently interfered with the accumulation of the cations in the root and their transport to the shoot, especially K^+. Potassium content of roots and shoots grown in Na_2SO_4 was lower than in NaCl and in the Hoagland controls, although the K^+ content in the external medium was the same in the three treatments. Even when the external osmolarity was due to K^+ salts, its concentration in the roots and the shoots was much lower in plants grown in K_2SO_4 than in those grown in KCl.

Table 2. Potassium and sodium content of sap expressed from roots and shoots of pea plants grown at various concentrations of different salts. Na^+ and K^+ as $\mu mol\ g^{-1}$ FW.

π_e — osmotic concentration of culture solution in mosmol.
Hoagland controls: π_e = 27; Roots: Na^+ - 1.1, K^+ - 21.1;
Shoots: Na^+ - 2.5, K^+ - 24.2; $\mu mol\ g^{-1}$ FW.

	NaCl			Na_2SO_4			KCl[*]			K_2SO_4[*]	
π_e	Na^+	K^+	π_e	Na^+	K^+	π_e	Na^+	K^+	π_e	Na^+	K^+
					R O O T S							
95	28.8	20.4	91	21.0	13.2	89	t	23.7	93	t	25.1
190	37.1	19.4	183	29.0	15.2	178	r a	34.4	185	r a	27.0
285	50.3	25.9	274	36.1	14.2	267	c	36.2	278	c	24.4
380	67.9	33.2	365	45.0	15.4	358	e s	53.3	370	e s	33.1
					S H O O T S							
95	14.9	27.5	91	16.2	17.8	89	t	68.2	93	t	22.4
190	16.0	37.2	183	17.2	17.4	178	r a	81.2	185	r a	30.0
285	24.2	39.5	274	23.8	18.5	267	c	91.9	278	c	34.7
380	36.2	54.1	365	27.0	19.6	355	e s	108.6	370	e s	54.6

[*]Sodium content of plants exposed to potassium salinity was less than in Hoagland controls

The most common salinity in nature is an excess of sodium which may appear with either chloride or sulphate anions. As shown above, plants exposed to sodium with sulphate anions, had the better osmotic adjustment. The height of these plants was less stunted than when chloride was the anion (Fig. 1).

The fresh weight, however, was higher in shoots exposed to Cl^- than in those exposed to SO_4^{--} (Fig. 2). This is apparently in accord with Strogonov's suggestion that succulence is induced by chloride whereas sulphate salts induce xeromorphism [8]. The changes in the FW/DW ratio, which were found here with increasing external salinity, conform with this idea (Fig. 3).

Dry weight of shoots, of plants exposed to NaCl, decreased linearly with the increase in external osmolarity. This was apparently due to increase in stomatal resistance and decrease in net photosynthesis (Fig. 4). In plants exposed to Na_2SO_4, in spite of better osmotic adaptation, the decrease in dry weight of shoots was identical to that caused by NaCl.

Incomplete osmotic adaptation is not apparently the decisive factor in the reduction of dry weight.

Fresh weight of shoots and height of plants exposed to either potassium salt, decreased linearly with the increase in external and internal osmolarity (correlation, r^2 of 0.98 for FW and of 0.97 for length of shoots).

Dry weight of shoots was higher in plants exposed to low concentrations of K_2SO_4 (< 150 mosmol) than in the Hoagland controls (Fig. 5). Such an increase was not observed in plants exposed to KCl. However, preliminary experiments showed that the stomatal resistance of plants exposed to 89 mosmol KCl was lower, and their rate of photosynthesis was twice as high as of those exposed to 93 mosmol K_2SO_4, even though the latter showed a large increase in dry weight.

The difference in dry weight could not be attributed to leaf area, as in plants grown in $\sim$ 90 mosmol KCl leaf area was almost twice as large as in plants exposed to similar osmolarity of K_2SO_4. However, plants of the latter treatment were the only ones which maintained a $\Delta\pi$ which was not lower than that of the Hoagland controls (Table 1).

The difference in the FW/DW ratio between the two potassium salt treatments (Fig. 3) was not expressed in the general FW values, which, when plotted against π_e, fall on the same line, decreasing with increase in the external osmolarity (a slope of 1, $r^2 = 0.98$).

The explanation of the fact that dry weight of plants grown in $\sim$ 90 mosmol KCl is lower than in that of plants grown in similar osmolarity of K_2SO_4 may be sought in differences in maintenance respiration, which was not measured in these experiments [6].

Preliminary experiments suggested that high levels of K^+ may benefit photosynthesis of Alaska peas. Indeed doubling the K content in the Hoagland control (from 6 to 12 mM) resulted in a decrease of 35% in stomatal resistance and an increase of 65% in the rate of photosynthesis.

In pea roots all three indicators of growth, FW, DW and length, decreased with increasing external osmolarity. The values for FW and length of roots decreased linearly with increasing osmolarity (correlation of 0.96 for FW and 0.95 for length). Dry weight values in plants from both sulphate salts were higher, and the slope of the line, when plotted against π_e, was steeper than in chloride salts. KCl was apparently much more damaging than K_2SO_4, as 89 mosmol of external KCl caused a very marked decrease in DW as compared to the controls (Fig. 6). However, further increases in KCl salinity had a very small effect.

The fresh weight and length of pea plants exposed to salinity do appear to be more directly affected by the osmotic component of the stress than dry matter production. The latter depends apparently on the interaction of several other factors. The preliminary results reported here indicate that the species of ion which induces salinity has a marked effect on osmotic adaptation and growth. These specific effects possibly cause changes in photosynthesis and the partitioning of photosynthetic products between anabolic and catabolic processes in the plant [6]. It seems important that these effects be investigated in greater detail.

REFERENCES

1. Epstein E. 1972. Mineral nutrition of plants: principles and perspectives. New York, London, Sydney, Toronto: Wiley & Sons.
2. Flowers TJ, Troke PF and Yoe AR. 1977. The mechanism of salt tolerance in halophytes. Ann Rev Plant Physiol 28: 89-121.
3. Gale J. 1975. Water balance and gas exchange of plants under saline conditions. In Poljakoff-Mayber, A. and Gale, J., eds. Plants in saline environment, Ecological Studies 15, pp. 168-185. Berlin, Heidelberg, New York: Springer-Verlag.
4. Greenway H and Rana Munns. 1980. Mechanism of salt tolerance in non-halophytes. Ann Rev Plant Physiol 31: 149-190.
5. Hsiao TC. 1973. Plant responses to water stress. Ann Rev Plant Physiol. 24: 519-570.
6. Schwarz M and Gale J. 1981. Maintenance respiration and carbon balance of plants at low levels of sodium chloride salinity. Jour Exp Bot 32: 933-941.
7. Séstak Z, Catsky J and Jarvis PG. 1971. Plant photosynthetic production. Manual of methods. The Hague: Junk Publ.
8. Strogonov BP. 1964. Physiological basis of salt tolerance of plants. IPST, Jerusalem.
9. Waisel Y. 1972. Biology of halophytes. New York, London: Academic Press.

LEGENDS FOR FIGURES

Fig. 1   Effect of NaCl concentrations on length of pea shoots (cm/plant).
Fig. 2   Fresh weight of shoots of pea plants as in Fig. 1 (mg/plant).
Fig. 3   FW/DW ratio of pea plants grown in the four different salinities.
Fig. 4   Effect of NaCl during growth of peas on stomatal resistance (r_s) in sec cm^{-1} (0), % inhibition of DW (Δ), and photosynthetic rate (X).
Fig. 5   Dry weight of pea shoots of plants grown at different levels of potassium salts salinities for 10 days (mg/plant).
Fig. 6   Dry weight of pea roots grown at different levels of potassium salts salinities for 10 days (mg/plant).
          π_e in mosmol.
          In Figs. 1, 2, 5 and 6:
          $0 - Na_2SO_4$; $\bullet - K_2SO_4$; $\Delta - NaCl$; $\blacktriangle - KCl$; salinities.

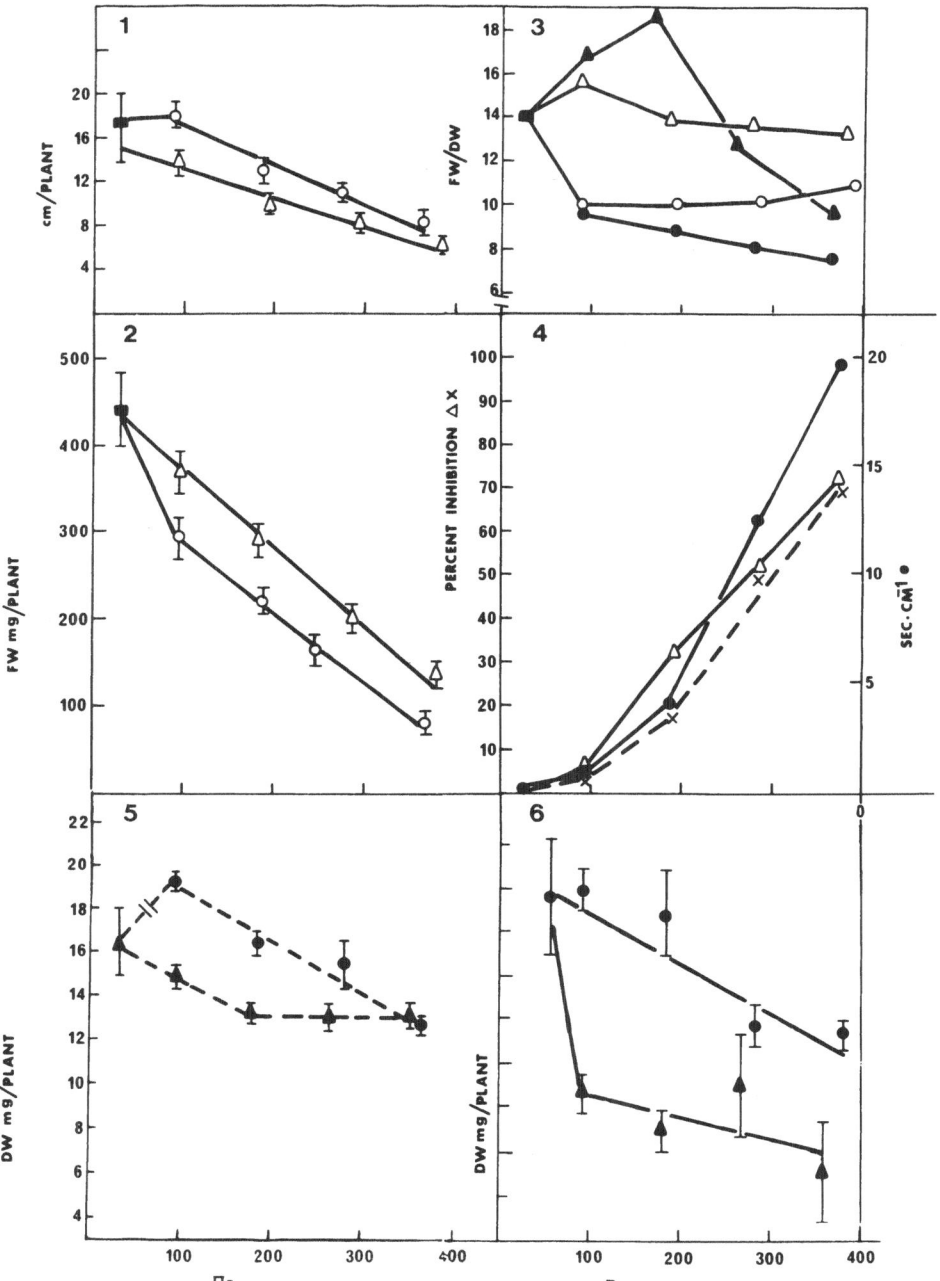

SALT TOLERANCE AND ROOT-ZONE AERATION IN *HELIANTHUS ANNUUS* (L.) - GROWTH AND STOMATAL RESPONSE TO SOLUTE UPTAKE.

P.E. Kriedemann, R. Sands and R.C. Foster
CSIRO Division of Soils, Glen Osmond, South Australia.

ABSTRACT

Sunflower plants were grown through their vegetative phase in temperature controlled (20-21°C) nutrient solution under greenhouse conditions. Cultures were either aerated continuously or flushed periodically with N_2. NaCl treatments (up to 200mM) were superimposed either during the course of such root-zone treatment, or else coincided with transfer from aerated to 'anoxic' conditions and *vice versa*.

Plants affected by root-zone 'anoxia' showed an early, but partly reversibly decline in laminar expansion and much reduced roots: tops ratio. The limited root system commonly failed to meet transpirational demand so that plants with 'anoxic' root-systems showed lower relative water content and wilted easily.

Interactive effects were apparent - tissue solute and especially Na^+ ion concentration were always stronger when salinisation coincided with root-zone 'anoxia'. However, root systems showed adaptive responses to poor aeration such as formation of cortical aerenchyma external to a well defined endodermis. Plants with such anatomical modifications, and probably other more subtle physiological adjustment, were able to sustain their Na^+ and Cl^- ion exclusion mechanisms better than non-adapted plants, and appeared less vulnerable to salt injury even when salinisation was accompanied by root-zone 'anoxia'.

INTRODUCTION

Irrigation agriculture makes extensive use of fine textured soils where crop root systems experience periodic deterioration in aeration status following each cycle of flood irrigation (8). Root function becomes impaired under such conditions (11) and energy dependent processes such as solute exclusion will be especially vulnerable (5). While salt tolerance in crop plants is generally lowered by aeration deficiency within the root-zone (13), some species e.g. *Helianthus annuus* (L.) show adaptation to waterlogging in the form of reduced tops growth and development of cortical aerenchyma external to vascular elements in hypocotyl and roots (3,10). Such modifications may well confer improved salt tolerance under conditions of poor root-zone aeration status, and experiments reported here were designed to test that proposition.

MATERIALS AND METHODS

Sunflowers (*Helianthus annus* L.) cv Grey Stripe were grown for periods up to 50 days in half-strength nutrient solution, under heated glasshouse conditions with root-zone temperature control (20°C). Root-zone aeration status was varied by continuous bubbling with compressed air versus periodic flushing of sealed culture vessels (3.22ℓ capacity) with N_2 gas. Redox potential (P_t reference electrode) ranged between +200 to +300 and +90 to +120 mv for aerated and 'anoxic' cultures respectively. Salt treatments were imposed by addition of 25 or 50 mM NaCl increments. Stomatal resistance was measured under greenhouse conditions with a Lambda steady state porometer (Model LI 1600); leaf growth data were derived from daily breadth measurements (B) subsequently converted to areas (A) according to a relationship of the form $A = 3.22B^{1.73}$ established empirically during the present experiments. Whole plant growth was derived from weight of harvested biomass - both fresh and oven dried (60°C forced draught). Tissue analyses (Na^+ and Cl^-) were based on a hot water extract of oven-dried and ground material using $AgNo_3$ conductimetric titration and atomic absorption spectroscopy for Cl^- and Na^+ ions respectively.

RESULTS

1. Leaf growth

Laminar expansion (area v/s time) is typically sigmoidal (e.g. Figure 1) and the initial log phase provides a sensitive index of foliar response to root-zone conditions. Relative leaf growth rate (RLGR in Figure 1) shows an abrupt reduction from 0.55 to 0.10 following transfer from aerated to 'anoxic' conditions compared with a decrease from 0.52 to 0.23 which would be attributable to normal ontogeny in control leaves (uppermost curve in Figure 1). Conversely, onset of aeration in previously 'anoxic' cultures (lowermost plot in Figure 1) does not restore RLGR to control levels (0.18 versus 0.23 respectively). Leaf growth response to root-zone 'anoxia' is heightened when

315

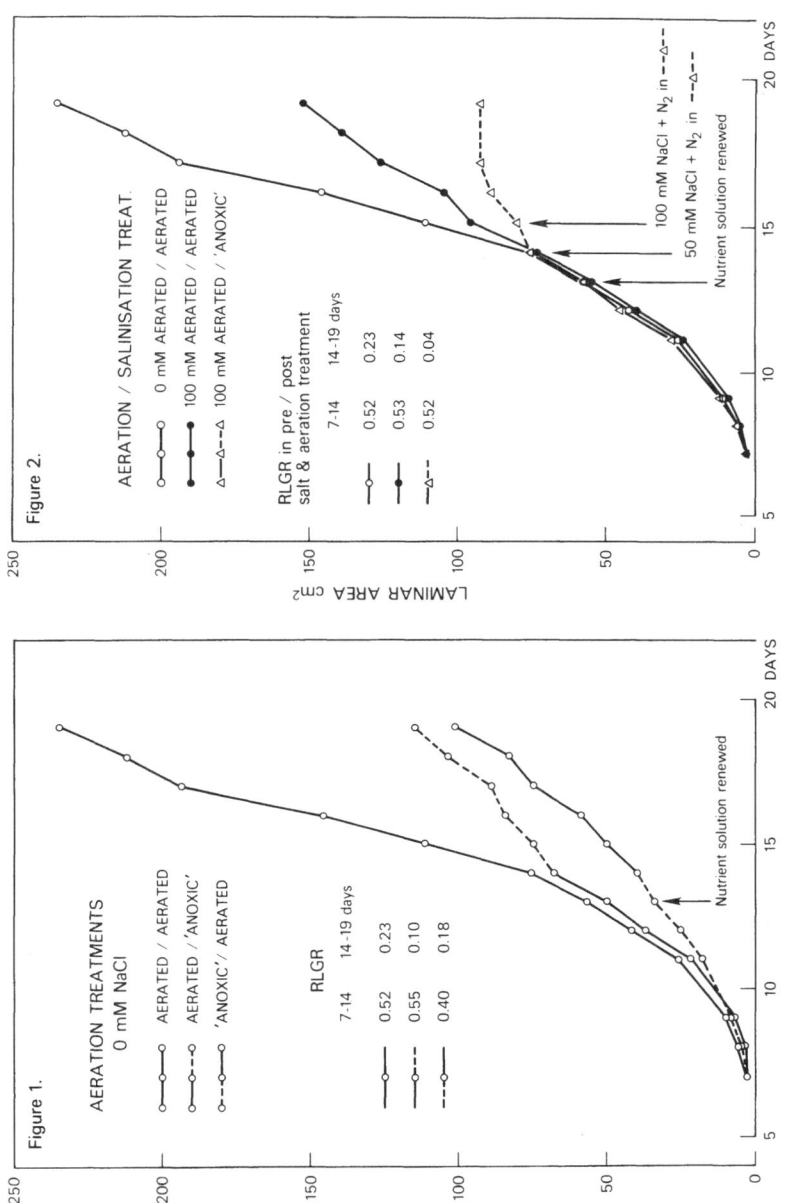

FIGURES 1 and 2. Leaf growth response to root-zone aeration status at 0 mM (Fig. 1) and 100 mM NaCl (Fig. 2). Relative Leaf Growth Rate (RLGR cm^2 cm^{-2} day^{-1}) is indicated. (Mean data from 2 representative replicates - parallel observations on other plants confirmed patterns.) Units on abscissa are coincident for Figures 1, 2 and 3; 5 days from start of leaf growth observations represents plant age of 19 days.

change in aeration status is accompanied by salinisation (see Figure 2). A progressive increase to 100 mM NaCl one day following nutrient solution renewal (to ensure that interpretation of solute effects was uncomplicated by provision of fresh nutrients) constrained laminar growth to 0.04 units under 'anoxia' compared with 0.14 in aerated cultures. Moreover, combined effects of salt plus 'anoxia' led to early cessation in leaf expansion (lowermost curve in Figure 2).

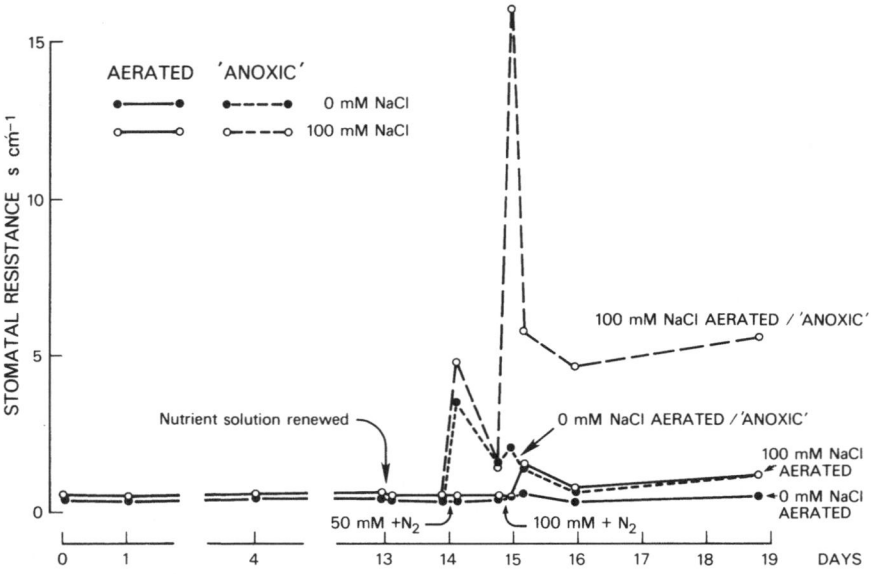

FIGURE 3. Timecourse of stomatal response to salinisation and root-zone aeration status. (Accompanying leaf growth responses shown in Figures 1 and 2; related data on laminar moisture in Table 1; mean values from 4 replicates).

2. Stomatal resistance

Short-term stomatal changes also reflect additive effects of aeration and salinity stresses as shown in Figure 3; sustained adjustments are summarised in Table 1 which provides more comprehensive data from the same experiment. Continuous aeration clearly offsets stomatal response to salinisation and only mild perturbation in r_s occurs following transfer from 50 to 100 mM NaCl; r_s showed no detectable response to 50 mM provided roots were aerated continuously. Previously aerated plants showed a strong stomatal response at the onset of 'anoxia' which was heightened by 50 and subsequently 100 mM NaCl (Figure 3). Prior exposure to root-zone 'anoxia' suppressed this initial response (right hand column, Table 1) although eventual steady state values showed no apparent dependence on previous aeration status. By day 19 for example (bottom line, Table 1) currently aerated plants had uniformly low resistance, with only minor increase under salinisation, whereas both aerated/ 'anoxic' and continuously 'anoxic' plants showed higher r_s; viz, a marginal increase at 0 mM but a substantial longer term increase due to salt addition. Changes in leaf moisture status (as % fresh weight in Table 1) accompanied these stomatal responses. Values were consistently lower in continuously 'anoxic' cultures, regardless of salinisation, which could derive from some symplastic pathway constraint. Moreover stomatal closing response from mid-morning to mid-afternoon provides a further indication of hydraulic limitation under 'anoxic' conditions - compare treatments on day 15 in Table 1. Aerated cultures maintain low resistance despite salinisation whereas 'anoxic' plants show partial closure.

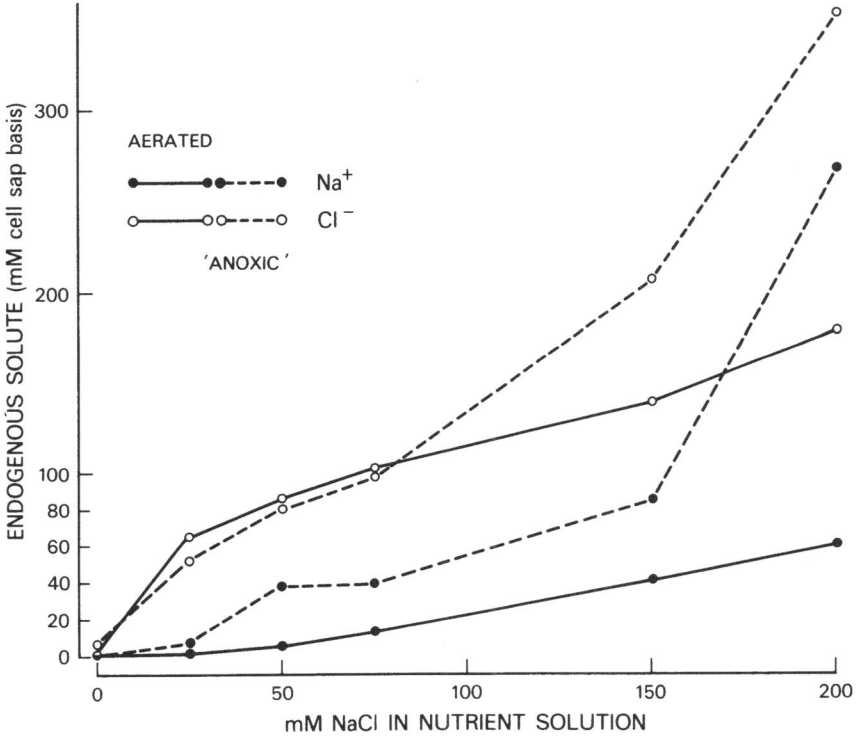

FIGURE 4. Laminar concentrations of Na^+ and Cl^- ions (mM cell sap basis) as a function of root-zone salinisation and aeration status. (Plant age 50 days; 'anoxia' applied on day 7; salinisation started on day 24).

TABLE 1. Stomatal and leaf moisture response to reciprocal changes in root-zone aeration status at the time of salinisation.

Day No.	Hours after salt.	Root-zone status:	Aerated		'Anoxic'		Aerated		'Anoxic'	
		NaCl (mM)	0	0	0	0	0	0	0	0
1	-	Stomatal resist.	.40	.50	.52	.57	.51	.58	.58	.62
14	-	Stomatal resist.	.39	.65	.37	.75	.45	.47	.47	.47
14		Leaf moisture	82.06		79.45		—		—	

Day No.	Hours after salt.	Root-zone status:	Aerated-Aerated		'Anoxic'-Aerated		Aerated-'Anoxic'		'Anoxic'-'Anoxic'	
14	2 (Noon)	NaCl (mM)	0	50	0	50	0	50	0	50
		Stomatal resist.	.37	.58	.49	.97	.53	4.80	.91	6.74
		Leaf moisture	82.61	83.58	78.51	80.24	83.08	80.79	79.92	79.53
		NaCl (mM)	0	100	0	100	0	100	0	100
15	20	Stomatal resist.	.46	.59	.70	1.70	1.65	1.49	.55	1.36
	23	Stomatal resist.	.44	.48	.81	4.50	2.07	16.00	.49	6.85
	27	Stomatal resist.	.60	1.60	.71	2.38	1.46	5.85	2.38	5.20
16	47	Stomatal resist.	.33	.83	.43	1.08	.72	4.70	1.90	1.92
		Leaf moisture	83.60	83.31	81.74	80.13	83.23	83.54	79.37	79.88
19	120	Stomatal resist.	.52	1.24	.54	1.01	1.21	5.64	1.08	6.69
		Leaf moisture	-	84.05	-	82.58	-	82.89	-	80.82

(Stomatal resistance s cm^{-1}; Leaf moisture as % fresh wt; Mean data 4 and 2 reps respectively)

3. Solute levels

Turgor-dependent processes such as leaf expansion and stomatal opening are clearly dependent on laminar solute levels and relevant data showing Na^+Cl^- ion concentration plus stomatal resistance, as a function of root-zone conditions, are given in Figure 4 and Table 2 respectively. Solute exclusion against high external concentration is no longer maintained when roots are deprived of continuous aeration, and stomatal resistance changes accordingly. Towards the end of this particular experiment (day 48 in Table 2) r_s was consistently higher due to root-zone 'anoxia', and especially at 150 and 200 mM where exclusion mechanisms against Na^+ Cl^- ions were breaking down (Figure 4). In particular, aerated plants maintained disproportionately low Na^+ relative to Cl^- within their leaf tissue over the entire concentration range, whereas 'anoxic' plants showed clear dysfunction with respect to Na^+ and Cl^- ion exclusion at 150 and 75 mM respectively. At 200 mM NaCl, plants accumulated almost equimolar concentrations throughout their aerial organs and especially in laminae where cell sap concentrations exceed NaCl levels in the root-zone bathing solution. Stomatal resistance (Table 2) matched this solute picture with a shift in r_s above 75 mM NaCl in 'anoxic' cultures. Both abaxial and adaxial leaf surfaces were similarly affected.

4. Adaptive responses

Morphological adjustments to root-zone O_2 deficiency include smaller aerial organs, a much reduced root biomass and formation of lysigenous aerenchyma in hypocotyl and lateral root cortex. Under the additional influence of salinisation, the endodermal layer can also become more prominent. The question then arose as to whether such modifications conferred any particular advantage with respect to solute exclusion under root-zone 'anoxia' as already inferred from stomatal responses in Table 1 where r_s increase following salinisation was minimised by preconditioning to 'anoxic' conditions. Accordingly, a series of 2 phase experiments were undertaken where well established seedlings were initially held in 'anoxic' nutrient solution which enabled adaptive responses to occur and then distributed between aerated versus 'anoxic' cultures in a reciprocal fashion at the start of salinisation, thereby yielding 4 basic treatments with respect to pre/post salinisation viz. 1. aerated/aerated, 2. 'anoxic'/aerated, 3. aerated/'anoxic', and 4. 'anoxic'/'anoxic'. If preconditioning under root-zone 'anoxia' subsequently confers improved salt tolerance under that condition compared with plants not so adapted, treatments 3 and 4 should provide the

TABLE 2. Long term stomatal adjustment to sustained aeration versus 'anoxia' during salinisation.

Plant age (days)	Time of day	Leaf Surface	Root Status	NaCl treatment (mM)					
				0	25	50	75	150	200
22	1400 hrs	Abaxial	Aerated	0.92	-	-	-	-	-
			'Anoxic'	0.84	-	-	-	-	-
38	1400 hrs	Abaxial	Aerated	0.69	0.51	0.60	0.49	0.93	-
			'Anoxic'	0.52	0.56	0.79	1.09	2.78	-
48	1200 hrs	Abaxial	Aerated	0.38	0.29	0.33	0.29	0.56	0.54
			'Anoxic'	0.35	0.34	0.41	1.55	4.72	5.35
		Adaxial	Aerated	0.60	0.54	0.40	1.42	1.31	1.36
			'Anoxic'	0.69	0.71	1.27	2.49	4.02	11.0

(Stomatal resistance s cm^{-1}; mean data 4-8 replicates; std. errors typically 6-10% of means)

TABLE 3. Na$^+$ and Cl$^-$ ion concentration in Sunflower plants as a function of root-zone aeration status and prior adaptation to 'Anoxia'

Root-zone aeration status during pretreatment v/s salinisation	Aerated Aerated		'Anoxic' Aerated		Aerated 'Anoxic'		'Anoxic' 'Anoxic'	
	Na$^+$	Cl$^-$	Na$^+$	Cl$^-$	Na$^+$	Cl$^-$	Na$^+$	Cl$^-$
Laminae	40	112	96	145	304	338	139	151
Petioles and Stem	76	126	126	132	180	195	147	135
Hypocotyl and Roots	77	55	77	76	61	61	67	51

(150 mM NaCl bathing roots; tissue solutes expressed as mM NaCl in cell sap: mean of duplicate determinations on four replicates; LSD for treatment effects 40.4/35.2 for Na$^+$ and Cl$^-$ ion concentration respectively).

critical comparison. Data provided in Table 3 on tissue solute levels, enable the inference that some adaptive response does indeed occur during phase 1. Leaves on non-adapted plants (Table 3) accumulated 2.2 times more Na^+ and Cl^- ion that in plants previously adapted to 'anoxia' while aerated specimens (both treatment 1 and 2) had Na^+ levels 87% lower. General concentrations of Na^+ and Cl^- within petioles and stems confirmed that distinction.

DISCUSSION

Salt tolerance threshold was lowered under root-zone 'anoxia' as evidenced by reduced leaf growth, accentuated stomatal closure and altered biomass distribution, but was most clearly reflected in solute uptake and especially Na^+ ion accumulation. Sunflower can therefore join the list of bean, barley, rice, tobacco, apple and citrus cited by West and Taylor (12,13) as established causes of aeration-dependent solute exclusion. However, sunflower roots showed adaptive responses to poor aeration status and salinisation such as aerenchyma formation within the cortex and external to a well defined endodermis. Plants with such anatomical features, and probably equipped with other more subtle physiological adjustments as well were able to sustain solute exclusion mechanisms and were less vulnerable to salt injury even when salinisation was accompanied by root-zone 'anoxia'. Significantly, Na^+ ion exclusion appears to have had an obligate need for root-zone aeration regardless of pretreatment, as reported previously (2,12,13,14). In the present experiments, selective discrimination against Na^+ relative to Cl^- ion was impaired by root-zone 'anoxia' regardless of aeration status during pretreatment (compare laminar levels of Na^+ and Cl^- in treatments 1 and 2 versus 3 and 4 in Table 3).

Increased stomatal resistance following salinisation appeared partly reversible, especially in aerated plants, but closing responses were accentuated under sustained 'anoxia' (Table 2) and especially when NaCl addition was accompanied by onset of root-zone 'anoxia' in previously aerated plants (e.g. day 14 in Table 1). However, two anomalies are noteworthy: 1. Abaxial versus adaxial leaf surfaces show distinctive stomatal behaviour reminiscent of the disparity between upper and lower surfaces on *Commelina communis* L. reported by Pemadasa (4), viz. adaxial stomata (day 48 in Table 2) show a stronger interaction with salinisation than is apparent with abaxial stomata. 2. Short-term partial stomatal closure in response to salinisation (aerated or 'anoxic' root-zone in Table 1) is not accompanied by any systematic change in leaf moisture. Sojka and Stolzy (7) report similar behaviour in sunflower following oxygen elimination from the root profile under greenhouse conditions; stomatal closure occurred even at optimal soil matric potential. In that case, increased root resistance at low soil oxygen tension did not account for higher stomatal resistance. A reduction in supply of root derived cytokinins from the waterlogged growing medium as noted for sunflower by (1) may well have been a contributing factor.

While growth effects due to salinisation showed no clear interaction with oxygen deprivation *per se* (as inferred from present experiments based on nutrient solution) shorter term functional responses described in the literature (6,9) reflect heightened salt sensitivity in waterlogged compared to aerated soil. This distinction between solution versus soil culture probably derives from other factors associated with root-zone anaerobiosis and especially metabolic end products generated by micro-organisms. Any breakdown in root tissue integrity within waterlogged saline soil is therefore likely to enhance anaerobiosis by providing respiratory substrates and thereby engender an autocatalytic decay of root tissue. The adaptive response to poor aeration status within the root-zone described above may have additional relevance in this context by contributing towards the maintenance of oxidative conditions at the soil/root interface.

REFERENCES

1. Burrows, W.J., and Carr, D.J. (1969) Effect of flooding the root system of sunflower plants on the cytokinin content in the xylem sap. *Physiol. Plant.* 22:1105-1112.
2. Cooper, A. (1982) The effects of salinity and waterlogging on the growth and cation uptake of salt marsh plants. *New Phytol.* 90:263-275.
3. Kawase, M. (1974) Role of ethylene in induction of flooding damage in sunflower. *Physiol. Plant.* 31:29-38.
4. Pemadasa, M.A. (1981) Abaxial and Adaxial stomatal behaviour and responses to fusicoccin on isolated epidermis of *Commelina communis* L. *New Phytol.* 89:373-384.
5. Rains, D.W. (1972) Salt transport by plants in relation to salinity. *Ann. Rev. Pl. Physiol.* 23:367-388.
6. Shalhevet, S., Maas, E.V., Hoffman, G.J., and Ogata, G. (1976) Salinity and the hydraulic conductance of roots. *Physiol. Plant.* 38:224-232.
7. Sojka, R.E., and Stolzy, L.H. (1980) Soil-oxygen effects on stomatal response. *Soil Science* 130:350-358.
8. Stolzy, L.H., and Fluhler, H. (1978) Measurement and prediction of anaerobiosis in soils. *IN*: 'Nitrogen in the Environment, Volume 1'. (eds. D.R. Neilson, and J.G. MacDonald). Academic Press. New York.
9. Wainwright, S.J. (1980) Plant in relation to salinity, *IN*: 'Adv. Bot. Res.' (Ed. H.W. Woolhouse). Academic Press, London and New York. pp 221-261.

10. Wample, R.L., and Reid, D.M. (1978) Control of adventitious root production and hypocotyl hypertrophy of sunflower (*Helianthus annuus*) in response to flooding. *Physiol. Plant.* 44:351-358.
11. Wenkert, W., Favsey, W.R., and Watters, H.D. (1981) Flooding responses in *Zeo mays* L. *Plant and soil.* 62:351-366.
12. West, D.W., and Black, J.D.F. (1978) Irrigation timing - its influence on the effect of salinity and waterlogging stresses in tobacco plants. *Soil Science.* 125:367-376.
13. West, D.W., and Taylor, J.A. (1980a) The response of *Phaseolus vulgaris* L. to root-zone anaerobiosis, waterlogging and high sodium chloride. *Ann. Bot.* 46:57-60.
14. West, D.W., and Taylor, J.A. (1980b) The effect of temperature on salt uptake by tomato plants with diurnal and nocturnal waterlogging of salinised root-zones. *Plant and Soil.* 56:113-121.

THE EFFECT OF HEAT AND SALINITY STRESS ON THE CARBON BALANCE OF *Xanthium strumarium*

MEIER SCHWARZ AND JOSEPH GALE, Department of Botany, Hebrew University of Jerusalem, Jerusalem, Israel, 91904

ABSTRACT

Photosynthesis and respiration of control and salinised (-0.5 MPa, NaCl) *Xanthium strumarium* plants were measured throughout the day at different temperatures. Maintenance respiration (R_m) was calculated.

With a rise in temperature between 16 and 28°C photosynthesis increased by 46% while in the salinised plants it decreased by 54%. Total respiration in the control plants increased with rising temperature ($Q_{10} \sim 1.75$) but decreased in the salinised plants, at temperatures above 18°. Salinity caused a two to threefold increase of R_m. Temperature also resulted in an increase in R_m. The effect was stronger in the control than in salinised plants.

These data support only partially the hypothesis that the increased sensitivity to salinity at high temperatures (reduced dry weight increment) is the result of a larger use of assimilate in R_m, for coping with the effects of heat. This occurs at a time when assimilation itself is reduced and R_m for coping with salt is increased. The results suggest, rather, that the adaptive increase in R_m at high temperatures may be reduced by salt thus increasing heat damage.

INTRODUCTION

There have been many reports showing that sensitivity to salinity increases at high temperature (1, 2, 3, 4, 5, 9, 14). However the mechanism of this response has not been clarified as specific effects of high temperature are often compounded with impaired water balance (7).

In a previous work we reported that low levels of salinity may reduce plant growth by increasing assimilate consumption for maintenance respiration (R_m) at the very time that carbon assimilation is reduced (16). It has also been shown that R_m increases with temperature (12, 13). In the present work we evaluated the relative contributions of the increase of R_m in response to temperature and the increase of R_m in response to salinity on the carbon balance of plants exposed to low level salinity. By "low" level of salinity is meant a level of salt in the root medium which brings about a reduction of growth (dry weight increment) without causing apparent signs of toxicity. At low levels of salinity osmotic adjustment may be complete, but stomata are partly closed and CO_2 uptake is impaired (6).

METHODS

Xanthium strumarium was chosen for this study as a representative C_3-glycophyte whose growth is reduced at low levels of salinity without other signs of toxicity. Seeds were germinated in vermiculite in a greenhouse and the seedlings were then transferred to aerated culture solution (15) in a growth chamber. pH was checked daily and the solutions changed weekly. Growth conditions were: 12 h light at 425 μmol $m^{-2}s^{-1}$ photon flux density (400-700 nm measured with a LI-188 Lambda quantum sensor). Day/night air temperatures were 25/18 ± 0.5°C and relative humidity 60 ± 5% throughout. At least 7 days after transfer to nutrient solution the osmotic potential (π) of the solution was decreased, at a rate of 0.05 MPa d^{-1}, by the addition of NaCl to a final π of -0.5 MPa. Osmotic potentials were checked with a cryoscopic osmometer.

Measurements of CO_2 exchange of control and salinised plants as a function of temperature were made 3-4 weeks after germination. During these measurements air and root temperatures (see Results) were controlled to within ± 0.5°C.

The gas exchange measurements were carried out in four parallel mini growth-chembers. Methods and data reduction were as previously described (16). All experiments were replicated 3 to 5 times. The duration of the light periods was only 5 hours per day in order to minimise changes in structural and storage material — which would otherwise tend to invalidate the calculation of maintenance respiration by this method.

RESULTS

An example of the daily course of CO_2 exchange of a control and of a salinised plant is shown in Fig. 1.

As seen in Fig. 1 net photosynthesis of the salinised plant was lower than that of the controls. This difference increased with the rise in

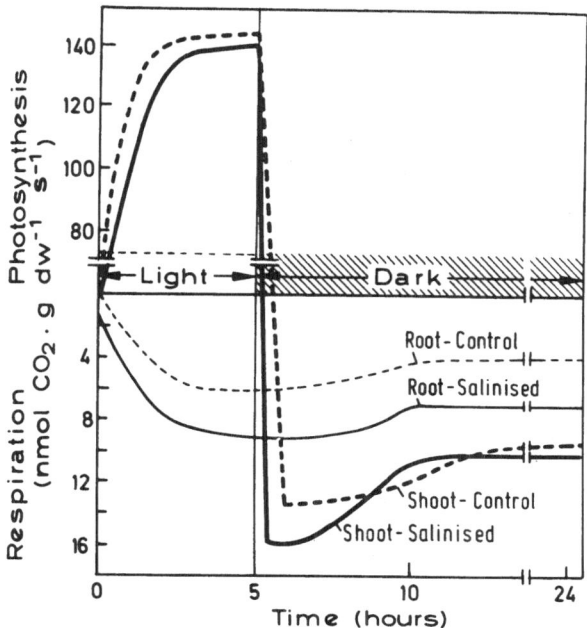

Fig. 1 An example of the daily course of uptake and efflux of CO_2 in a control and a salinised *Xanthium strumarium* plant.
Salinised plant grown with -0.5 MPa NaCl in the nutrient solution. Temperatures: day — shoots 33°, roots 29°; night — shoots 31°, roots 27°C. Light period 5 h at 610 µ mol $m^{-2}s^{-1}$ (400-700 nm). Dark period 19 h. Note the different scales used for photosynthesis and for respiration.

temperature (Fig. 2). Root respiration was higher in the salinised plant than in the control throughout the 19 h dark period. The difference in respiration rate of the tops was much smaller. This pattern recurred in all replicates of the experiment.

There was a small increase of net-photosynthesis of the control plants with temperature up to about 28°C ($Q_{10} \sim 1.4$) — Fig. 2. On the other hand net-photosynthesis of the salinised plants decreased very considerably with rising temperature. However it should be noted that in these experiments the light period lasted only 5 hours (in 24). Part of the reduced photosynthesis in the salinised plants was due to the longer time required to achieve the maximum rate of photosynthesis, at the onset of the light period.

Net 24 hours respiration rates of whole representative control

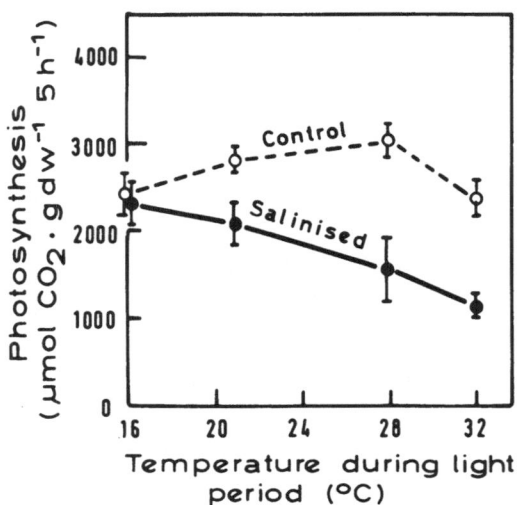

Fig. 2 Effect of salinity and temperature on the net photosynthesis of *Xanthium strumarium* plants during a 5 hour light period. Salt and light levels as in Fig. 1.

and salinised plants as functions of temperature are shown in Fig. 3. The temperatures shown are the average for root and top during the dark period. These values were chosen as the dark period continued for 19 hours and was dominant in determining the rate of respiration.

As shown in Fig. 3 total dark respiration of the control (non-salinised) plants increased with temperature ($Q_{10} \sim 1.75$). In the salinised plants total dark respiration also rose initially with temperature but between 18 and 23°C there was a 50% decrease.

As previously described (15) R_m was calculated from the intercept of total daily respiration plotted as a function of total daily photosynthesis. Results are given in Fig. 4.

As seen in Fig. 4, in this set of experiments R_m was increased by a factor of 3 in the salinised plants. However the response to temperature was lower in the salt treated than in the control plants (an increase of 160 μmol $CO_2 \cdot$g dry $wt^{-1}d^{-1}$ for a rise in temperature of from 15 to 29°C, vs. 250 μmol for the controls) and the rate response to temperature was smaller (Q_{10} of 1.3 vs. 1.9).

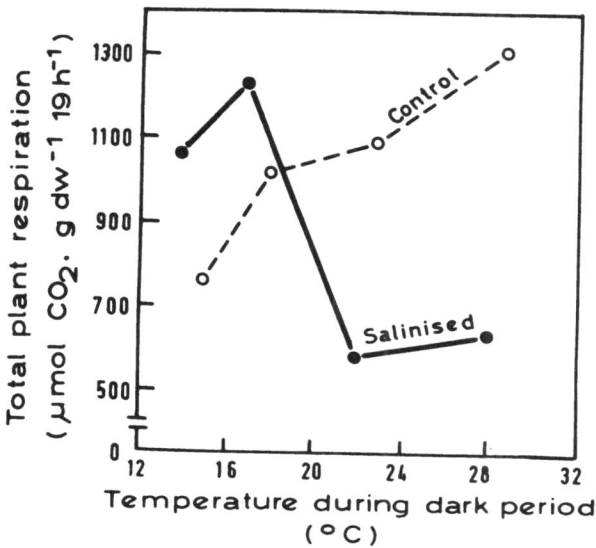

Fig. 3 Examples of the effect of temperature on the total respiration (CO_2 efflux) of control and salinised *Xanthium strumarium* plants. Salt level as in Fig. 1. See text for note on temperature used. Light intensity during light period — 610 μ mol $m^{-2}s^{-1}$ (400-700 nm).

DISCUSSION

Most of the difference in rate of respiration (per unit dry weight) of the control and salinised plants was found in the roots. Total respiration (mol CO_2/plant/day) was also greater as the roots in the saline plants constituted a higher percentage of the plant (33.2 ± 1.3% by dry weight) than in the controls (24.3 ± 0.7%).

Lambers *et al.* (10, 11) have suggested that respiration via the cyanide insensitive pathway may be induced when use of assimilates is inhibited. However in parallel experiments (Schwarz and Gale in preparation) we found that CO_2 supplementation during the growth period can, to a large extent, overcome growth inhibition caused by low level salinity. This appears to indicate that for *Xanthium strumarium*, salinity does not inhibit use of assimilate for growth. Furthermore, as noted above (Methods) growth was reduced to a minimum in these experiments, by having only a 5 h day with a 19 h night, thus minimising the possibility of surplus assimilate.

In our previous report it was shown that low level salinity imposes

330

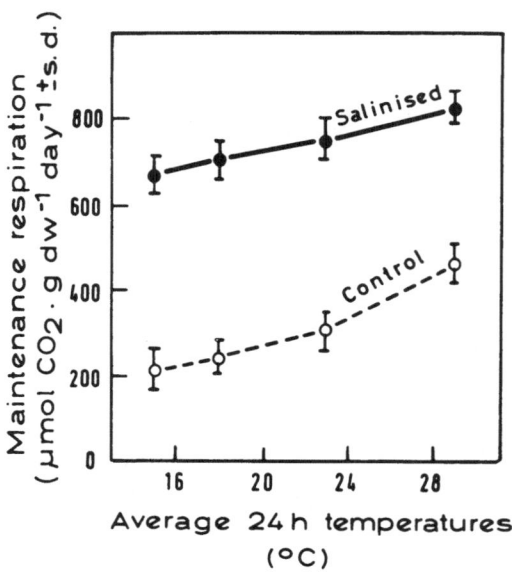

Fig. 4 Maintenance respiration of control and salinised *Xanthium strumarium* plants as a function of temperature. Salt level as in Fig. 1.

a burden on the carbon balance of *Xanthium* (16). Extra assimilate is used for maintenance at the very time when photosynthesis is reduced. This finding is further supported by the present results (Figs. 2 and 4). The initial hypothesis in the present work was that high temperatures increase sensitivity to salinity by inducing loss of assimilated carbon, by way of a large increase in R_m. However the data shown in Fig. 4 give only very partial support to this idea. As shown above the extra R_m in the salinised plants, resulting from the increase of temperature from 15 to 29°C, was only 64% of that found for the controls. High temperature did put a small extra burden on the carbon balance of the salinised plants. However, quantitatively this effect would be secondary to the decrease in assimilation (Fig. 2) and to the increase in salt induced R_m (Fig. 4).

If the increase of R_m with temperature of the control plants is indeed a reflection of adaptive processes (8), then the reduction of this increase under saline conditions may be an important factor contributing to the greater sensitivity to salt reported at high temperatures.

REFERENCES

1. Ashi, S. M. and Powers, W. L., 1938. Salt tolerance of plants at various temperatures. Plant Physiol. 12: 767-789.
2. Bernstein, L. and Ayers, A. D., 1953. Salt tolerance of six varieties of green beans. Am. Soc. Hort. Sci. Proc. 57: 243-248.
3. Berry, J. A. and Raison, J. K., 1981. Responses of macrophytes to temperature, in Physiological Plant Ecology I, Responses to the Physical Environment (ed. O. L. Lange), pp. 277-338. Springer Verlag, New York.
4. Brouwer, R., 1963. Some physiological aspects of growth factors in the root medium on growth and dry matter production. Jaarb. Inst. Biol. Scheikundig, 11-30.
5. Brouwer, R., 1979. Root functioning, in Environmental Effects on Crop Physiology (ed. J. S. Landsberg). Academic Press, New York.
6. Gale, J., 1975. Water balance and gas exchange of plants under saline conditions, in Plants in Saline Environments (eds. A. Poljakoff-Mayber and J. Gale). Springer Verlag, Berlin.
7. Gale, J., 1975. The combined effect of environmental factors and salinity on plant growth, in Plants in Saline Environments (eds. A. Poljakoff-Mayber and J. Gale). Springer Verlag, Berlin.
8. Gale, J., 1982. Evidence for essential maintenance respiration of leaves of *Xanthium strumarium* at high temperatures. J. Expt. Bot. 33: 471-476.
9. Kemp, P. R. and Cunningham, G. L., 1981. Light, temperature and salinity effects on growth, leaf anatomy and photosynthesis of *Distichlis spicata greene*. Am. J. Bot. 68: 507-516.
10. Lambers, H., Blacquiere, T., and Stuiver, C. E. E., 1980. Energy metabolism of *Plantago coronopus* L. as affected by NaCl. Abstract Meeting FESPP, Santiago de Compostela, Spain.
11. Lambers, H., Posthumus, F., Stulen, I., Lanting, L., van de Dijk, S. J. and Hofstra, R., 1981. Energy metabolism of *Plantago lanceolata* as dependent on the supply of mineral nutrients. Physiol. Plant. 51: 85-92.
12. McCree, K. J., 1974. Equations for the rate of dark respiration of white clover and grain sorghum, as function of dry weight, photosynthetic rate and temperature. Crop Science 14: 509-514.
13. McCree, K. J. and Van Bavel, C. H. M., 1980. Respiration and crop production, in Environmental Effects on Crop Physiology (ed. J. J. Landsberg). Academic Press, New York.
14. Meiri, A. and Shalhevet, J., 1973. Crop growth under saline conditions, in Ecological Studies, Analysis and Synthesis (ed. B. Yaron). Springer Verlag, Berlin.
15. Schwarz, M., 1968. Guide to Commercial Hydroponics. Keter Publ., Jerusalem.
16. Schwarz, M. and Gale, J., 1981. Maintenance respiration and carbon balance of plants at low levels of sodium chloride salinity. J. Exp. Bot. 32: 933-941.

COMPENSATORY MECHANISMS IN THE ENERGY METABOLISM OF SEEDLINGS UNDER ROOT ANAEROBIOSIS

P. HOFFMANN; E.-M. WIEDENROTH
Humboldt-University, Section of Biology, DDR - 1040 Berlin, Reinhardtstr. 4

ABSTRACT

Oxygen deficiency in the root medium of *Triticum aestivum* seedlings leads to changes in root morphology.

The gas exchange of the shoots is influenced only to a small extent, whereas in the roots respiration is gradually decreased and substituted by fermentation. These consequences are partially compensated by exudation and degradation of the intermediates in toxic concentrations and an increased substrate decomposition to ensure the ATP level. Compensatory mechanisms are discussed in a general concept of regulatory mechanisms valid for normal development as well as under resistance physiological load, ensuring the energy state of the organism. Under normal conditions cells will work following the overflow concept. Under stress conditions gradually compensation mechanisms become effective, starting with short term effects (by changing redox state and/or phosphorylation activity) over processes which need longer times (by changes of pool sizes, substrate availability etc.) to limited changes in the realization of the genetic program (from de novo synthesis of participating enzymes to variations in organelle numbers).

1. INTRODUCTION

Oxygen deficiency is a widespread limiting factor for plant growth in modern agriculture. In this paper the reactions of wheat seedlings on hypoxia or anoxia in the rhizosphere are discussed with special regard to energy metabolism. Compensatory adaptations on various levels have been found, expecially in morphology, anatomy, catabolic pathways, assimilate distribution, including exudation, and gas exchange of the root as well as the shoot.

The distinct reactions of photosynthesis to changes in the activity of root metabolism stress the dependence of the shoot on the root as the only source of water, minerals and certain rhizogen metabolites which can be stored in the above-ground parts only to a small extent. The looser: relations in the opposite directions (shoot → root) support a hypothesis that generally an excess of photosynthates is translocated into the root which is wastefully respired up to the extent of demand for growth and maintenance there. Sink capacity of the root therefore seems to be of little importance for photosynthesis. Even under stress the energy balance is assumed to be the general regulatory principle of interrelations between root and shoot.

2. MATERIAL AND METHODS

The experiments were carried out with seedlings of *Triticum aestivum* L. cv. Hatri and Kaspar, resp. Plants were cultivated in KNOP solution, growing in a growth room under fluorescent tubes in continuous light (25°C; 30 $W.m^{-2}$) Oxygen deficiency was induced by flushing the nutrient solutions with nitrogen continuously (12 l $N_2.h^{-1}.6$ l solution^{-1}). Gas exchange was measured by IRGA and oxygen sensitive electrodes. For the determination of assimilate distribution the ^{14}C-technique was used (13). The biochemical parameters concerning the fermentation were measured according to POSKUTA and HORLACHER (unpublished). Pyridinucleotide amounts were calculated on the basis of the recycling method (8; 9), adenylate contents according to the modified luciferin-luciferase test (10).

3. RESULTS AND DISCUSSION

3.1. Morphology

Under the defined conditions of hydroculture on oxygen free nutrient solution a general retardation of biomass accumulation of wheat seedlings was found (about -20 to -30 % in 8 d old seedlings). Root anaerobiosis acts on roots and shoots nearly to the same extent, the root/shoot ratio therefore is little affected (0,35). On the other hand, the consumption of endosperm reserves decreased indicating that growth retardation may be interpreted as slowing the speed of early stages of ontogenesis.

Furthermore O_2-deficiency in the root medium decreased the water content of the shoot (on 84-87 %) as well as the root (on 90 %) relatively to the control. So the results support our hypothesis that the water content of the seedling shoots is a sensitive but inspecific indicator of stress in the root medium (11).

Cultivation of wheat seedlings on oxygen free nutrient solution affected the development of the roots of different order in different strength. Longitudinal growth of the first three (so called seminal) roots is retarded to a higher extent than those of the adventitious ones. Furthermore these adventitious roots appeared 2-3 days earlier under oxygen deficiency than in the control, but the formation of lateral roots is retarded whilst the distribution of root hairs remained unaffected.

According th these morphological reactions the whole structure of the root systems is changed under anaerobiosis to enable the plant, in connection with other adaptations, to overcome periods of oxygen deficiency in the rhizosphere.

3.2. Gas exchange of roots

For 8-10-d-old wheat seedlings cultivated on nutrient solution the critical oxygen concentration below which the O_2-uptake rate starts to decrease varies from 1.5 - 3 mg O_2.l^{-1} according to the physiological state of the plants, but only at values lower than 0.8 mg O_2.l^{-1} the decrease of root respiration becomes important (fig. 1; 14). Following the hypothesis of LAMBERS (5) that under hypoxic conditions at first only the cyanide-resistant, non-phosphorylating pathway of electron transport will be affected, the initial relatively slow decrease in O_2-uptake may only be due to blocking this energetically inefficient pathway. We suggest therefore that only below 10 % of saturation-concentration oxygen deficieny will retard also the activity of the ATP-yielding cytochrome chain.

On the basis of alternating measurement of O_2-uptake from the ambient solution by CLARK electrodes and the CO_2-output into the gaseous phase by IRGA one can calculate the respiratory quotient even of the roots of intact plants. This RQ was found to vary between 2 and 3 in oxygen-saturated solution and normal air, respectively, indicating that even under normoxic conditions in young roots fermentation processes take place. This pheno-menon was described already by RUHLAND and RAMSHORN (7). From the RQ values measured one can calculate that fermentation participates nearly 75 % in the dissimilation on the whole.

Recently it has been shown in our laboratory that an oxygen concentration of 10 % saturation level in the nutrient solution lowered CO_2-release as

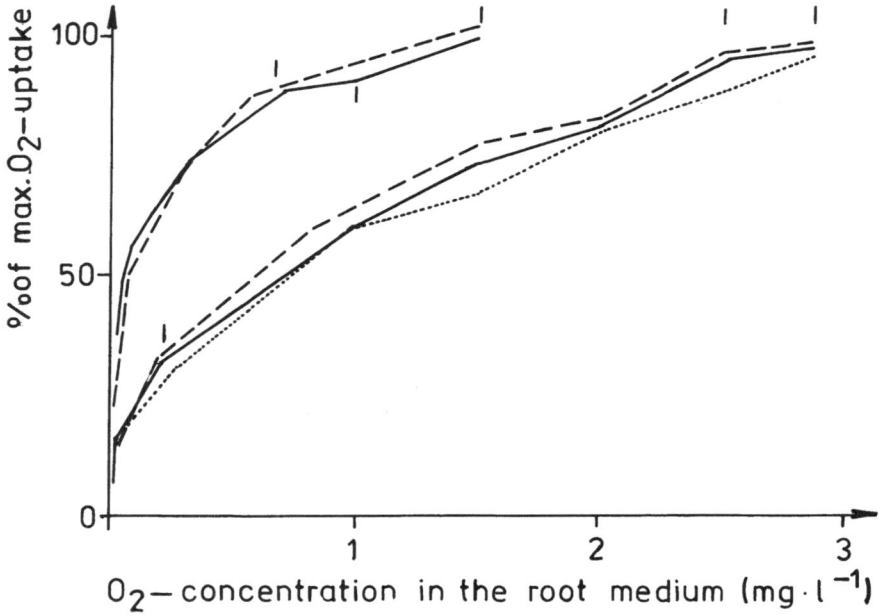

.Fig. 1 : Gas exchange of roots in dependence on the O_2-concentration in the
root medium.

well as O_2-uptake of wheat seedlings to the same extent by nearly 50 %. The
RQ therefore remained unchanged supporting the hypothesis mentioned above
that hypoxia only retards the activity of the alternative oxidase without
influencing the energy yield of the cytochrome pathway and therefore no
change to enhanced fermentation is induced.

Strong anoxia, on the other hand, blocking the O_2-uptake from the ambient
medium completely, decreased the CO_2-release from the roots of intact plants
by 50 % and of isolated roots by 65 %. Under restituted normoxia gas-exchange
rates reach nearly the initial values before O_2-deficiency. It may be concluded from these data that

- in intact plants a certain internal O_2-transport via the intercellular
  system enables the root to higher CO_2-relase, corresponding to that under
  hypoxia;

- in isolated roots only the same amount of glucose is dissimilated by
  fermentation as is respired aerobically under normal conditions;

- therefore the low energy yield of fermentation seems to be sufficient to maintain the living structure of root tissue (maintenance respiration) during temporary anaerobiosis.

These observations support the hypothesis (4) that the shoot generally exports more carbohydrates into the roots than required there under normoxic conditions and that the amount of photosynthates translocated therefore does not determine the activity of root metabolism immediately (15).

3.3. Gas exchange of shoots

The apparent photosynthesis is not or only a little influenced by root anaerobiosis up to 17 hours. On the other hand, a longer time of oxygen deficiency in the rhizosphere during the first days stimulated the net

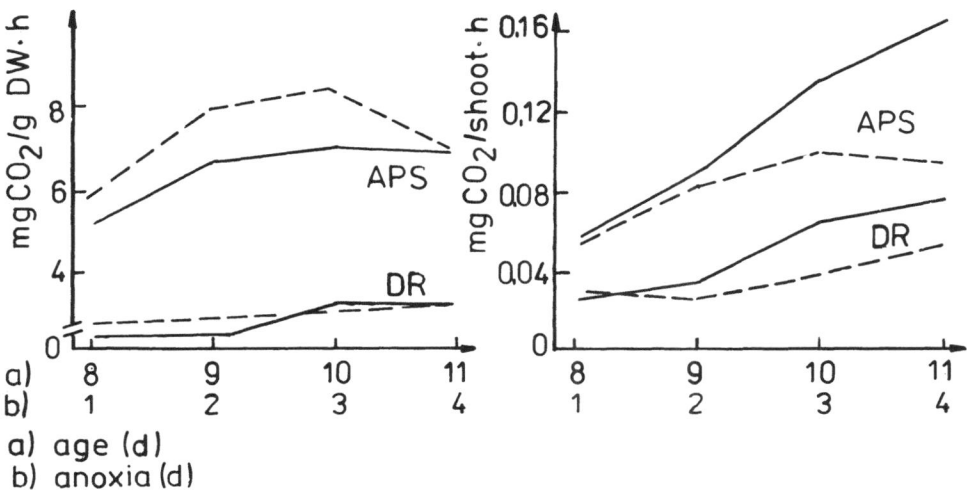

Fig. 2 : Time course of apparent photosynthesis (APS) and dark respiration (DR) of shoots in wheat seedlings (cv. Caspar) under root anaerobiosis. Solid line = control; broken line = O_2-deficiency.

photosynthesis rate, compensating in some degree the stress- induced growth retardation of biomass accumulation of the shoot (fig. 2; 14). Dark respiration of the shoot is less sensitive against root anaerobiosis than apparent photosynthesis, but is strongly retarded by a combination of a dark period with the lack of oxygen in nutrient solution.

The results mentioned above underline that O_2-deficiency-induced changes in root metabolism cannot influence shoot metabolism immediately by means of a sink/source interrelation because

- roots possess a certain pool of carbohydrates, and

- under normoxia shoots export a surplus of photoassimilates, so that even a higher demand of the roots does not influence the matter balance of the shoots. Compensating changes in shoot metabolism therefore occur only after longer periods of O_2-deficiency in the root medium.

3.4. Assimilate distribution

Already a 24-hour-period of strong root anaerobiosis caused drastic changes in the distribution of labelled photoassimilates. O_2-deficiency after $^{14}CO_2$-application induced an increase of ethanol and a decrease of the socalled starch fraction in roots as well as in shoots, and an enhancement of exudation from the roots into the ambient medium. The incorporation of labelled carbon into structure components is not clearly dependent on the experimental procedures used. The fact that flushing the root medium with nitrogen has a stronger effect than simple sealing the root containers indicates that the changes in assimilate distribution are really due to oxygen deficiency and not caused by CO_2-enrichment in the nutrient solution.

O_2-deficiency before $^{14}CO_2$-application followed by normal aeration after labelling induced the corresponding distribution pattern in the shoot only, whereas the root behaved indifferently and no enhanced exudation could be noted. It may be suggested that the root as the directly affected organ immediately responds to restituted normoxia, and the preceding root anaerobiosis after-affects the shoot only.

The distribution of labelled compounds is highly dependent on the time between $^{14}CO_2$-application and its measurement.

3.5. Biochemical parameters

3.5.1. Fermentation. It has recently been shown by HORLACHER and POSKUTA (unpubl.) that the ethanol concentration in the root tissue markedly increased after 24 hours of root anaerobiosis, this is accompanied by a high exudation of ethanol into the root medium. Both were followed by an increase of ADH activity, which caused an enhanced destruction of ethanol during the next 24 hours. The enhanced exudation of toxic metabolites of fermentation into the surrounding medium has been known up to now for halophytes only.

Triticum seedlings are not able to avoid ethanol accumulation by transferring it into the shoot. Also the increase of the ADH in the shoot is very small.

The LDH-activity in root tissue is only to a small extent stimulated by anaerobiosis; in shoots no influence was observed. Therefore lactate fermentation is uniportant in these seedlings.

During the whole time of O_2-deficiency studied the amount of glucose remains constant in the shoot; in the root the content is even higher, at the expense of starch (13).

3.5.2. <u>Pyridinnucleotides and adenylates</u>. Inhibitor experiments suggest that the ADP/ATP ratio may be one of the more limiting factors for root respiration and not only the amount of carbohydrates translocated from the shoot. This is in accordance with our data on the pyridine nucleotide content in wheat seedlings after a 16-hour-period of darkness, expressed in terms of the reduction charge (HOFFMANN et al., unpubl.).

The catabolic as well as the anabolic reduction charge of the shoot decreased drastically in darkness as a consequence of the continuous final oxidation of the reduced pyridine nucleotides maintaining the ATP level essential for normal metabolism.

Oxygen deficiency and darkness acts more drastically, also on the ATP content (6).

Even under these conditions, after repeated aeration of only 90 min., the action of compensatory mechanisms can be observed.

So a whole cascade of overflow mechanisms in roots and shoots stabilizes the energy metabolism of plants allowing them to overcome various stress conditions without being forced to adapt the fundamental metabolic pathways genetically fixed over the long period of phylogenesis.

REFERENCES

1. Armstrong W. 1975. Waterlogged soils. In : Ethevington J.R. (Ed.) Environment and plant ecology. pp. 181-218, John Wiley and Sons, London-New York-Sidney-Toronto.
2. Chirkova T.V. 1978. Some regulatory mechanisms of plant adaptation to temporal anaerobis. In : Hook D.D. and Crawford R.M.M. (Eds.) Plant Life in Anaerobic Environments pp. 137-154, Ann Arbor Science Publ., Ann Arbor, Mich. 1978.

3. Grineva G.M. 1975. Regulatsiya Metabolizma in Rastenii pri Nedostatke Kisloroda (Regulation of Plant Metabolism under Oxygen Deficiency). Nauka, Moskva.

4. Lambers H. 1979. Energy Metabolism in Higher Plants in Different Environment. Thesis. Groningen.

5. Lambers H. 1980. The physiological significance of cyanide resistant respiration in higher plants. Plant, Cell and Environment 3, 239-302.

6. Miginiac-Maslow M., Hoarau A. 1979. The adenine nucleotide levels and the adenylate energy charge values of different *Triticum* and *Aegilops* species. Z. Pflanzenphysiol. 93, 387-394.

7. Ruhland W., Ramshorn K. 1938. Aerobe Gärung in aktiven pflanzlichen Meristemen. Planta 28, 471-514.

8. Slater T.F., Sawyer B. 1962. A colorimetric method for estimating the pyridine nucleotide content of small amounts of animal tissue. Nature 193, 454-456.

9. Smith P.J.C. 1961. Assay of pyridine nucleotide coenzymes and associated dehydrogenases with phenazine methosulphate. Nature 190, 84-85.

10. Strehler B., Thother J. 1982. Firefly luminescence in the study of energy transfer mechanisms. I. Substrate and enzyme determination. Arch. Biochem. Biophys. 40, 28-41.

11. Wiedenroth E.-M. 1976. Chlorophyllbildung und Gaswechsel von Weizen-Keimpflanzen (*Triticum*) under dem Einfluss variierter Temperatur- und Sauerstofbedingungen im Wurzelbereich. Tagungsber. Akad. Landwirtschaftswiss. DDR 143, 297-309.

12. Wiedneroth E.-M., Poskuta J. 1978. Photosynthesis, photorespiration, respiration of shoots and respiration of roots of seedlings as influenced by oxygen concentration. Z. Pflanzenphysiol. 89, 217-225.

13. Wiedenroth E.-M., Poskuta J. 1981. The influence of oxygen deficiency in roots on CO_2 exchange rates of shoots and distribution of ^{14}C-photoassimilates of wheat seedlings. Z. Pflanzenphysiol. 103, 459-467.

14. Wiedenroth E.-M. 1981 a. The distribution of ^{14}C-photoassimilates in wheat seedlings under root anaerobiosis (Eds.) Structure and Function of Roots. pp. 389-393. M. Nijhoff/Dr. W. Junk Publ., The Hague-Boston-London.

15. Wiedenroth E.-M. 1981 b. Relation between photosynthesis and root metabolism of cereal seedlings influenced by root anaerobiosis. Photosynthetica 15, 575-591.

CHLOROPLAST DEVELOPMENT AND PHOTOSYNTHESIS IN ECHINOCHLOA (BARNYARD GRASS) AND RICE: O_2 AND FLOODING STRESS

ROBERT A. KENNEDY, DELMAR VANDER ZEE AND CONNIE S. BOZARTH
DEPARTMENT OF HORTICULTURE
WASHINGTON STATE UNIVERSITY
PULLMAN, WASHINGTON 99164-6414, U.S.A.

ABSTRACT

Primary leaf plastid development and photosynthetic activity in *Echinochloa* (barnyard grass) were monitored during a 5 day time course of germination and growth in a nitrogen atmosphere. After 8 hours imbibition in the light but without oxygen, the proplastids were spheroidal and contained a few rudimentary stromal thylakoids. However, during the next 126 hours of anaerobiosis, plastid size and stromal thylakoid numbers increased markedly. Under anaerobiosis plastid morphology was also characterized by the development of prolamellar bodies, previously reported only in etiolated tissues, while changes in starch grains and plastoglobuli indicated carbohydrate and lipid metabolism, respectively. Upon exposure to air the arrested development of chloroplasts was reversed. Compared to etioplasts when exposed to light, however, there was a lag for the first 18 hours before the plastids developed rapidly into typical, C_4 mesophyll and bundle sheath chloroplasts.

Chlorophyll synthesis and development of photosynthetic capability began slowly when seedlings were transferred from N_2 to air. Whereas etiolated seedlings exhibited net CO_2 fixation after approximately 12 hours after exposure to light, anaerobically-grown seedlings did not show a positive CO_2 gas exchange until about 55 hours after transfer from N_2 (light) to air (light). The kinetics for CO_2 uptake more closely resemble those of RuBP carboxylase, the C_3 cycle enzyme, than they do PEP carboxylase. Upon transfer to air (light), "anaerobic" shoots had about 60% of their ultimate PEPcase, but less than 20% of their RuBPcase activity. The predominance of PEPcase over RuBPcase in shoots grown without oxygen also occurs in rice, a C_3 plant. Thus, when rice and oryzicola were transferred to air after germination in an N_2 (light) environment, the early-labeled $^{14}CO_2$ fixation products in both grasses were predominantly

C_4 acids. Based on $^{14}CO_2$ labeled intermediates, however, oryzicola appeared capable of developing carbon cycle capability faster than rice. Such a head start may, in part, explain oryzicola's ability to emerge and outcompete rice in field situations.

INTRODUCTION

Echinochloa crus-galli (barnyard grass) is the most serious rice field weed world-wide. In California, *E. crus galli* var. *oryzicola* frequently reduces rice yields by up to 40%. Like rice, oryzicola has the unusual ability to germinate without oxygen, producing a 2-3 cm unpigmented shoot after 7 days germination in the light. If oxygen is supplied during anaerobic growth, the shoots green up, chloroplasts develop and photosynthesis proceeds with only a slight lag when compared to air/dark (etiolated) controls. We have been interested in *oryzicola* as a system to study development of the photosynthetic apparatus and the role of oxygen in chloroplast development. In the field, oryzicola routinely germinates and emerges through 25 cm of water. Therefore, we have also been interested in the CO_2 fixation capability of oryzicola under aquatic environments and have compared the early-labeled photosynthetic products and carboxylase enzymes to those of rice.

MATERIALS AND METHODS

Seeds of *Echinochloa crus-galli* var. *oryzicola* and rice *Oryza sativa* cv. S201) were obtained from the Rice Experiment Station, Biggs, California. Seeds were germinated as before (2), flushing continuously with 99.995% N_2 or air at 25 C. Except for dark controls, irradiance was 75 to 100 uE m^{-2} sec^{-1}.

For transmission electron microscopy (TEM), plumule tissues within emerged coleoptiles were fixed as described previously (4) using 2% glutaraldehyde in 0.1 M PO_4 buffer (pH 6.8), post-fixed in 2% O_sO_4 and stained en bloc with 1% uranyl acetate.

Carboxylase enxyme assays (5) and $^{14}CO_2$ incorporation experiments (3) were conducted as published earlier. Chlorophyll was assayed as described by Arnon (1).

RESULTS

Chloroplast ultrastructure: N_2/light. The general cytology of plumule cells in oryzicola seedlings germinated for 8 hours without oxygen was characterized by cells with dense cytoplasm and little vacuolation. Organelles observed include numerous spherosomes and many discrete but rudimentary mitochondria and chloroplasts (Fig. 1). After 8 hours anoxia the chloroplasts (proplastids) were spheroidal in shape, they had an intact outer membrane and contained two to four short thylakoid membranes, dense stroma and many ribosomes. No differentiation could be seen between bundle sheath and mesophyll chloroplasts at this stage of development and no starch was observed.

By 96 hours anoxia, considerable chloroplast development had taken place, especially with respect to internal membranes (Fig. 2). At this time the chloroplasts were generally elongate in shape and contained many peripheral membranous vesicles just inside the plastid envelope. These vesicles appeared to be continuous with the inner envelope membrane (insert). This stage of development was also characterized by extension of the thylakoid membranes. Starch grains were commonly observed within the chloroplasts. Perhaps the most distinguishing feature of the plastids at 96 hours was the prolammellar body (PLB), a dense cluster of interconnecting tubular membranes arranged in a para-crystalline array.

By 134 hour anoxia (Fig. 3), starch grains had disappeared. Peripheral membrane vesicles were more common and for the first time plastoglobuli were seen, most frequently in close association with the PLB or peripheral membrane vesicles.

FIGURE 1. Plumule cells imbibed 8 hr under anoxia. Proplastid (PP), spherosomes (S), cell wall (CW), mitochondria (M). Note the short stromal thylakoids (ST). Bar = 1 um.
FIGURE 2. Plastids from 96 hr anoxia with pleomorphic shapes. Cytoplasm of cup-shaped plastid (C), extended stromal thylakoids (ST), prolamellar body (PLB), peripheral membrane vesiculation (MV). Bar = 1 um. Inset 2a shows connection between peripheral membrane vesicles and inner envelope membrane of plastid (arrow heads). Bar = 0.25 um.
FIGURE 3. Section through membrane rich area of 134-hr anoxic plastids. Peripheral membrane vesicles (MV), plastoglobuli (PG), perforated shaped plastid (C). Bar = 1 um. Inset 3a shows continuity between peripheral membrane vesicles and membrane invagination into the stroma (arrow head). Bar = 0.25 um.
FIGURE 4. Plastids (P) at 4 hr of air showing thylakoid (T) formation from peripheral membrane vesicles (arrows). Bar = 1 um.
FIGURE 5. Plastid at 8 hr of air showing large starch grains (SG), plastoglobuli (PG), and grana stacking initials (arrows). Bar = 1 um.
FIGURE 6. Margin of plastid shows simpler peripheral membranes (arrow head) at 8 hr air than seen earlier under anoxia. Thylakoid (T), chloroplast envelope (CE). Bar = 0.25 um.
FIGURE 7. Early evidence of dimorphic chloroplasts at 36 hr of air in bundle sheath (BS) and mesophyll cells (MC). Bar = 2 um.

344

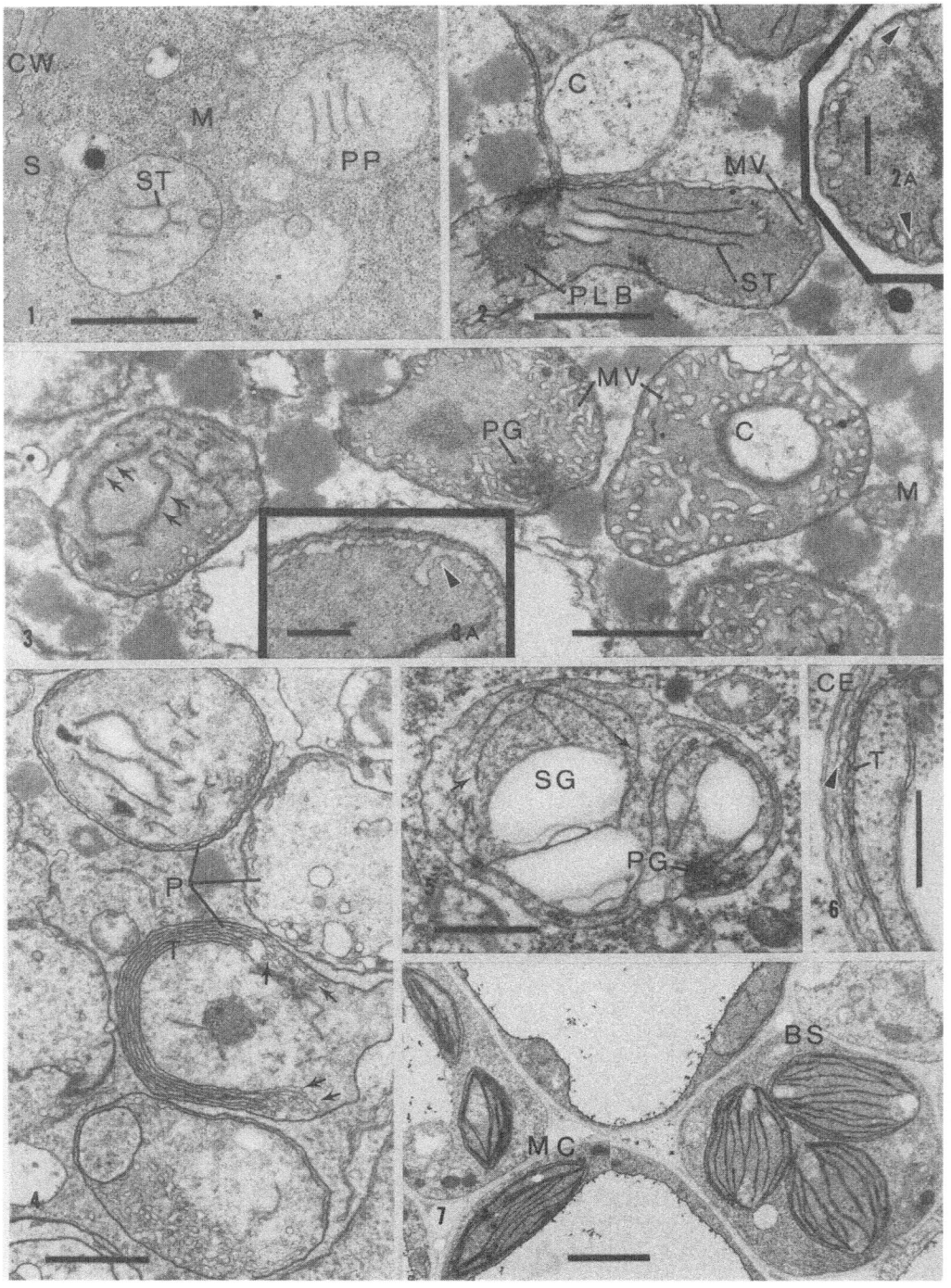

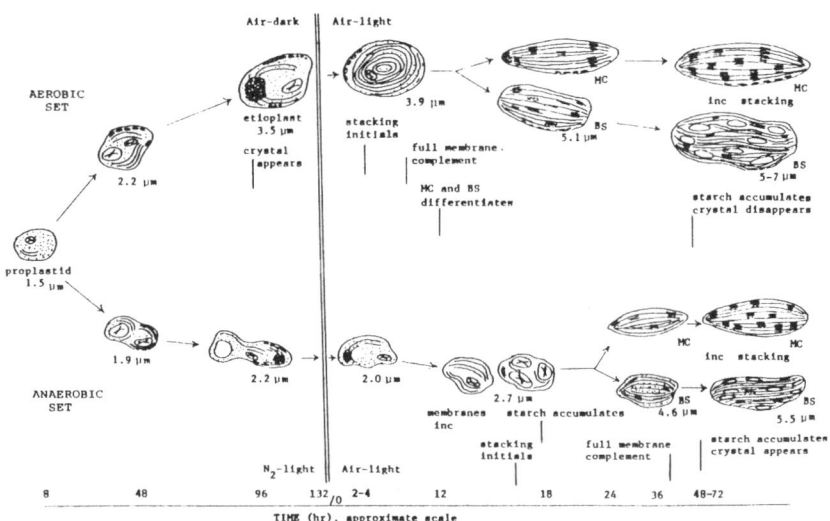

FIGURE 8. Summary of proplastid to chloroplast development under air—dark and N_2-light pretreatments followed by light and air post-treatments, respectively. Approximate size (length) of plastids is given and major events are indicated by vertical lines corresponding to the time scale. Mesophyll chloroplast (MC), bundle sheath chloroplast (BS). Plastids not necessarily drawn to scale.

Chloroplast ultrastructure: air/dark. After 4 hours germination, chloroplasts in etiolated (air/dark) seedlings contained many starch grains (Fig. 4) and showed an extensive development of thylakoid membranes which seemed to originate from peripheral membrane vesicles. As under anaerobic conditions, cup-shaped chloroplasts were frequently seen. Although not shown, by this stage in development etioplasts already contained a PLB, which was not observed until much later in the absence of oxygen. Chloroplast development after 8 hours germination was marked by continued starch build-up and thylakoid formation (Fig. 5). Whereas stacking initials were never seen in "anoxic plastids" (cf. Fig. 3), after only 8 hours germination in air/dark stacking initials were seen in etioplasts (Fig. 5). As under anoxia, peripheral membrane vesicles were common within developing plastids.

Finally, by 36 hours the chloroplasts in the mesophyll and bundle sheath cells were highly developed, contained large starch grains, an extensive grana and stroma thylakoid system and plastoglobuli (Fig. 7).

As expected of C_4 grass chloroplasts, a peripheral reticulum was present, but at this stage of development little differentiation between mesophyll and bundle sheath chloroplasts was observed, i.e., mesophyll chloroplasts were only slightly more "granal" than were bundle sheath chloroplasts. Figure 8 summarizes the differences in timing and nature of chloroplast development observed in oryzicola under anaerobic and aerobic conditions.

Chlorophyll synthesis. Chlorophyll (Chl) formation in greening oryzicola shoots germinated in air (dark) or anoxia (light) for $5\frac{1}{2}$ days then exposed to light or air, respectively, is shown in Figures 9 and 10. As expected, the Chl a/b ratio dropped faster and chorophyll synthesis occurred much earlier in etiolated tissue (Fig. 9) as compared to that grown without O_2 (Fig. 10). In the latter there was an 18 to 24 hour lag in the synthesis of chlorophyll, although by 96 hours total Chl and the Chl a/b ratio in the two tissues were similar and equal to that of plants grown normally.

CO_2 assimilation: N_2. Development of CO_2 fixation capability in both anaerobic and dark grown seedlings paralleled Chl synthesis. In etiolated plants, net CO_2 fixation occurred approximately 12 hours after exposure to light, it increased rapidly thereafter and leveled off between 36 to 48 hours (Fig. 11). Although "anoxic seedlings" exhibited a lag in Chl synthesis, they seemed to display an even greater lag in net CO_2 fixation (Fig. 12). Compared to etiolated tissue, which exhibited net CO_2 fixation when Chl levels reached approximately 0.25 mg/g fresh wt. (12 hours), N_2-grown seedlings showed not net CO_2 fixation until total Chl levels were three times those of air (dark) grown plants and 50 to 55 hours after exposure to air. None-the-less, by day 5 the photosynthetic rate in both plants was approximately 40 umol/g fresh wt/hr.

Figure 13 shows activity of RuBP and PEP carboxylase in air (dark) and N_2 (light) grown seedlings. Initially, the activity of PEP carboxylase was greater than that of RuBP carboxylase whether the seedlings were grown in air or N_2. And in either case, PEP carboxylase activity increased rapidly upon exposure to light or air, respectively. RuBP carboxylase activity on the other hand increased quickly when etiolated seedlings were transferred to light, but did not increase similarly when N_2-grown seedlings were exposed to air. Instead, a 35 to 40 hour lag was observed before RuBP carboxylase activity increased, and even then, RuBP carboxylase activity of N_2-grown seedlings was less than 20% of that of etiolated seedlings 96 hours after transfer to ambient conditions.

Early labeled photosynthetic intermediates reflected the developmental
pattern of carboxylase enzyme activity. When 6 day old, N_2-grown seedlings
were exposed to $^{14}CO_2$ for 30 seconds, C_4 acids contained nearly 40% of the
total radioactivity (Table 1). None-the-less, about 8% of the ^{14}C was
located in intermediates of the C_3 pathway (PGA and sugar phosphates).
Rice, a C_3 plant, also had heavy labeling of the C_4 acids malate and
aspartate, with less radioactivity located in C_3 cycle products. In both
plants, after 48 hours exposure to O_2 the percent label in C_3 cycle intermediates increased substantially. This occurred in rice even more than
oryzicola, possibly owing to the lag in RuBP carboxylase synthesis in oryzicola, discussed above.

CO_2 assimilation: flooding. When oryzicola seeds were germinated
under water to more realistically simulate their germination and growth in
rice fields, the seedlings initially were unpigmented, as when grown
anaerobically, and greened up when they were between 20-30 mm in height.
Initial carboxylase enzyme activities were also similar to anaerobically-
grown seedlings (Table 2). At first PEP carboxylase activity was much
greater than that of RuBP carboxylase, but it increased very little during
the 3 week time course examined. Again as in anaerobic seedlings, RuBP
carboxylase activity was near zero initially, but it increased very rapidly
after an initial lag in synthesis. Interestingly, in seedlings grown under

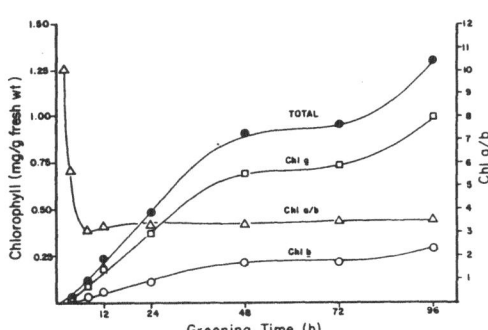
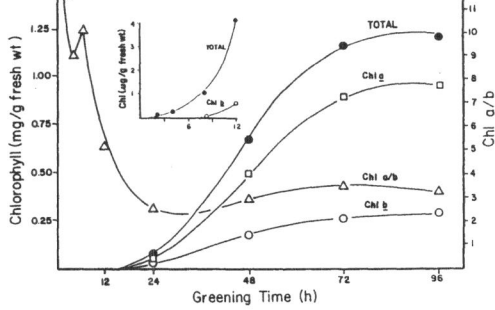

FIGURE 9. Time course of Chl formation
and Chl a/b ratios in greening oryzicola shoots germinated in air-dark for
5 1/2 days, then exposed to light.

FIGURE 10. Time course of Chl formation and Chl a/b ratios in greening
oryzicola shoots germinated under
anoxic-light conditions for 5 1/2
days, then exposed to air. Expanded
scale in inset shows chlorophyll concentration in ug/g fresh weight.

flooded conditions, RuBP carboxylase activity eventually was much greater than PEP carboxylase activity, whereas the opposite was true in N_2-grown seedlings. These differences between plants grown under N_2 versus flooding is also seen in photosynthetic products. While N_2-grown seedlings had a majority of label in C_4 acids (Table 1), water grown seedlings exhibited a much higher percentage of ^{14}C in C_3 cycle products (PGA, sugars and sugar phosphates) and comparatively little radioactivity in C_4 acids (Table 3).

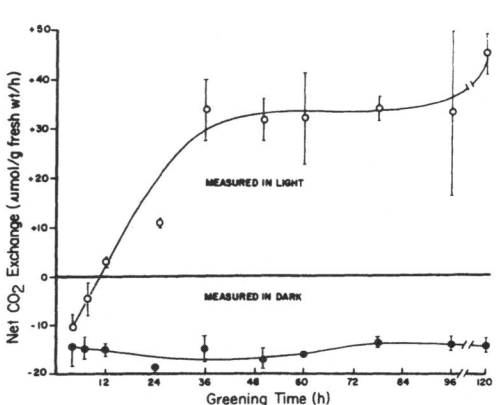

FIGURE 11. Net CO_2 exchange in etiolated seedlings after exposure to light, measured in light and dark with IR gas analysis. Points are means of five repetitions; bars represent SE if greater than 0.5.

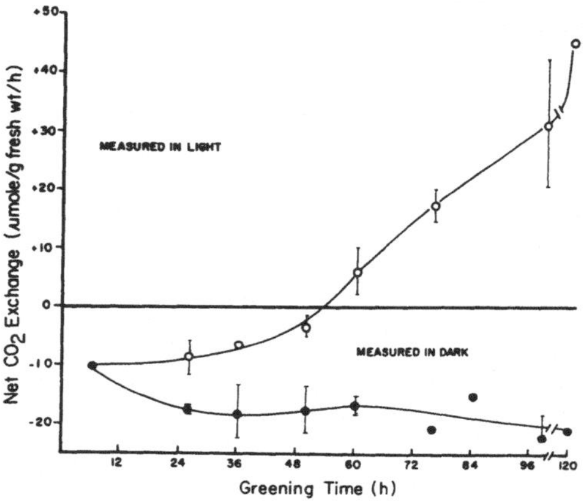

FIGURE 12. Net CO_2 exchange in anoxic-germinated seedlings after exposure to air, measured in light and dark with IR gas analysis. Points are means of five repetitions; bars represent SE if greater than 0.5.

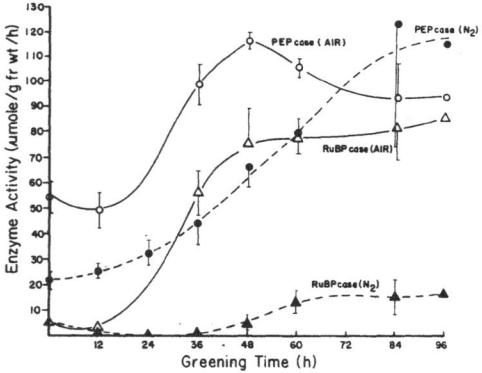

FIGURE 13. Enzyme activity of PEP and RuBP carboxylase in oryzicola shoots during greening following air-dark or N_2-light germination for 5 1/2 days. Points are means of three experiments; bars represent SE. Curves appeared the same when expressed on a dry weight or protein basis.

Table 1. Early labeled photosynthetic products in oryzicola and rice grown anaerobically. Seedlings were germinated for 6 days in N_2, then exposed to $^{14}CO_2$ for 30 seconds, or transferred to air for 48 hours prior to the exposure to $^{14}CO_2$. Killing of the tissue and separation of labeled products was as given in Methods.

| | Photosynthetic Products (% of total) | | | | | | |
	Malate	Aspartate	Glycerate	Glutamate	Alanine	PGA	Sugars/S-P
Oryzicola							
N_2- 0 hr O_2	27.9	10.1	7.6	12.8	2.2	1.6	6.0
N_2-48 hr O_2	9.8	7.7	15.4	5.4	6.7	6.9	26.2
Rice							
N_2- 0 hr O_2	11.0	16.6	4.3	28.0	-	2.4	-
N_2-48 hr O_2	15.9	9.1	6.9	2.6	5.9	8.3	37.2

Table 2. PEP and RuBP carboxylase enzyme activity in oryzicola seedlings germinated and grown for varying periods of time under water.

| | Enzyme Activity (umol/g fresh wt/hr) | |
	PEP carboxylase	RuBP carboxylase
5 days	11	6
7 days	13	26
10 days	8	56
15 days	13	56

Table 3. Early labeled photosynthetic products in oryzicola seedlings germinated and grown under flooded conditions for 5 or 7 days. Experimental conditions were as given in Table 1 except that the $^{14}CO_2$ incorporation period was 60 seconds.

| | Photosynthetic Products (% of total) | | | | | | |
	Malate	Aspartate	Glycerate	Glutamate	Alanine	PGA	Sugars/S-P
5 days	11.9	3.0	9.0	6.0	3.9	23.5	42.6
7 days	5.5	4.1	4.1	5.5	3.3	10.1	55.9

DISCUSSION

In the present work we have studied the role of oxygen in the development of photosynthesis and the photosynthetic apparatus in barnyard grass, a
rice field weed able to germinate and grow anaerobically. The results are
interesting for several reasons. First, the system employed provides a
unique alternative to studying chloroplast development. Most of what we know
about the development of chloroplasts has come from studies using etioplasts,
which may not reflect the sequence of chloroplast development naturally.
Second, the extent to which the chloroplasts developed in the total absence
of oxygen is surprising. Previous studies on the effects of anaerobiosis on
cell structure have often indicated just the opposite, a disintegration of
organelles (see 4). In contrast, the complex development of chloroplast
structure and photosynthetic function in oryzicola indicates that considerable
anabolic activity is possible without oxygen.

Other results are also interesting and warrant further study. For
example, even though the seedlings were grown in the light, prolamellar
bodies were frequently seen in "anoxic chloroplasts." Except for a few
exceptions (4), PLB's have previously only been seen in etioplasts,
quickly dispersing in the light. The PLB's shown here were formed in
seedlings continuously grown in the light, but without oxygen. The present
studies also suggest a close correlation of inner envelope vesiculation to
increases in stromal membranes.

Last, synthesis and/or increases in carboxylating enzyme activities
differ when comparing seedlings grown in a truly anaerobic environment to
those from a low oxygen (flooded) environment. In the former, PEP carboxylase
activity rapidly increased to control levels when exposed to air, whereas
RuBP carboxylase did not. Under water, the opposite occurred; PEP carboxylase activity did not increase as seedlings greened up during their
first 7-10 days growth, while RuBP carboxylase activity increased sharply.
These results suggest that the anaerobic system may be a profitable system
to study development of photosynthetic structure and function, but it may
not truly reflect the situation in the field.

REFERENCES

1. Arnon, D. I. 1949. Copper enzymes in isolated chloroplasts, polyphenol oxidase in Beta vulgaris. Plant Physiol. 24:1–15.
2. Kennedy, R. A., Barrett, S. C. H., Vander Zee, D., Rumpho, M. E. 1980. Germination and seedling growth under anaerobic conditions in Echinochloa crus-galli (barnyard grass). Plant, Cell and Environment 3:243–248.
3. Kennedy, R. A. Laetsch, W. M. 1973. Relationship between leaf development, carboxylase enzyme activities and photorespiration in the C_4-plant Portulaca oleracea L. Planta 128:149–154.
4. Vander Zee, D., Kennedy, R. A. 1982. Plastid development in seedlings of Echinochloa crus-galli under anoxic germination conditions. Planta 155:1–7.
5. Williams, L. E., Kennedy, R. A. 1978. Photosynthetic carbon metabolism during leaf ontogeny in Zea mays L.: Enzyme studies. Planta 142:269–274.

EFFECT OF HIGH LIGHT AND HIGH LIGHT STRESS ON COMPOSITION, FUNCTION AND STRUCTURE OF THE PHOTOSYNTHETIC APPARATUS

H.K. LICHTENTHALER, R. BURGSTAHLER, C. BUSCHMANN, D. MEIER, U. PRENZEL, A. SCHÖNTHAL

Botanisches Institut (Pflanzenphysiologie), Universität Karlsruhe, Kaiserstraße 12, D-7500 Karlsruhe, F.R.G.

ABSTRACT

The differences in chloroplast ultrastructure, pigment composition and variable fluorescence of plants grown at low light and high light were determined, and the effect of a white light stress on the photosynthetic apparatus of LL and HL-plants and of sun and shade leaves was studied.

1. HL-chloroplasts possess fewer thylakoids per chloroplast, lower grana stacks, a higher proportion of exposed photosynthetic membranes and a higher percentage of the photosystem I chlorophyll a/ß-carotene-proteins CPI and CPIa than LL-chloroplasts. They are characterized by higher values for the ratio chlorophyll a/b and lower values for the ratio chlorophyll a/ß-carotene (a/c), xanthophylls/ß-carotene (x/c) and chlorophylls/carotenoids. HL and sun leaves possess a smaller photosynthetic unit size, as seen from lower values for the ground fluorescence fo and the maximum fluorescence fp.

2. A 45 min exposure of LL and HL-radish plants to a water filtered strong white light (SL: PAR 1000 μE $\cdot$ m^{-2} $\cdot$ sec^{-1}) does not decrease the degree of thylakoid stacking or the ratio of appressed to exposed thylakoid membranes. There occurs, however, a gradual adaptation response, as seen from an increased ß-carotene proportion, increasing a/b values together with decreasing values for a/c, x/c and a+b/x+c. Similar changes in prenylpigment ratios were also found in Cissus leaves and in leaves of Fagus. Upon prolonged irradiation the level of photosynthetic pigments per leaf area decreases. At light + heat stress (SL + IR: strong white light without water filter) this pigment degradation proceeds faster.

Exposure to strong white light causes photo-inhibition of the photosynthetic apparatus and decreases ground fluorescence and variable fluorescence, before any decrease in chlorophyll content can be detected. It is assumed that high light stress causes the fast activation or accumulation of quencher molecules (e.g. ß-carotene or others) which protect the photosynthetic apparatus by quenching excitation energy.

1. INTRODUCTION

The high light adaptation response of leaves and plants, as found in full sun light (sun leaves) and in plants grown at high light intensity (HL-leaves), is bound to special changes in the morphology, physiology, composition and structure of leaves and of chloroplasts (1-7). At high quanta fluence rates the photosynthetic apparatus (sun-type of HL-chloroplast) is built and adapted for high rates of photosynthetic quantum conversion (2-5, 7). As compared to the shade-type (low light) chloroplasts of shade leaves and LL-plants, the HL-chloroplast is characterized by a different pigment and pigment protein composition (7-10). In contrast, the LL-chloroplasts exhibit a higher stacking degree of thylakoids and also possess a higher proportion of the light-harvesting chlorophyll a/b-proteins (6-9).

The formation of either HL-chloroplasts or LL-chloroplasts demonstrates the capacity of growing plants for an ontogenetic adaptation of their photosynthetic apparatus to the incident light quantity. Light intensity also regulates the number of chloroplasts per mesophyll cell with a higher chloroplast number in sun and HL-leaves (5, 10) than in shade or LL-leaves. The question rises whether the HL or LL-photosynthetic apparatus of fully developed, green leaves can be adapted to a new light regime. This would require different mechanisms a) for the reorganization of HL-chloroplasts, when exposed to low quanta fluence rates, and b) for the adaptation of LL and HL-chloroplasts to a very strong light intensity. In the latter case a sudden exposure to high light would represent a high-light stress which might cause a photoinhibition of photosynthesis (12) and a photodestruction of chlorophylls and of certain pigment proteins (e.g. LHCPs), a process which might be overlapped by a new accumulation of pigment proteins (e.g. chlorophyll a/ß-carotene-proteins).

Because of the differences in prenylpigment ratios between HL and LL-chloroplasts, changes in the ratios of photosynthetic pigments are good indicators for changes in the reorganization of chloroplasts. In order to get information on the adaptation possibility of developed chloroplasts and/or their photodestruction, we determined the prenylpigment ratios of fully green leaves before and after exposure to a very strong white light, and compared these with the fluorescence induction kinetics. We also looked for spontaneous changes in chloroplast ultrastructure upon strong light exposure.

2. MATERIALS AND METHODS

Seedlings were germinated and grown for three days in the dark on peat (25°C, 65% relative humidity) and then illuminated. The plants were kept either under high light (HL: 15 klux; 67 $W \cdot m^{-2}$; PAR 250 $\mu E \cdot m^{-2} \cdot sec^{-1}$) or low light (LL: 1.5 klux; 7 $W \cdot m^{-2}$; PAR 26 $\mu E \cdot m^{-2} \cdot sec^{-1}$) growth conditions provided by Osram lamps HQI-E 400/DV using 5 cm running water as IR-filter. The light source used for strong white light exposure (SL) consisted of a Osram lamp HQI-E 1000 W/N (SL: up to 75 klux, 230 $W \cdot m^{-2}$; PAR 1000 $\mu E \cdot m^{-2} \cdot sec^{-1}$). For simultaneous light + heat stress the same lamp was used without water filter (SL + IR: up to 80 klux; 560 $W \cdot m^{-2}$; PAR 1100 $\mu E \cdot m^{-2} \cdot sec^{-1}$).

The leaf pigments were extracted using acetone and light petrol. Chlorophylls and the total carotenoid content were measured in diethylether using the new extinction coefficient of Lichtenthaler and Wellburn 1981 (13). The levels of individual carotenoids were determined after TLC separation on silicagel plates (1), and also by high performance liquid chromatography (8, 9).

The fluorescence induction curves were registered using a special apparatus with a Helium-Neon-laser as described (2, 9).

Leaf segments were fixed in buffered 5% glutardialdehyde (pH 7.4) and postfixed and poststained as previously described (2, 5). The degree of thylakoid stacking and the ratio of appressed to exposed thylakoid membranes were determined by measuring the length of stacked (appressed) and of exposed membranes (stroma thylakoids + end grana membranes) with a milage tracer.

3. RESULTS AND DISCUSSION

3.1. Differences between HL and LL-chloroplasts

The ultrastructure of chloroplasts from plants grown at high light is quite different from that of LL-plants (Fig. 1). LL-chloroplasts possess more appressed membranes, as seen by broader and higher grana stacks, higher values for the ratio appressed to exposed thylakoid membranes and a larger number and range of thylakoids per granum (Table 1). Further differences in chloroplast ultrastructure and composition are summarized in Figure 2.

A particular characteristic of LL-chloroplasts is their higher proportion of the light-harvesting chlorophyll a/b-proteins, the LHCPs (7) which, as carotenoids, mainly contain lutein and neoxanthin and exhibit chlorophyll a/b ratios between 1.1 and 1.3 (8, 9). ß-carotene is preferentially located in the chlorophyll a-proteins of photosystem I (CPI and CPIa) and in the pigment-protein CPa (7) which is thought to be the reaction center of photosystem II (Fig. 3).

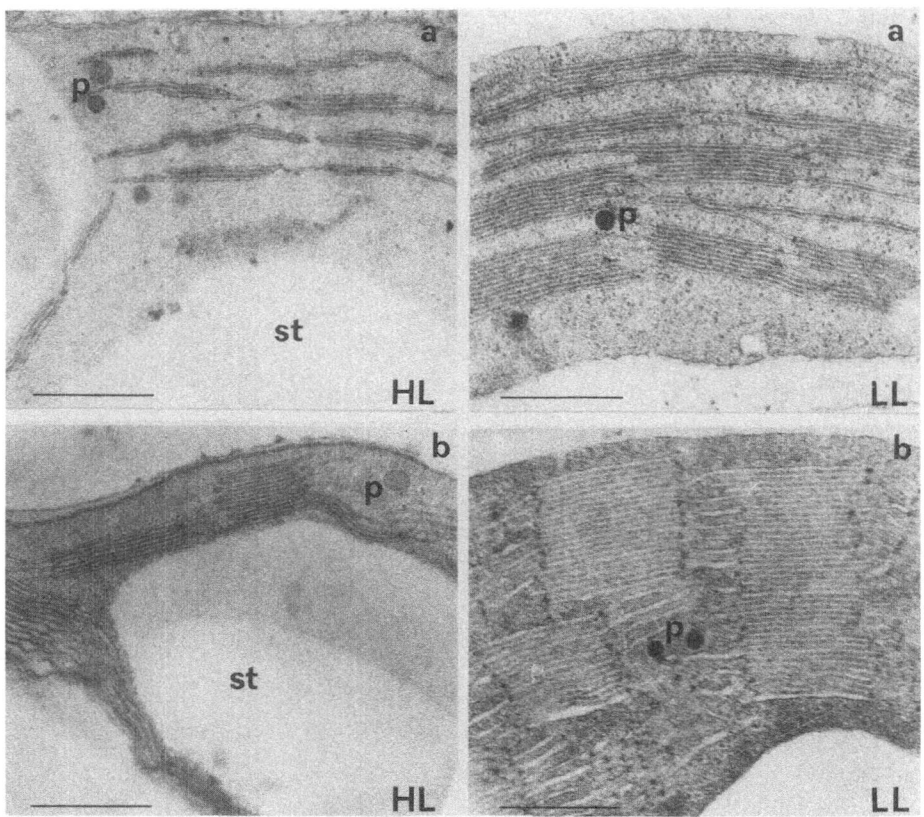

FIGURE 1. Ultrastructure of chloroplasts a) from the cotyledons of 8 d old radish plants and b) from leaves of 28 d old wheat plants grown in continuous white light of high or low intensity. p = plastoglobuli; st = starch; bar = 0.5 μm.

HL-chloroplasts contain a higher proportion of the chlorophyll a/ß-carotene-proteins (CPI, CPIa) and either the same amounts (17) or a slightly higher percentage of the chlorophyll a/ß-carotene-protein CPa (7). They are thus characterized by a higher proportion of ß-carotene and chlorophyll a which is documented by higher chlorophyll a/b ratios and lower values for the ratio xanthophylls/ß-carotene (x/c) and the ratio chlorophyll a/ß-carotene (a/c). Since the a/c ratio is rather low in the chlorophyll a-proteins (CPIa : 12; CPI : 9; CPa : 4 - 8) and very high in the LHCPs (60 - 180), the a/c ratio proves to be a very efficient parameter for the determination of changes in the arrangement of pigment proteins. A further feature of HL-chloroplasts is in most cases a higher amount of total carotenoids on a chlorophyll basis which is documented by lower values for the ratio chlorophylls/carotenoids (a+b/x+c) than in LL-chloroplasts.

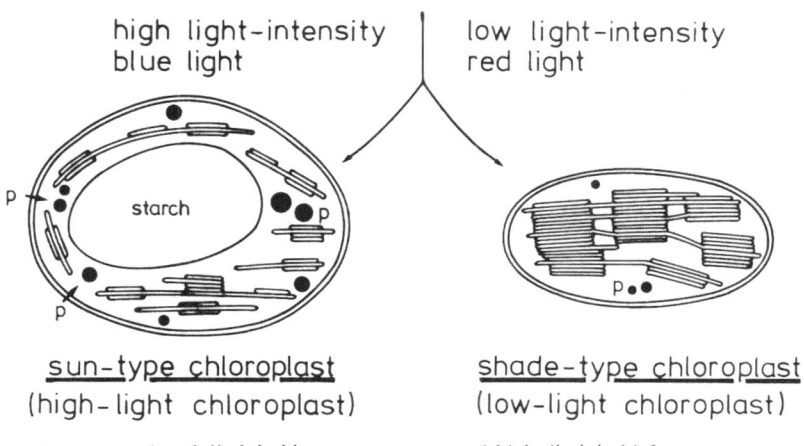

etioplast

high light-intensity
blue light

low light-intensity
red light

sun-type chloroplast
(high-light chloroplast)

shade-type chloroplast
(low-light chloroplast)

1. low amounts of thylakoids per chloroplast section

2. narrow grana stacks

3. few thylakoids per granum

4. low stacking degree

5. more exposed chloroplast membranes

6. higher proportion of chlorophyll a-proteins CPI, CPIa

7. large starch grains

8. many plastoglobuli (p)

9. lower chlorophyll content

1. high thylakoid frequency per chloroplast section

2. broad grana stacks

3. high grana stacks with many thylakoids

4. high stacking degree

5. more appressed thylakoid regions

6. higher amounts of light-harvesting chlorophyll-proteins LHCP s

7. no starch

8. few plastoglobuli (p)

9. more chlorophyll per chloroplast

FIGURE 2. Development of sun-type and shade-type chloroplasts from etioplasts indicating the ontogenetic adaptation response of the photosynthetic apparatus to high or low light intensity. The scheme is based on the data of (2, 5,6,7,11).

Since the increase in thylakoid stacking of LL-radish chloroplasts is paralled by an identical increase in the LHCP amounts (7, 17), it is assumed that the LHCPs participate in thylakoid stacking, as indicated in Figure 4. An involvement of LHCPs in thylakoid stacking is also suggested by proteolysis experiments with chloroplast membranes (21). In more recent works using different methods it had been shown that stacked membrane regions predominantly contain photosystem II units (18, 19, 20) with a full complement of LHCPs (16, 20). In contrast, exposed membranes mainly contain photosystem I units; their photosystem II units appear to possess less LHCPs per PSII core (14, 16, 20)

Table 1. Differences in the ratio of appressed to exposed thylakoid membranes and in the average number (and range) of thylakoids per granum in chloroplasts from sun and shade leaves (Fagus) and from the leaves of HL and LL-seedlings.

	ratio of appressed to exposed thylakoids	number (range) of thylakoids per granum	
Fagus sylvatica [a]			
sun leaf	1.3	4	(2-30)
shade leaf	4.7	9.5	(2-60)
Raphanus sativus (8 d old) [b]			
HL-cotyledon	1.2	2.7	(2-6)
LL-cotyledon	1.8	4.5	(2-17)
Triticum aestivum (28 d old)			
HL: 4th-leaf	1.2 [c]	8	(2-14) [d]
LL: 3rd-leaf	2.7 [c]	16	(2-45) [d]
Zea mays (16 d old; mesophyll-chloroplasts)			
HL: 3rd-leaf	1.2 [c]	7.6	(2-40) [d]
LL: 2nd-leaf	3.3 [c]	16.8	(2-80) [d]

based on a) 25 median chloroplast sections with 500 grana, b) 40 median sections with 1300 grana, c) 10 median sections, d) 20 median sections with 200 grana.

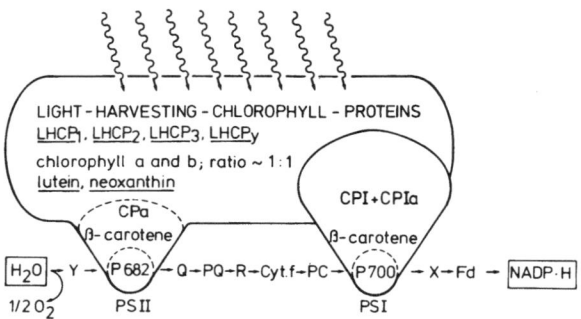

FIGURE 3. Hypothetical model for the organization of the chlorophyll — carotenoid-proteins in the thylakoid membrane. The preferential location of lutein, neoxanthin and chlorophyll b in the LHCPs and of ß-carotene with the chlorophyll a-proteins CPI, CPIa and CPa is indicated (after 7).

as shown in Figure 4. There is some information that the PSI units may occur in two forms which differ in the amount of antenna chlorophyll (16). Whether the two P 700-containing chlorophyll a-proteins CPI and CPIa, which are different in their pigment composition (9), derive from different PSI units, has yet to be investigated. Since CPIa exhibits a lower chlorophyll a/b ratio, it seems to possess a larger antenna than CPI (9).

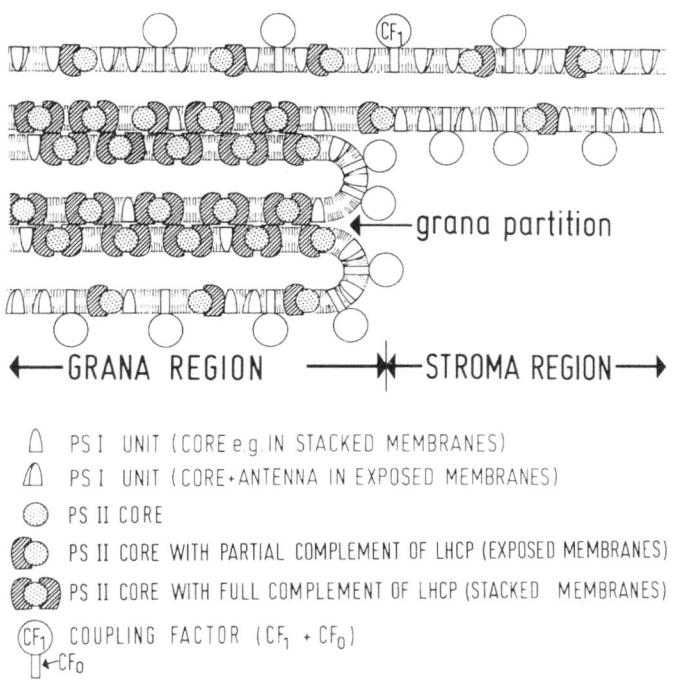

←—GRANA REGION ——————✳—STROMA REGION—→

◠ PS I UNIT (CORE e.g. IN STACKED MEMBRANES)
◠ PS I UNIT (CORE+ANTENNA IN EXPOSED MEMBRANES)
◯ PS II CORE
◖◗ PS II CORE WITH PARTIAL COMPLEMENT OF LHCP (EXPOSED MEMBRANES)
◖◗ PS II CORE WITH FULL COMPLEMENT OF LHCP (STACKED MEMBRANES)
(CF₁) COUPLING FACTOR (CF₁ + CF₀)
∏←CF₀

FIGURE 4. A model for the arrangement of photosystem I units and photosystem II units (with different amounts of LHCPs in appressed (grana) and in exposed membranes (stroma region, end grana membranes), indicating the participation of the light-harvesting chlorophyll a/b-proteins (LHCPs) in the stacking of thylakoids; after Lichtenthaler et al. 1982 (7), including the results of (14-21).

3.2. Effects of high light stress

When plants which are grown at low or medium light intensity, are placed under high light growth conditions, they will gradually adapt their photosynthetic apparatus to the new light regime. This also applies to HL-plants when transferred to an even higher quanta fluence rate. The time course for a full accomplishment of the adaptation response may be a few days, a week, or perhaps longer; it largely depends on the plant type and on the age and physiological state of the leaves.

How is this adaptation response performed and how fast is it detectable? Are the old thylakoids broken down and replaced by newly formed ones? Is there a rearrangement of thylakoids by incorporation of newly formed membrane components? Several processes may occur simultaneously.

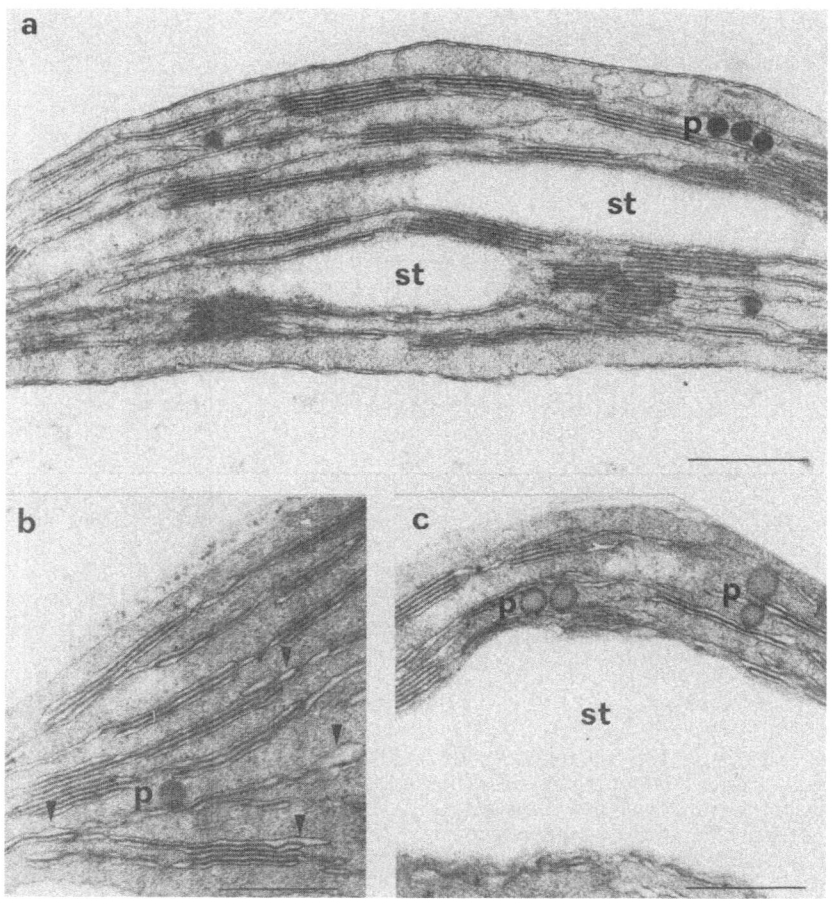

FIGURE 5. Ultrastructure of chloroplasts from the cotyledons of 11 d old HL-radish seedlings (3 d dark growth + 8 d light/dark cycles of 14/10 hours).
a) control; b) and c) after a 90 min exposure to a very strong white light (SL: 75 klux; 230 W · m⁻²; PAR 1 mE · m⁻² · sec⁻¹; water filter).
p = plastoglobuli; st = starch; arrows = swollen thylakoids; bar = 0.5 μm.

Since HL-chloroplasts possess fewer photosynthetic membranes per chloroplast section and a lower proportion of the light-harvesting chlorophyll a/b-proteins than LL or medium-light chloroplasts, one can assume that a partial destruction of membranes and pigment proteins (e.g. LHCPs) will occur at high light stress. In fact, pigments possess a higher turnover rate at HL than at LL-conditions (22). Another possibility would be a new formation of photosystem I units which move in lateraly into the appressed regions of the grana stacks (Fig. 4), and decrease the degree of thylakoid stacking by partially separating the granalphotosystem II

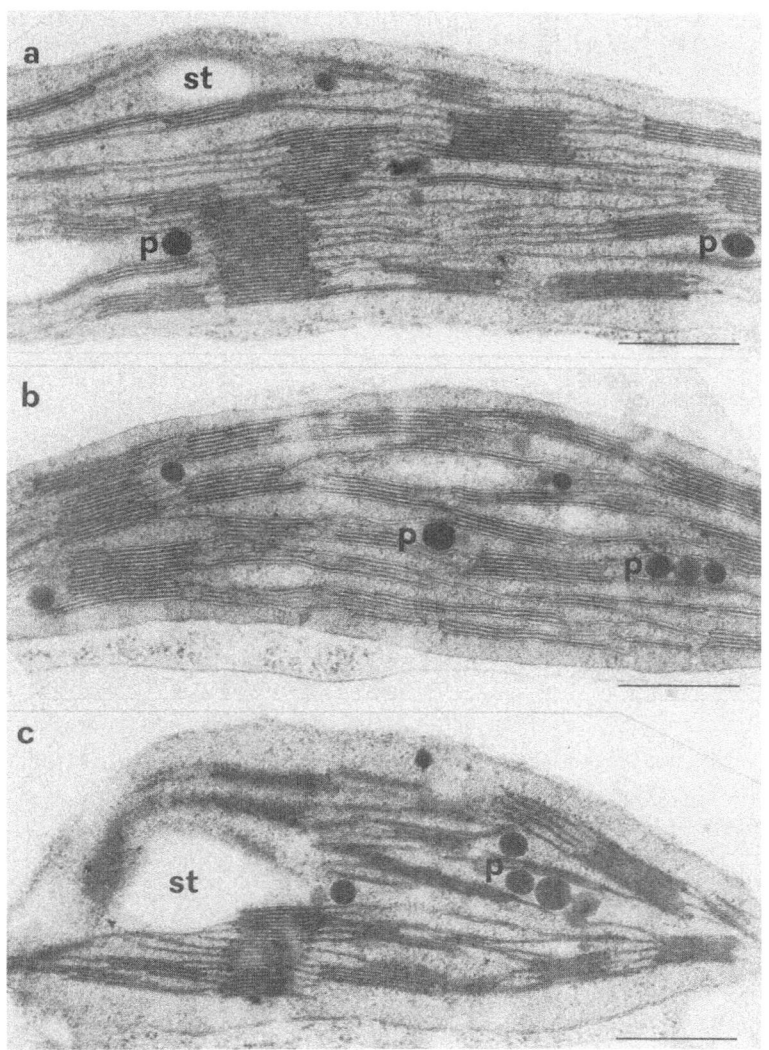

FIGURE 6. Ultrastructure of chloroplasts from the cotyledons of 27 d old LL-
radish plants (3 d dark growth + 24 d light/dark cycles of 14/10 hours).
a) control; b) after a 45 min and c) after a 90 min exposure to a very strong
white light (SL: 75 klux; 230 W · m^{-2}; PAR 1 mE · m^{-2} · sec^{-1}; water filter).
p = plastoglobuli; st = starch; bar = 0.5 μm.

units. A decrease of the total thylakoid content per chloroplast could also be
achieved by a chloroplast division.

Table 2. Change in the prenylpigment ratios and in the percentage composition of carotenoids (weight %) of cotyledons from low light radish plants during exposure to a water-filtered, strong white light (SL: 75 klux; 230 $W \cdot m^{-2}$; PAR 1 $mE \cdot m^{-2} \cdot sec^{-1}$).

	LL-plant (control)	+ SL 30 min	+ SL 70 min	+ darkness 70 min
a) prenylpigment ratios				
a/b	2.65	2.83	2.94	2.72
a/c	16.9	15.6	13.7	17.5
x/c	5.3	3.5	3.0	4.9
a+b / x+c	7.4	6.7	5.4	7.1
b) % composition of carotenoids				
β-carotene	16	22	25	17
lutein	53	48	47	52
zeaxanthin	0	5	11	0
violaxanthin	15	5	3	16
neoxanthin	13	12	11	13
others	3	4	3	2

3.2.1. Ultrastructure. Exposure of LL-radish plants to a very strong white light (SL: PAR 1 $mE \cdot m^{-2} \cdot sec^{-1}$) for 45 min does not change the average height and width of the grana or the stacking degree of thylakoids (Fig. 5a,b). At longer exposure times (90 min) we observed besides fully normal appearing chloroplasts also an increased number of those with more diffuse thylakoids (Fig. 5c) which seems to indicate a beginning thylakoid breakdown. In HL-plants (PAR 250 $\mu E \cdot m^{-2} \cdot sec^{-1}$) which had been treated with light stress for 45 min no significant change in the stacking degree or grana width could be observed; however, the osmiophilic character of the plastoglobuli was decreased and many thylakoids were swollen (Fig. 6), apparently indicating a photostress situation.

It had been postulated that upon illumination of restacked thylakoid systems there may occur a lateral movement of photosystem I units into the stacked grana regions (23). This process seems to be restricted to this particular artificial system. Our ultrastructural studies of chloroplasts from darkened and illuminated LL and HL-plants do not provide any indication for such a lateral movement in the natural system.

3.2.2 Pigment content. When green radish seedlings (grown on peat) are submitted to a high light stress (PAR 1 $mE \cdot m^{-2} \cdot sec^{-1}$, water filter), rather fast changes in the carotenoid composition and in the prenylpigment ratios of LL and HL-plants occur, no matter whether the total chlorophyll and carotenoid content increases slightly or remains the same. Violaxanthin is de-epoxidized

Table 3. Changes in pigment content and carotenoid composition of green leaves of Cissus antarctica upon exposure to a 3 h high-light stress (SL: 75 klux; 230 W $\cdot$ m^{-2}; PAR 1 mE $\cdot$ m^{-2} $\cdot$ sec^{-1}) and a 20 min high-light + heat stress (SL + IR: 80 klux; 560 W $\cdot$ m^{-2}; PAR 1.1 mE $\cdot$ m^{-2} $\cdot$ sec^{-1}). Mean of 3 extracts with 3 leaves per each condition.

	initial value	SL 3 h	decrease	SL + IR 20 min	decrease
a) pigment content (μg/100 cm^2 leaf area)					
chlorophylls	3693	3255	(12%)	2525	(32%)
chlorophyll a	2875	2532		2174	(24%)
chlorophyll b	818	723		351	(57%)
carotenoids	635	571	(10%)	533	(16%)
ß-carotene (c)	196	199	(0%)	177	(10%)
xanthophylls (x)	439	372	(15%)	356	(19%)
b) prenylpigment ratios					
a/b	3.5	3.5		6.2	
a/c	14.7	12.7		12.3	
x/c	2.24	1.87		2.01	
a+b / x+c	5.8	5.7		4.7	
c) % composition of carotenoids					
ß-carotene	30.8	34.9		33.2	
lutein + zeaxanthin + antheraxanthin	44.9	51.5		54.3	
violaxanthin	14.1	4.3		4.1	
neoxanthin	10.2	10.3		8.3	

to zeaxanthin indicating that the violaxanthin-zeaxanthin-cycle (24, 25) is fully operative. Simultaneously the percentage of ß-carotene increases and that of lutein declines (Table 2). Parallel to this, we also observed an increase in the chlorophyll a/b ratios and a decrease in the values for a/c, x/c and a+b/x+c with increasing strong light exposure. These changes in the prenylpigment ratios are indicators for an adaptation of the photosynthetic apparatus by new formation of the chlorophyll a/ß-carotene-proteins. They proceed at best under slight water stress conditions, when the stomata are almost closed and the main photosynthetic CO_2-fixation is excluded. NADPH and ATP, then not needed for CO_2-fixation, are apparently used for the synthesis of the isoprenoid pigments.

When leaves, in which the photosynthetic electron transport is partially blocked by diuron (short immersion in 10^{-4}M solution of DCMU), are submitted to a photo-stress treatment, there occurs, up to 1 h, a strong increase in the percentage of ß-carotene by new formation. This points to a protective function of ß-carotene which may possibly act as a quencher.

Similar changes in carotenoid composition were also found in photo-stressed Cissus leaves. After a 3 h strong-light stress the chlorophyll and carotenoid

Table 4. Level of photosynthetic pigments, prenylpigment ratios and percentage of individual carotenoids in sun leaves of Fagus before and after exposures to a very strong white light (SL). 1) values obtained using an IR-filter of 4 cm running water (SL: 50 klux; 155 W · m^{-2}; PAR 700 µE · m^{-2} · sec^{-1}) and 2) without water filter (SL + IR: 50 klux; 350 W · m^{-2}; PAR 700 µE · m^{-2} · sec^{-1}). The initial value was taken at dawn before sunrise.
Mean of 3 repetitions with 3 leaves per extract and condition.

		1) SL		2) SL + IR	
	initial value	1.5 h	5 h	2 h	5 h
a) pigment content (µg/100 cm^2 leaf area)					
chlorophylls	4382	4716	2940	4536	2020
chlorophyll a	3363	3619	2256	3456	1571
chlorophyll b	1019	1097	684	1080	449
carotenoids	673	798	507	710	326
ß-carotene (c)	240	275	188	229	108
xanthophylls (x)	433	523	319	481	218
b) prenylpigment ratios					
a/b	3.3	3.3	3.3	3.2	3.4
a/c	14.0	13.1	12.0	15.1	14.5
x/c	1.8	1.9	1.7	2.1	2.0
a+b / x+c	6.5	5.9	5.8	6.4	6.2
c) % composition of carotenoids					
ß-carotene	35.8	34.4	37.3	32.6	33.0
lutein + zeaxanthin	42.7	48.7	44.1	52.7	53.8
antheraxanthin	1.6	3.4	2.5	2.3	2.0
violaxanthin	12.5	5.9	8.4	3.8	2.2
neoxanthin	7.4	7.6	7.7	8.6	8.0

content per leaf area begins to decline (Table 3). The decrease in the carotenoid level is due to the loss of xanthophylls, while the ß-carotene level is not influenced. A faster rate of pigment destruction proceeds in a combined light and heat stress (SL + IR, Table 3). The degradation of chlorophylls proceeds faster than that of carotenoids and the destruction of chlorophyll b and xanthophylls faster than that of chlorophyll a and ß-carotene respectively, indicating a preferential degradation of the light-harvesting chlorophyll a/b-xanthophyll-proteins.

Fagus leaves were submitted to a light stress treatment of a lower intensity (PAR 700 µE · m^{-2} · sec^{-1}). An exposure of up to 2 h under photosynthetic conditions (full water supply, continuous stream of fresh air) had little influence on the pigment levels (e.g. Table 4; 1.5 h SL). The differences in prenylpigment ratios between sun and shade leaves remained the same, a/b 3.3 and 2.8, a/c 14 and 20.7, x/c 1.8 and 2.7, and a+b/x+c 6.5 and 7.9 for sun and shade leaves.

Prolonged exposure to this more moderate light stress decreased the chlorophyll and carotenoid content (Table 4; 5 h SL). Small changes in the ratios a/c and a+b/x+c indicate that only a minor adaptive rearrangement of the photosynthetic apparatus occured. The induction of adaptive responses in Fagus leaves apparently needs a higher quanta fluence rate than the applied intensity. At a prolonged combined light + heat stress the rate of pigment degradation is larger (Table 4; SL + IR: 5h). It is of interest that in sun and in shade leaves the degree of violaxanthin to zeaxanthin transformation was always higher under combined light + heat stress than under light stress conditions alone.

3.2.3. <u>Variable fluorescence</u>. The differences in the photosynthetic activity between sun and shade leaves and HL and LL-leaves are also visible in the fluorescence induction kinetics (Kautsky effect) (2, 5, 7). The fluorescence decrease fd parallels the onset of oxygen evolution (Fig. 7). The fluorescence decrease ratio R_{fd} corresponds to the rate of oxygen evolution. Sun leaves exhibit higher values for the fluorescence decrease ratio R_{fd} than shade leaves (Fig. 8a). The ground fluorescence fo and the maximum fluorescence are lower in sun than in shade leaves, and it takes a longer time period to reach the fluorescence maximum, but the final steady-state fluorescence fs is reached earlier (Fig. 8a,b). By repeating the 5 sec kinetic (2nd flash after a 0.1 sec dark interval) a fast rise of the fluorescence to a broad maximum with a small dip at the beginning is observed. The repetition of the flashes up to 100 times results in a low and constant fluorescence which equals the fs-values of the slow kinetic. Usually, the variable fluorescence kinetics are measured after a dark adaptation of 15 min (Fig. 8a, b). When the dark period is shortened to 1 or 1.5 min, one obtains principally the same kinetics, however, a smaller fluorescence yield (Fig. 8c).

After a 1 h light stress treatment both sun and shade leaves exhibit a decreased ground fluorescence (Fig. 9a, b). The variable part of the fluorescence (fluorescence rise and decrease signals) is missing; the fluorescence kinetic passes from fo directly to the level of the steady state fluorescence fs. After a longer regeneration period the variable fluorescence signal appears (Fig. 9b). Such a regeneration does, however, not occur, when the leaves had been treated with a combined light + heat stress (SL + IR; Fig. 9c).

A light stress treatment of HL or LL-radish plants affects the fluorescence kinetics in a similar way as in sun and shade leaves of the beech. A decrease of the maximum fluorescence fp and of the ground fluorescence fo is found in the primary and secondary leaves of HL and LL-plants (Table 5). The variable part of

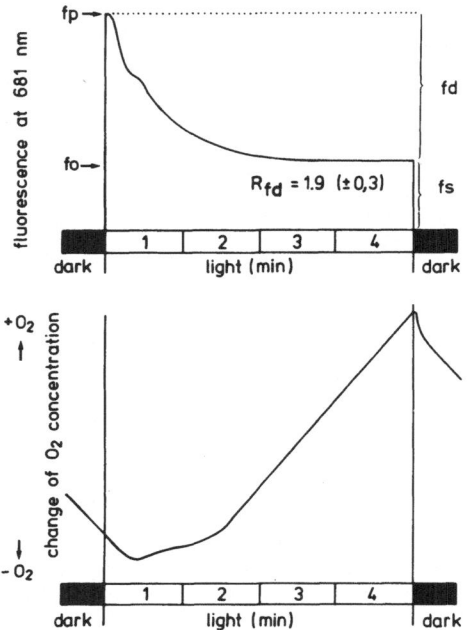

FIGURE 7. Variable fluorescence (fluorescence rise and decrease, Kautsky effect) and oxygen evolution upon illumination of 20 min darkened cotyledons from 10 d old radish seedlings. The fluorescence decrease fd parallels the oxygen evolution. The fluorescence decrease ratio Rfd = fd/fs is a measure for the potential photosynthetic quantum conversion of a leaf. fo = ground fluorescence; fp = maximum fluorescence; fs = steady state fluorescence.

the fluorescence kinetic (Kautsky effect) is, however, not lost.

The results of the fluorescence kinetics indicate that high light stress causes a photoinhibition of the photosynthetic apparatus which seems to be more effective in the fully developed sun and shade leaves of the beech than in the primary and secondary leaves of radish seedlings. The fluorescence induction curve in photo-stressed sun and shade leaves of the beech is similar to that of diuron-treated leaves, except that the whole fluorescence signal is much lower. With its small height it resembles very much the fluorescence kinetics obtained after incubation of leaves with halogenated naphthoquinones which are known to block the photosynthetic electron transport and which, in addition to this, are effective quenchers of fluorescence (26). Based on this comparison we assume that strong-light stress either activates existing quencher molecules or

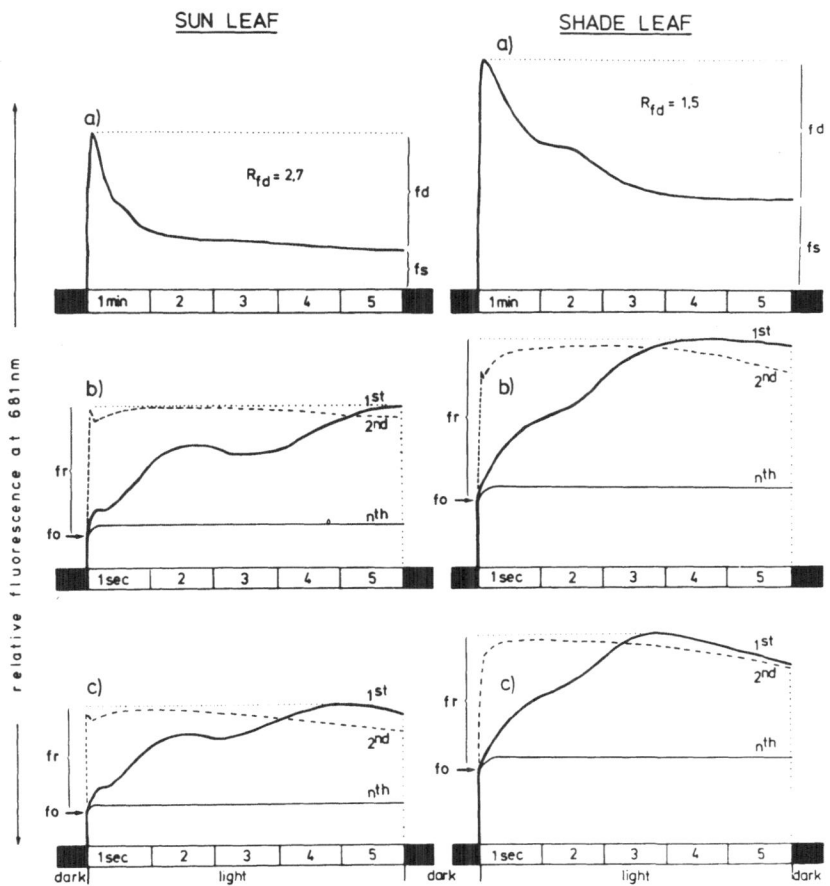

FIGURE 8. Fluorescence induction kinetics (variable fluorescence, Kautsky effect) in sun and shade leaves of Fagus sylvatica. a) min-range: fluorescence decrease during onset of photosynthesis, b) and c) fast fluorescence rise signal: 1st induced by the first 5 sec-flash of a darkened leaf; 2nd = second 5 sec-flash after a 0.1 sec dark interval; nth = fluorescence after about 50 to 100 5 sec-flashes.
In a) and b) the leaves were dark adapted for 15 min; in c) the fluorescence measurements were performed directly, after a maximum dark period of only 1 or 1.5 min.
fd = fluorescence decrease; fs = steady state fluorescence; fr = fluorescence rise above fo; fo = ground fluorescence.
R_{fd} = fluorescence decrease ratio, defined as the ratio of fluorescence decrease fd to the steady state fluorescence fs (R_{fd} = fd/fs).

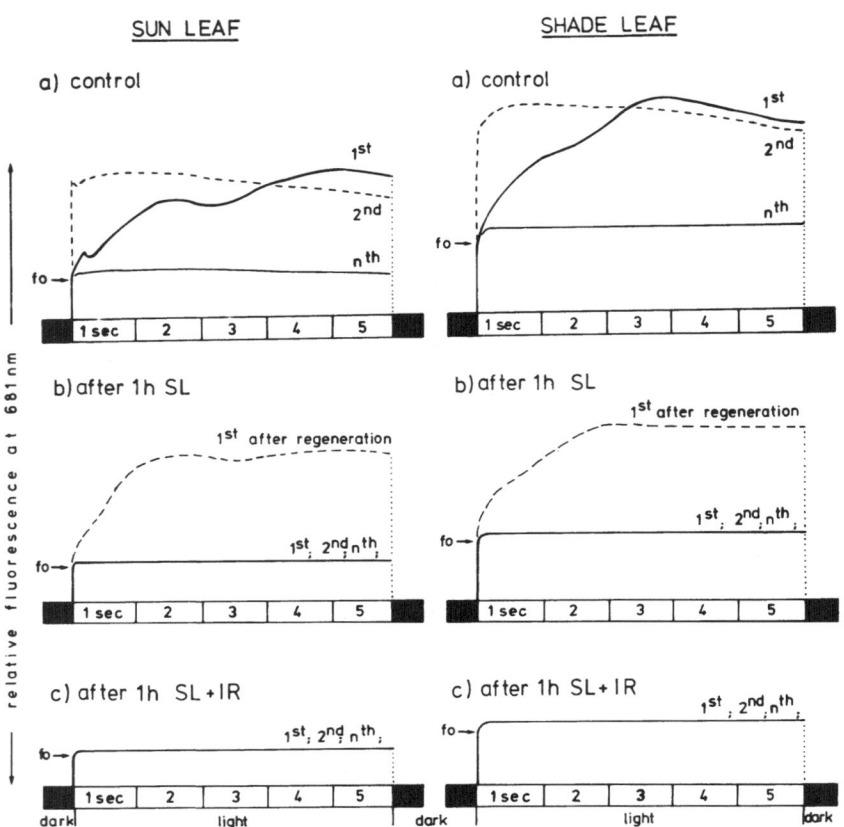

FIGURE 9. Fluorescence induction kinetics (Kautsky effect) of sun and shade leaves of Fagus sylvatica under stress conditions.
a) before treatment (control), b) after 1 h strong-light stress (SL: 50 klux; 155 W · m⁻²; PAR 700 uE · m⁻² · sec⁻¹) and after a regeneration period of 10 h darkness + 12 h low light, and c) after 1 h light + heat stress (SL + IR: 50 klux; 350 W · m⁻²; PAR 700 μE · m⁻² · sec⁻¹). 1st, 2nd, and nth as in Figure 8.

induces a fast new formation of quenchers which consume the excess excitation energy. The nature of such a quencher (or quenchers) is not known. Carotenoids, in particular ß-carotene, are known to protect excited chlorophylls against photooxidation by quenching triplet states of chlorophyll and also singlet oxygen (27, 28). The fast increase in ß-carotene levels upon exposure of leaves to light stress may indicate that ß-carotene is one possible candidate.

A photoinhibition of chloroplasts and a loss of the variable fluorescence pattern were firstly described by Kok (12) assuming an inhibition site close to the antenna of photosystem II. More recent investigations with cucumber point t

Table 6. Decrease of maximum fluorescence (fp) and of ground fluorescence (fo) in the primary and secondary leaves of LL-radish plants and HL-radish plants upon exposure to a very strong high light (SL: 75 klux; 230 W·m^{-2}; 1 mE· m^{-2}· sec^{-1}; with water filter). Fluorescence values given in relative units. Mean values of 3 leaves per each condition.

| | LL-leaves | | | | HL-leaves | | | |
| | primary | | secondary | | primary | | secondary | |
	fp	fo	fp	fo	fp	fo	fp	fo
initial value	6.2	1.3	5.2	1.1	3.7	0.7	4.8	0.9
+ 15 min SL	4.9	1.1	4.5	1.0	3.2	0.4	2.9	0.5
+ 60 min SL	4.5	1.0	3.4	0.8	2.9	0.4	3.0	0.5
% decrease of initial	29%	23%	35%	27%	22%	43%	38%	44%

a possible attack on the oxidizing (donor) site of photosystem II (29, 30). Photoinhibition which is similar in sun and shade plants (31) can apparently be diminished by CO_2 (32), and is hindered by photorespiration (33). From some of our data it appears that the degree of photoinhibition also depends on the aperature of the stomata and is more pronounced under water stress conditions.

ACKNOWLEDGEMENTS

This work was sponsored by a grant from the Deutsche Forschungsgemeinschaft. We wish to thank Mrs. Gabriele Ihrig, Mrs. Waltraud Meier and Mr. Bernhard Volk for assistance.

REFERENCES

1. Lichtenthaler HK. 1979. Z. Naturforsch. 34c, 936.
2. Lichtenthaler HK, Buschmann C, Döll M, Fietz H-J, Bach T, Kozel U, Meier D, Rahmsdorf U. 1981. Photosynth. Res. 2, 115.
3. Boardman NK. 1977. Rev. Plant Physiol. 28,355.
4. Wild A. 1979. Ber. Dtsch. Bot. Ges. 92, 341.
5. Lichtenthaler HK. 1981, in Photosynthesis VI (G. Akoyunoglou ed), p. 273. Philadelphia, Balaban Intern. Sciences Services.
6. Meier D, Lichtenthaler HK. 1981. Protoplasma 107, 195.
7. Lichtenthaler HK, Kuhn G, Prenzel U, Buschmann C, Meier D. 1982. Z. Naturforsch. 37c, 464.
8. Lichtenthaler HK, Prenzel U, Kuhn G. 1982. Z. Naturforsch. 37c, 10.
9. Prenzel U, Lichtenthaler HK. 1982, in Biochemistry and Metabolism of Plant Lipids (J.F.G.M. Wintermans and P.J.C.Kuiper eds.), in press. Amsterdam, Elsevier/North Holland Biomedical Press.
10. Butterfaß T. 1979. Patterns of Chloroplast Reproduction. Wien, Springer.
11. Buschmann C, Meier D, Kleudgen HK, Lichtenthaler HK. 1978. Photochem. Photobiol. 27, 195.
12. Kok B, Gassner EB, Rurainski HJ. 1965. Photochem. Photobiol 4, 215.
13. Lichtenthaler HK, Bach TJ, Wellburn AR. 1982, in Biochemistry and Metabolism of Plant Lipids (J.F.G.M. Wintermans and P.J.C. Kuiper eds.). Amsterdam,

Elsevier/North Holland Biomedical Press.
14. Staehelin LA, Giddings TH, Badami P, Kryzmowski WW. 1978. in Light Transducing Membranes, p. 335. New York, Academic Press.
15. McDonnel A, Staehelin LA. 1980. J. Cell Biol. 84, 40.
16. Staehelin LA. 1981, in Photosynthesis III (G.Akoyunoglou ed.), p. 3. Philadelphia, Balaban Intern. Sciences Services.
17. Lichtenthaler HK, Kuhn G, Prenzel U, Meier D. 1982. Physiol Plant., in press
18. Andersson B, Anderson JM. 1980. Biochim. Biophys. Acta 593, 427.
19. Anderson JM. 1981. FEBS Letters 124, 1.
20. Arntzen CJ. 1978. Current Topics in Bioenergetics 8, 111.
21. Carter DP, Staehelin LA. 1980. Arch. Biochem. Biophys. 200, 374.
22. Grumbach KH, Lichtenthaler HK. 1981. Photochem. Photobiol 35, 209.
23. Barber J. 1980. FEBS Letters 118, 1.
24. Hager A. 1975. Ber. Dtsch. Bot. Ges. 88, 45.
25. Yamamoto HY. 1979. Pure and Appl. Chem. 51, 639.
26. Pfister K, Lichtenthaler HK, Burger G, Musso H, Zahn M. 1981. Z. Naturforsch. 36c, 645.
27. Mathis P, Kleo J. 1973. Photochem. Photobiol. 18, 343.
28. Wolff C, Witt HT. 1969. Z. Naturforsch. 24b, 1031.
29. Critchley C, Smillie RM. 1981. Austral. J. Plant Physiol. 8, 133.
30. Critchley, C. 1981. Plant Physiol. 67, 1161.
31. Powles SB, Thorne SW. 1981. Planta 152, 471.
32. Powles SB, Osmond CB, Thorne SW. 1979. Plant Physiol. 64, 982.
33. Osmond CB. 1981. Biochim. Biophys. Acta 639, 77.

MULTIPLE EFFECTS OF HEAVY METAL TOXICITY ON PHOTOSYNTHESIS

F. VAN ASSCHE and H. CLIJSTERS

Department SBM, Limburgs Universitair Centrum, Universitaire Campus,
B-3610   Diepenbeek, Belgium

1. ABSTRACT

Photosynthetic activity of many species of green algae and higher plants
is inhibited by accumulation of toxic amounts of heavy metals.

Evidence is presented that zinc toxicity impairs photosynthetic electron
transport, mainly at the oxidizing side of PS 2, inducing a decrease of
photophosphorylation capacity and NADPH production. In addition Ribulose
1,5 biphosphate carboxylase activity is directly inhibited. Both effects
can regulate photosynthetic CO_2 fixation. CO_2 diffusion from the ambient air
to the chloroplast is reduced by structural changes to the stomata and spongy
parenchyma of the poorly expanding leaf. They are considered to be secondary
influences of zinc toxicity, the primary site of action being merely at the
metabolic level.

These experimental results are discussed in relation to effects of other
heavy metals on photosynthesis, already described in literature.

2. INTRODUCTION

In industrial areas, heavy metals occur locally in high concentrations in
the environment inducing a reduction of plant biomass production and of the
nutritional quality of crops (11). Laboratory experiments indicated that
photosynthesis is highly sensitive to heavy metals, more so than respiration
(13,18,41). Photosynthetic CO_2 fixation in higher plants was inhibited after
uptake of Tl, Pb, Cd, Ni (10), Zn (41), Cd (3), Ni, Cr, Co (2), Cr (4) and
Al (36). When supplied to the nutrient medium of green algae, Zn, Cd, Hg (12),
Cr (47) and Cu (11) also reduced photosynthesis.

In these studies to assess their effect on photosynthesis in a first approach
heavy metals were added to isolated systems *in vitro*, such as intact chloroplasts or chloroplast fragments, enzymes, and epidermal strips. Careful extra-
polation of these data to the situation *in vivo* provided valuable information

on the interaction of heavy metals with photosynthesis.

In more recent studies plants were cultivated on various substrates, containing heavy metals at concentrations similar to those found in contaminated soils or polluted water. After uptake and assimilation of the pollutants by the plants their effects on metabolism were studied in isolated systems.

We report on the effects of the assimilation of toxic amounts of Zn on photosynthesis in *Phaseolus vulgaris L.*. These results are compared with the effects of other heavy metals on photosynthesis in higher plants and green algae.

3. MATERIAL AND METHODS

3.1. Plant Material

Dwarf beans, *Phaseolus vulgaris L.* c.v. Limburgs Vroege, were cultivated on vermiculite in a growth chamber under a photon flux density of 200 $\mu E \ m^{-2} \ s^{-}$ during 12 h per day. Day and night temperatures were 25°C and 21°C respectively and relative humidity was kept at 80%.

Optimal plant growth was obtained by supplying 50 ppm zinc sulphate to the nutrient solution, which contained all the essential macro-elements (control plants). Growth inhibition by toxicity was induced by increasing the zinc content level to 200 ppm (for details, see 41). All measurements were performed on fully expanded primary leaves during the third week after sowing; at this moment leaf zinc content reached values of about 200 and 400 ppm of the dry matter in the control and treated plants respectively.

3.2. Photosynthetic electron transport and photophosphorylation

Type C chloroplasts (14) were isolated as described previously (32). Electron transport and related ATP production were estimated in an assay, combining measurements on an oxygen electrode (Rank Brothers) and on a LKB Luminometer.

Electron transport reactions were measured polarographically at 22°C in the following reaction mixtures (final volume 2 ml) :
- Photosystem 1+2 (PS 1+2) : aerated Tricinebuffer pH 7,8 0,01 M; K_2HPO_4 2 mM NaCl 2 mM; $MgCl_2$ 5 mM; and methylviologen (MV) 0,25 mM (32). Stimulated electron flow was obtained by adding 1 μmol ADP.
- Photosystem 1 (PS 1) : The same aerated coctail as for PS 1+2, including 3-(3,4-dichlorophenyl)-1,1 dimethyl urea (DCMU) 0,25 μM; NaN_3 0,25 mM; 3,6 diaminodurene (DAD) 0,1 mM and L-Ascorbic acid 1 mM (6).

- Photosystem 2 (PS 2) : Tricinebuffer 0,01 M pH 7,8, saturated with nitrogen;
2 mM K_2HPO_4; 5 mM $MgCl_2$; 1,6 mM DAD and 4,8 mM $K_3Fe(CN)_6$ (35). Eventually
3 mM semicarbazide (SC) were added to the reaction mixture (49).
Chlorophyll content, determined according to Bruinsma (9), was about 50 µg
for the PS 1+2 reaction mixture, 25 µg for PS 1 and PS 2 reactions.
White light of 200 µE $m^{-2} s^{-1}$ was provided from a slide projector to the
reaction cuvette. Photophosphorylation was measured in essentially the same
reaction mixtures after addition of the specific bioluminescence reagents
(ATP-monitoring reagent, LKB 1250-121, and 5×10^{-3} M ADP). Illumination and
temperature conditions were as above. ATP production in the light was
derived from a standard curve and corrected with an internal standard,
5×10^{-9} mol ATP, added at the end of the reaction. Full description of the
assay procedure and manipulations will be published in detail elsewhere.

3.3. Enzyme activities

These were measured on two types of extracts :

a) on a leaf-homogenate, prepared at 4° C by homogenising 2 g of fresh leaf
   tissue for 30 s with a Virtis blendor in Tris-HCl buffer 0,1 M pH 8,
   containing 1 mM EDTA; 1 mM dithiothreitol (DTT); 10 mM $NaHCO_3$; 20 mM $MgCl_2$.
   The slurry was squeezed through eight layers of cheesechloth and the
   filtrate was assayed for enzyme activity.

b) on a chloroplast stroma protein enriched fraction, prepared according to
   Walker (44). In this method precooled leaves were mixed with the same homogenizer for 3 s in an extraction buffer consisting of 50 mM 2-(N-morpholino)
   ethane sulphonic acid (MES) pH 8,0; sorbitol 0,33 M; $MgCl_2$ 5 mM; Na ascor-
   bate 2 mM; K_2HPO_4 10 mM; cystein 1 mM; polyethyleneglycol (PEG) 0,6 % and
   EDTA 1 mM. The tissue fresh weight/extraction volume ratio was 1:5. The
   slurry was filtered through 2 layers of miracloth and cotton wool and the
   filtrate centrifuged for 30 s at 1000 g in glass tubes, with acceleration
   and braking time not exceeding 15 s. The pellet was gently resuspended and
   washed another two times in the same medium, until the protein content of
   the supernatant was beyond the detection limit. The pellet was resuspended in 1 ml of a tris-HCl buffer pH 8,0 containing 1 mM (DTT); 1 mM
   EDTA, and osmotically shocked for 1 min. After centrifugation of the
   broken chloroplasts at 1000 g, the supernatant was used for the enzyme
   assays.

Ribulose biphosphate carboxylase (RUBPC; E C 4.1.1.39) was measured according to McLilley (29). The chloroplast stroma preparation was preincubated

for 15 min. at 25° C with 20 mM $MgCl_2$; 10 mM $NaHCO_3$ (27); 1 mM NADPH; and
0,5 mM Phosphogluconate (25). Ribulosebiphosphate oxygenase (RUBPO) activity
was estimated polarographically on the leaf homogenate only according to
Lorimer (27). Phosphoribulokinase (EC 2.7.1.19) was measured according to
Racker (33), phosphoglyceratekinase (EC 2.7.2.3) according to Bradbeer (7)
and NADPH- dependent glyceraldehydephosphatedehydrogenase (EC 1.2.1.13) as
described by Wu and Racker (48).

Protein content of the samples was determined with a commercial Biorad
reagent according to Bradford (8).

4. RESULTS

Under photophosphorylation stimulating conditions (addition of ADP), PS 1+2
electron transport activity of chloroplasts from plants having received toxic
zinc doses, was reduced to about half of the activity rate of the controls
(table 1). A significantly lower photophosphorylation activity was also observed
but the ATP/2 electron ratio was not significantly affected by the zinc treatment.

Table 1. Photosynthetic electron transport (A), photophosphorylation (B)
and ATP/2 electron ratio (C) measured on type C chloroplasts from
Phaseolus vulgaris L., grown on optimal and toxic concentration of
zinc in the substrate

		Zinc nutrition	
		Optimal (control)	Toxic
PS 1+2 ($H_2O \rightarrow MV$)	A	255±37	135[*]±25
	B	189±26	87[*] ±40
+ ADP, phosphate	C	1,5±0,3	1,3 ±0,6
PS 1 (HAsc/DAD⁻ $\rightarrow$ MV)	A	694±109	620 ±231
	B	267±62	149[*]±75
	C	0,73±0,14	0,53[*]±0,24
PS 2 ($H_2O \rightarrow DAD^+/K_3fe(CN)_6$)	A	827±156	555[*]±161
	B	286±70	194[*]±85
	C	0,69±0,14	0,70 ±0,24
($H_2O/SC \rightarrow K_3Fe(CN)_6$)		502±152	1012[*]±184

A : µeq. mg chl^{-1}.h^{-1}
B : µmol ATP.mg chl^{-1}.h^{-1}
C : ATP/2 electrons
[*] : significant difference (p=0,05; t-test); mean value of 10 replications
± standard error.

No significant difference was found between the two levels of zinc nutrition for the electron transport related to PS 1 alone, but ATP production at this site was significantly lower in plants with zinc toxicity. Both effects resulted in a significantly lower ATP/2 electrons ratio (table 1). PS 2 electron transport activity and ATP production were equally inhibited by high zinc levels; ATP/2 electrons ratios were identical for both types of plants (table 1). Removal of DAD from the reaction mixture decreased the PS 2 electron transport rate in control chloroplast, even when semicarbazide was supplied. Addition of the latter compound however induced a very high PS 2 activity rate in chloroplasts from plants receiving toxic zinc levels.

Table 2. Enzyme activities (mU/mg protein) as a function of zinc nutrition in *Phaseolus vulgaris L*. The measurements were made in a leaf homogenate (A) or in a chloroplast stroma protein enriched fraction (B).

	Zinc nutrition	
	Optimal (control)	Toxic
A.		
RUBP Carboxylase	342±45	261 *±35
RUBP-Oxygenase	69±8	71 ±7
RUBPC/RUBPO	5,0±0,9	3,7 *±0,6
Phosphoribulokinase	955±123	636*±119
Phosphoglyceratekinase	3379±1125	3124 ±391
NADPH-glyceraldehydephosphate dehydrogenase	686±234	500 ±140
B.		
RUBP Carboxylase	554±155	427 ±145
Phosphoribulokinase	1210±201	1060 ±251
Phosphoglyceratekinase	5089±611	4559 ±927
NADPH-glyceraldehydephosphate dehydrogenase	666±267	715 ±320
Mean Activity $\frac{B}{A}$ ×100	134	155

* : see table 1.

In the same plants RUBPC was significantly inhibited in the leaf homogenate (table 2A). The oxygenase activity was not affected. Consequently zinc toxicity resulted in a significant decrease of the RUBPC/RUBPO activity ratio. Phosphoribulokinase and NADPH-glyceraldehydephosphate dehydrogenase were inhibited to the same extent as RUBPC. No significant effect was observed on the highly active phosphoglycerate kinase.

In the chloroplast stroma protein enriched fraction the mean specific activity increased 34 % in the controls, but 55 % in the intoxicated plants (table 2 B). Only for RUBPC there was a strong tendency to inhibition (77 % of the control activity). This inhibition was not significant due to the high variability of the data as a result of a relatively more compex extraction procedure.

5. DISCUSSION

In a previous paper (41) we showed that the basal photosynthetic electron transport rate was not inhibited by toxic amounts of Zn; even an increase in basal PS 1 related partial electron flow was observed. From table 1 however it is clear that under conditions supporting photophosphorylation severe inhibition is observed for PS 1+2-activity. In these conditions mainly PS 2 is inhibited. Recovery of this activity by semicarbazide suggests that the site of Zn action is at the oxidizing side. Semicarbazide introduces electrons at that side in a similar way as diphenylcarbazide. It can only donate electrons when the electron transport is inhibited (49); therefore no effect of semicarbazide is observed on chloroplasts from control plants.

Cd was shown to inhibit photosynthetic electron transport at the same site in tomato; recovery of the inhibition by feeding high amounts of Mn indicated interaction of this heavy metal with the Mn-containing water splitting enzyme (3). It is worth mentioning that addition of Zn (39, 41) or Cd (43) to isolated chloroplasts inhibited electron transport at the same site *in vitro*, and application of the same artificial electron donors could overcome this effect in a similar way. Moreover PS2-inhibition by Zn and Cd at the water splitting side was also found in *Euglena gracilis* (12) in a quasi *in vivo* approach, where artificial electron transport dyes were taken up by the living alga.

In general PS 2 activity proves to be very sensitive to heavy metals. *In vitro* addition of Hg interfered close to the reaction center of PS 2 (20, 31) in chloroplasts, but results with *Euglena gracilis* suggested an effect similar to that of Zn and Cd (12). The *in vitro* inhibitory activity of Pb was also Mn-indenpendent and directly related to the PS 2 reaction center (30). Similar to these observations the PS 2 inhibition by Cu in *Ankistrodes-mus falcatus* (38) and by Cr in *Lemna minor* (4) was not restored by diphenyl-carbazide. For Cd no interaction was found at this site *in vitro* (26).

The general sensitivity of PS 2 electron transport to heavy metals
provides good arguments for the use of variable fluorescence for monitoring
the effect on plants of soil and air pollution by these elements (1).

PS 1 activity with MV as an electron acceptor is generally more resistant
to heavy metal toxicity. The minor inhibitory effect we found for Zn
(table 1) was also observed in *Euglena* for Zn, Cd, and Hg (12) and in
Lemna, grown in high concentrations of Cr (4). Pb salts however inhibit
PS 1 activity of isolated chloroplasts at several sites (18, 19, 46) and in
Anacystis nidulans Hg interferes with the electron transport between PS 2
and PS 1 by substitution for Cu in plastocyanin (21).

At the reducing side of PS 1 however, the NADPH-oxidoreductase is
severely inhibited by Cd, Zn and Hg in *Euglena* (12) and by Cr in *Lemna* (4).
For chloroplasts of higher plants the reports are rather scarce, probably
due to the technical difficulties for measuring this activity, although
interference of Hg at this site was also demonstrated for these organelles
in vitro (20).

The photophosphorylation activity related to PS 2 electron transport is
inhibited by Zn (table 1), but this metal does not affect coupling of this
partial reaction of photophosphorylation as the ATP/2 electrons ratio is
not changed (table 1). However a small but significant reduction of this
ratio by Zn is observed for the electron transport between PS 2 and PS 1.
Uncoupling by this heavy metal seems to occur at that site, and could account
for the small but not significant decrease of the ATP/2 electrons ratio observed for the overall PS 1+2 electron transport. Pb (16), Cd (3) and Hg (20) also
inhibited photosynthetic energy transfer.

It is clear that heavy metals can reduce photosynthetic NADPH- and
ATP-production. Decrease of ATP production can occur when oxidative phospho-
rylation is enhanced (13). Dark respiration was stimulated but net photo-
synthesis reduced in excised *Acer saccharinum* leaves, treated with Cd (22).
Pb (24) and Cd (23) increased dark respiration in Soja. Alternatives for
NADP reduction were also suggested (15).

Shortage of ATP and NADPH could affect the photosynthetic CO_2 reduction
in vivo, but significant inhibition of the specific activity of several
Calvin-cycle enzymes was also observed in leaf homogenates from zinc treated plants. The soluble protein content of these leaves however was higher
(41). This extra protein might be extra-chloroplastic : a comparable
increase of soluble protein was observed in Soja after treatment with Cd

(23) and Pb (24), and it was associated to stimulated activity of extra-chloroplastic senescence enzymes. The inhibitory effect observed on specific enzyme activity could therefore be overestimated.

This possible artefact should be overcome by preparing a fraction, enriched in chloroplast stroma protein. In this enriched fraction the mean relative increase of specific enzyme activities was proportionally higher for treated (+ 54 %) than for control plants (+ 34 %) (table 2B), suggesting that relatively more (extra-chloroplastic) protein was eliminated in the extract from the former plants.

The strong tendency to inhibition by Zn of RUBPC in the enriched fraction (table 23) is worth some further consideration. The RUBPC/RUBPO activity ratio, independent of the protein content, decreased after zinc application (table 2A); a significant increase of the CO_2 compensation *in vivo* was previously observed (42). These results confirm an effect of zinc on the RUBPC/RUBPO enzyme complex. RUBPC/RUBPO requires CO_2 and Mg for full activation (27). Uptake of heavy metals induces metal substitution in the metallo-proteins, impairing enzyme activity; substitution of Zn and Mn by Cd was proven to be the inhibitory mechanism of Cd toxicity for carbonic anhydrase (13) and for the PS 2 water splitting enzyme (3) respectively. *In vitro* substitution of Mg by Mn (28) or Co and Ni (45) decreased the RUBPC/RUBPO activity ratio. The question arises whether the observed shift in RUBPC/RUBPO activity (table 2B) is due to a partial substitution of Mg by Zn *in vivo* as a result of Zn accumulation in the leaf.

As Zn, Cd, Pb and Hg are powerful SH-antagonists (40), they also affect enzyme activity in this way. The *in vivo* inhibition of the SH-enzyme NADPH-oxidoreductase by heavy metals is already discussed (12). *In vitro* RUBPC- and phosphoribulokinase-activities were strongly affected by supplying Cd (13) or Pb (18).

From these data it is clear that several modes of action exist by which heavy metals interfere with proteins, that are related to photosynthesis.

It was suggested that heavy metals may impair leaf transpiration and CO_2 fixation by decreasing leaf conductance to CO_2 diffusion as a result of stomatal closure. A direct effect on *in vivo* stomatal regulation was postulated for Cd, Tl, Ni and Pb (10) or for Al (36) in experiments with isolated epidermis strips, floating on heavy metal solutions and showing stomatal closure in these conditions (5, 37). Cd inhibited net photosynthesis and transpiration and decreased stomatal (g'_s) and internal (g'_i) conductance to CO_2 diffusion in excised leaves (22). In intact plants however stomata are primarly regulated by internal CO_2 concentration, which is a result of both CO_2 fixing and releasing processes. ATP and NADPH are required for stomatal function, and the production of both coenzymes is directly affected by heavy metals. The assumption, that inhibition of net photosynthesis by Al was exclusively at the stomatal level, was later refuted because its effect on chloroplast activity (17). In conditions of zinc toxicity we measured a premature loss of stomatal regulation (42), possibly reflecting early onset of senescence as observed for Cd (29) and Pb (24). It has yet to be proven that heavy metals directly interfere with K^+ fluxes in the guard cells.

Zinc toxicity strongly decreased leaf expansion, resulting in a more compact leaf structure (42). These structural changes were also related to the observed increase of g'_i and g'_s. G'_i consists of a structural and a metabolic component, and the latter was recently shown to be regulatory (50). We therefore believe that changes of leaf architecture and stomata are rather side effects of heavy metals, the primary site of action being merely at the metabolic level.

Heavy metal stress has multiple effects on photosynthesis metabolism. Therefore algae and higher plants show a general response to toxic concentrations, although the degree of toxicity is different from one metal to another, and some of them are even indispensable micro-nutrients. Environmental pollution may induce additive or synergistic effects, as mixed contamination by several heavy metals usually occurs.

6. ACKNOWLEDGEMENTS

Skilful technical assistance of Mrs. C. Bogaert-Vanherle is gratefully acknowledged. This research was supported by the "Instituut voor Aanmoediging van het Wetenschappelijk Onderzoek in Nijverheid en Landbouw (I.W.O.N.L.)", Brussels.

7. REFERENCES

1. Arndt U (1974) The Kautsky-effect : A method for the investigation of the actions of air pollutants in chloroplasts. Environm Pollut 6:181-194.
2. Austenfeld FA (1979) Nettophotosynthese der Primär und Folgeblätter von *Phaseolus vulgaris* L. unter dem Einfluss von Nickel, Kobalt und Chrom. Photosynthetica 13 (4):434-438.

3. Baszynski T Wadja L Krol M Wolinska D Krupa Z and Tukendorf A (1980). Photosynthetic activities of cadmium-treated tomato plants. Physiol Plant 48:365-370.

4. Baszinsky T Krol M and Wolinska D (1981) Photosynthetic apparatus of *Lemna minor* L. as affected by chromate treatment. In Akoyunoglou, G ed pp 111-122. Balaban Int Sci Serv Philadelphia.

5. Bazzaz FA Carlson RW and Rolfe GL (1974) The effect of heavy metals on plants : I : Inhibition of gas exchange in sunflower by Pb, Cd, Ni and Tl. Environm Pollut 7:241-246.

6. Böger P and Kunert KJ (1978) Phytotoxic action of paraquat on the photosynthetic apparatus. Z Naturforsch 33C:688-694.

7. Bradbeer JW (1969) The activities of the photosynthetic carbon cycle enzymes of greening bean leaves. New Phytol 68:233-245.

8. Bradford MM (1976) A Rapid and sensitive method for the quantitation of microgram quantities of protein utilising the principle of protein - dye binding. Anal Biochem 72:248-256.

9. Bruinsma J (1963) The quantitative analysis of chlorophylls a and b in plant extracts. Photochem Photobiol 2:241-250.

10. Carlson RW, Bazzaz FA and Rolfe GL (1975) The effect of heavy metals on plants. II Net photosynthesis and transpiration of whole corn and sunflower plants treated with Pb, Cd, Ni and Tl. Environm Res 10:113-120.

11. Cottenie A, Dhaese A and Camerlynck R (1976) Plant quality response to uptake of polluting elements. Qual Plant Pl Fds Hum Nutr XXVI:293-319.

12. De Filippis LF, Hampp R and Ziegler H (1981) The effects of sublethal concentrations of zinc, cadmium and mercury on *Euglena*. II Respiration photosynthesis and photochemical activities. Arch Microbiol 128:407-411.

13. Ernst WHO (1980) Biochemical aspects of cadmium in plants. In JO Nriagu, ed Cadmium in the environment, part 1. pp 639-653. J. Wiley & Sons.

14. Hall DO (1972) Nomenclature for isolated chloroplasts. Nature New Biol 235:125-126.

15. Halliwell B (1978) The chloroplast at work. A review of modern developments in our understanding of chloroplast metabolism. Prog Biophys Molec Biol 33:1-54.

16. Hampp R, Ziegler H and Ziegler I (1973) Die Wirkung von Bleiionen auf die $^{14}CO_2$-Fixierung und die ATP-Bildung von Spinatchloroplasten. Biochem Physiol Pfl 164:126-134.

17. Hampp R and Schnabl H (1975) Effect of aluminium ions on $^{14}CO_2$-fixation and membrane system of isolated spinach chloroplasts. Z Pflanzenphysiol 76(4):300-306.

18. Höll W and Hampp R (1975) Lead and Plants. Residue Rev 54:79-113.

19. Homer JR, Cotton R and Evans EH (1979) The effects of lead on whole-leaf photosynthesis determined by fluorescence measurements. Biochem Soc Transact 7:1259-1260.

20. Honeycutt RC and Krogmann DW (1972) Inhibition of chloroplast reactions with phenylmercuric acetate. Plant Physiol 49:376-380.

21. Kleinen Hammans JW, Rabou LPLM and Pietersen HQ (1976) Participation of pigment complexes in uptake and incorporation of mercury ions by *Anacystis*. Photosynthetica 10 (4):440-446.

22. Lamoreaux RJ and Chaney WR (1978) The effect of cadmium on net photosynthesis, transpiration and dark respiration of excised Silver Maple leaves. Physiol Plant 43:231-236.

23. Lee KC Cunningham BA Paulsen GM Liang GH and Moore RB (1976A) Effects of cadmium on respiration rate and activities of several enzymes in Soybean seedlings. Physiol Plant 36:4-6.

24. Lee KC, Cunningham BA, Chung KH, Paulsen GM and Liang GH (1976B) Lead effects on several enzymes and nitrogenous compounds in soybean leaf. J. Environm Qual 5(4):357-359.

25. Lendzian KJ (1978) Activation of ribulose-1,5-biphosphate carboxylase by chloroplast metabolites in a reconstituted spinach chloroplast system. Planta 143:291-296.

26. Li EH and Miles CD (1975) Effect of Cadmium on photoreaction II of chloroplasts. Plant Sci Lett 5:33-40.

27. Lorimer GH Badger MR and Andrews TJ (1977) D-Ribulose-1,5-biphosphate carboxylase-oxygenase. Improved methods for the activation and assay of catalytic activities. Anal Biochem 78:66-75.

28. Lorimer GH and Miziorko HM (1981) RuBP Carboxylase : The mechanism of activation and its relation to catalysis. In Akoyunoglou G ed Photosynthesis IV, Regulation of Carbon Metabolism pp 3-16. Balaban Int Sci Serv Philadelphia.

29. McLilley R and Walker DA (1974) An improved spectrophotometric assay for ribulose biphosphate carboxylase. Biochem Biophys Acta 358:226-229.

30. Miles CD Brandle JR Daniel DJ Chu-Der O Schnare PD and Uhlik DJ (1972) Inhibition of photosystem II in isolated chloroplasts by Lead. Plant Physiol 49:820-825.

31. Miles D Bolen P Faraq S Goodin R Lutz J Moustafa A Rodriguez B and Weil C (1973) Hg^{++} - A DCMU independent electron acceptor of photosystem II. Bioch Biophys Res Commun 50(4):1113-1119

32. Oben G and Marcelle R (1975) The effects of CCC and GA on some biochemical and photochemical activities of primary leaves of bean plants. In R Marcelle ed Environmental and biological control of photosynthesis. pp 211-216. Junk, The Hague.

33. Racker E (1957) The reductive pentose phosphate cycle I. Phosphoribulokinase and ribulose diphosphate carboxylase. Arch Biochem Biophys 69:300.

34. Reeves SG and Hall DO (1978) Photophosphorylation in chloroplasts. Biochim Biophys Acta 463:275-297.

35. Saha S, Ouitrakul R Isawa S and Good NE (1971) Electron transport and photophosphorylation in chloroplasts as a function of the electron acceptor 246(10):3204-3209.

36. Schnabl H und Ziegler H (1974) Der Einfluss des Aluminiums auf den Gasaustausch und das Welken von Schnitt-pflanzen. Ber Deutsch Bot Ges 87:s.13-20.

37. Schnabl H und Ziegler H (1975) Uber die Wirkung von Aluminiumionen auf die Stomatabewegung von *Vicia Faba* Epidermen. Z Pflanzenphysiol 74:394-403.

38. Shioi Y Tamai H and Sasa T (1978) Inhibition of photosystem II in the green alga *Ankistrodesmus falcatus* by Copper. Physiol Plant 44:434-438.

39. Tripathy BC and Mohanty P (1980) Zinc inhibited electron transport of photosynthesis in isolated barley chloroplast. Plant Physiol 66:1174-1178.

40. Valle BL and Ulmer DD (1972) Biochemical effects of mercury, cadmium and lead. Ann Rev Biochem 41:91-128.

41. Van Assche F Clijsters H and Marcelle R (1979) Photosynthesis in *Phaseolus vulgaris* L., as influenced by supra-optimal zinc nutrition. In Marcelle R et al eds. Photosynthesis and plant development pp 175-184. The Hague : Junk.

42. Van Assche F Ceulemans R and Clijsters H (1980) Zinc mediated effects on leaf CO_2 diffusion conductances and net photosynthesis in *Phaseolus vulgaris* L. Photosynth Res 1:171-180.

43. Van Duijvendijk-Matteoli MA and Desmet GM (1975) On the inhibitory action of cadmium on the donor side of Photosystem II in isolated chloroplasts. Biochim et Biophys Acta 408:164-169.

44. Walker DA (1972) Chloroplasts (and Grana) : Aqueous (including high carbon fixation ability). In San Pietro A ed Meth in Enzymol XXIII, pp 211-220.

45. Wildner GF and Henkel J (1979) The effect of divalent metal ions on the activity of Mg^{++} depleted ribulose-1,5-diphosphate oxygenase. Planta 46:223-228.

46. Wong D and Govindjer (1976) Effects of lead ions on photosystem I in isolated chloroplasts : Studies on the reaction center P 700. Photosynthetica 10(3):241-254.

47. Wium-Andersen S (1974) The effect of chromium on the photosynthesis and growth of diatoms and green algae. Physio Plant 32:308-310.

48. Wu R and Racker E (1959) Regulatory mechanisms in carbohydrate metabolism. III Limiting factors in glycolysis of ascites tumour cells. J Biol Chem 234,1029.

49. Yamashita T and Butler WL (1969) Inhibition of the Hill Reaction by tris and restoration by electron donation to photosystem 2. Plant Physiol 44: 435-438.

50. Zima J Sestak Z Catsky J and Ticha I (1981) Ontogenetic changes in leaf CO_2 uptake as controlled by photosystem and carboxylation activities and CO_2 transfer. In Akuyunoglou G ed Photosynthesis VI. pp 23-31. Balaban Int Sci Serv Philadelphia.

INDEX